Ernst Schering Foundation Symposium
Proceedings 2007-2
Organocatalysis

Ernst Schering Foundation Symposium
Proceedings 2007-2

Organocatalysis

M.T. Reetz, B. List, S. Jaroch, H. Weinmann
Editors

With 200 Figures

Springer

Series Editors: G. Stock and M. Lessl

Library of Congress Control Number: 2007943075

ISSN 0947-6075

ISBN 978-3-540-73494-9 Springer Berlin Heidelberg New York

This work is subject to copyright. All rights are reserved, whether the whole or part of the material is concerned, specifically the rights of translation, reprinting, reuse of illustrations, recitation, broadcasting, reproduction on microfilms or in any other way, and storage in data banks. Duplication of this publication or parts thereof is permitted only under the provisions of the German Copyright Law of September 9, 1965, in its current version, and permission for use must always be obtained from Springer-Verlag. Violations are liable for prosecution under the German Copyright Law.

Springer is a part of Springer Science+Business Media
springer.com

© Springer-Verlag Berlin Heidelberg 2008

The use of general descriptive names, registered names, trademarks, etc. in this publication does not emply, even in the absence of a specific statemant, that such names are exempt from the relevant protective laws and regulations and therefor free for general use. Product liability: The publisher cannot guarantee the accuracy any information about dosage and application contained in this book. In every induvidual case the user must check such information by consulting the relevant literature.

Cover design: design & production, Heidelberg
Typesetting and production: le-tex publishing services oHG, Leipzig
21/3180/YL – 5 4 3 2 1 0   Printed on acid-free paper

Preface

Chemical synthesis is one of the key technologies that form the basis of modern drug discovery and development. For the rapid preparation of new test compounds and the development of candidates with often highly complex chemical structures, it is essential to use state-of-the-art chemical synthesis technologies. Due to the increasing number of chiral drugs in the pipeline, asymmetric synthesis and efficient chiral separation technologies are steadily gaining in importance. Recently a third class of catalysts, besides the established enzymes and metal complexes, has been added to the tool kit of catalytic asymmetric synthesis: organocatalysts, small organic molecules in which a metal is not part of the active principle.

Despite considerable efforts to explore and extend the scope of asymmetric organocatalytic reactions in recent years, their use in medicinal and process chemistry is still rather low. This is even more surprising as the field was pioneered by the medicinal chemistry laboratories of Schering AG and Hoffmann La Roche in the late 1960s and early 1970s by using proline as asymmetric catalyst in a Robinson annulation to obtain steroid CD ring fragments, a process now referred to as the Hajos–Parrish–Eder–Sauer–Wiechert reaction.

In an effort to increase the awareness within the community of medicinal and process chemists, and to learn more about recent progress in this rapidly evolving field, the Ernst Schering Foundation enabled us to organize a symposium on 'Organocatalysis,' which took place in Berlin, Germany, from 18 to 20 April 2007. The proceedings of this symposium are detailed in this book.

S.C. Pan and B. List's paper spans the whole field of current organocatalysts discussing Lewis and Brønsted basic and acidic catalysts. Starting from the development of proline-mediated enamine catalysis—the Hajos–Parrish–Eder–Sauer–Wiechert reaction is an intramolecular transformation involving enamine catalysis—into an intermolecular process with various electrophilic reaction partners as a means to access α-functionalized aldehydes, they discuss a straightforward classification of organocatalysts and expands on Brønsted acid-mediated transformations, and describe the development of asymmetric counteranion-directed catalysis (ACDC).

Impressive applications of chiral amine organocatalysts in natural product synthesis come from the D. Enders' laboratory. Enders and colleagues elegantly employ the different modes of enamine and iminium activation in domino reactions leading to highly functionalized cyclohexenes. Using dihydroxyacetone acetonide the amine organocatalyst functions like an artificial aldolase eventually leading to carbohydrates, sphingolipids, and carbasugars. In the following chapter, M. Christmann reports examples for organocatalytic key steps in the total synthesis of the terpene alkaloid and telomerase inhibitor UCS1025A.

More examples for 'applied organocatalysis' are presented by H. Gröger, who gives an overview of organocatalytic methods already applied on a technical scale. Based on case studies, he shows several examples that satisfy the criteria of a technically feasible process such as high catalyst activity and stability, economic access, sustainability, atom economy, and high volumetric productivity.

Another privileged class of Lewis basic organocatalysts are nucleophilic carbenes, which have been proven to be extremely versatile for different transformation, albeit strongly depending on the electronic and steric nature of the catalyst. F. Glorius and K. Hirano apply N-heterocyclic carbenes (NHC) for a conjugate umpolung of α,β-unsaturated aldehydes into homoenolates, which are then reacted with aldehydes or ketones to γ-butyrolactones or β-lactones. A review on N-heterocyclic carbenes as a class of organocatalysts beyond the well-known Stetter reaction with diverse modes of action is provided by K. Zeitler.

Complementary to the great variety of Lewis basic organocatalysts, Brønsted acids gain increasing importance and significantly expand the scope of organocatalytic processes. M. Rueping and E. Sugiono demonstrate the versatility of chiral BINOL-phosphate-based Brønsted acids in the asymmetric hydrogenation of quinolines, benzoxazines and pyridines and they highlight the potential of this method in alkaloid total synthesis. The first radical reactions catalyzed by a chiral organic additive are realized in a piperidone and dihydroquinolinone synthesis by T. Bach and coworkers. An organocatalyst derived from Kemp's acid interacts with the secondary lactam or amide functionality of the substrates through double hydrogen bond contacts. This is an addition to the covalent involvement of amine organocatalysts, and allows for optimizing the stereochemical outcome by using nonpolar solvents, such as trifluorotoluene. Hydrogen bonding networks in chiral thiourea catalysis are also elegantly used by A. Berkessel in the kinetic resolution of oxazolones and oxazinones.

Various aspects of organocatalysis with larger molecules are also covered in this book. Possible benefits from immobilization approaches for organic catalysts are pointed out by M. Benaglia. Apart from catalyst recycling or simplified workup procedures, catalyst immobilization can be additionally advantageous in terms of catalyst development and optimization. The use of soluble supports, such as polyethylene glycol, often allows the direct transfer and application of already optimized reaction conditions.

In the field of enzyme catalysis, some of the major drawbacks, such as narrow substrate scope, or low selectivity and thermostability, can be successfully addressed by using directed evolution. In contrast to a rational design which uses site-specific mutagenesis, these studies utilize genomic technologies like saturation mutagenesis and gene shuffling to create powerful, tailor-made proteins as large molecule organocatalysts. An even more effective strategy to enhance enzyme catalysis is the symbiosis of rational design and randomization, as applied in CASTing (combinatorial active-site saturation test) in combination with iterative saturation mutagenesis which was introduced by M.T. Reetz.

Although organocatalysis is still in its infancy compared to metal-catalyzed processes or enzyme-mediated transformations, there has been tremendous progress within the last few years. New reactions have been developed and applied for technical processes. Novel types of catalysts are constantly introduced expanding the scope of organocatalytic methodology. Moreover, increasing mechanistic insights will help to further improve known catalytic transformations and to exploit reactivity. The Ernst Schering Foundation workshop offered a broad overview on organocatalytic processes, mechanisms and possible applications and provided an outlook for the future establishment of organocatalysis as a third important strategy in asymmetric catalysis, complementing metal- and biocatalysis not only in academia, but also in industry. The editors would like to acknowledge the generous support of the Ernst Schering Foundation, which allowed us to set up this exciting workshop. We trust that the readers will share the enthusiasm and excitement in the rapidly expanding field of asymmetric organocatalysis.

Manfred T. Reetz
Benjamin List
Stefan Jaroch
Hilmar Weinmann

Contents

New Concepts for Organocatalysis
S.C. Pan, B. List   1

Biomimetic Organocatalytic C–C-Bond Formations
D. Enders, M.R.M. Hüttl, O. Niemeier 45

Organocatalytic Syntheses of Bioactive Natural Products
M. Christmann 125

Asymmetric Organocatalysis on a Technical Scale:
Current Status and Future Challenges
H. Gröger 141

Nucleophilic Carbenes as Organocatalysts
F. Glorius, K. Hirano 159

N-Heterocyclic Carbenes: Organocatalysts Displaying
Diverse Modes of Action
K. Zeitler 183

New Developments in Enantioselective Brønsted Acid Catalysis:
Chiral Ion Pair Catalysis and Beyond
M. Rueping, E. Sugiono 207

Chiral Organocatalysts for Enantioselective
Photochemical Reactions
S. Breitenlechner, P. Selig, T. Bach 255

Organocatalysis by Hydrogen Bonding Networks
A. Berkessel 281

Recoverable, Soluble Polymer-Supported Organic Catalysts
M. Benaglia 299

Controlling the Selectivity and Stability of Proteins
by New Strategies in Directed Evolution:
The Case of Organocatalytic Enzymes
M.T. Reetz 321

List of Editors and Contributors

Editors

Reetz, M.
Max-Planck-Institut für Kohlenforschung, Kaiser-Wilhelm-Platz 1,
45470 Mülheim an der Ruhr, Germany
(e-mail: reetz@mpi-muelheim.mpg.de)

List, B.
Max-Planck-Institut für Kohlenforschung, Kaiser-Wilhelm-Platz 1,
45470 Mülheim an der Ruhr, Germany
(e-mail: list@mpi-muelheim.mpg.de)

Jaroch, S.
Bayer Schering Pharma AG, 13342 Berlin, Germany
(e-mail: stefan.jaroch@bayerhealthcare.com)

Weinmann, H.
Bayer Schering Pharma AG, 13442 Berlin, Germany
(e-mail: hilmar.weinmann@bayerhealthcare.com)

Contributors

Bach, T.
Lehrstuhl für Organische Chemie I, Technical University Munich,
Lichtenbergstr. 4, 85747 Garching, Germany
(e-mail: thorsten.bach@ch.tum.de)

Benaglia, M.
Dipartimento di Chimica Organica e Industriale –
Universitá degli Studi di Milano, Via C. Goli 19, 20133 Milan, Italy
(e-mail: maurizio.benaglia@unimi.it)

Berkessel, A.
Department of Organic Chemistry, University of Cologne,
Greinstraße 4, 50939 Cologne, Germany

Breitenlechner, S.
Lehrstuhl für Organische Chemie I, Technical University Munich,
Lichtenbergstr. 4, 85747 Garching, Germany

Christmann, M.
Institute of Organic Chemistry, RWTH Aachen University,
Landoltweg 1, 52074 Aachen, Germany
(e-mail: christmann@oc.rwth-aachen.de)

Enders, D.
Institute of Organic Chemistry, RWTH Aachen University,
Landoltweg 1, 52074 Aachen, Germany
(e-mail: Enders@rwth-aachen.de)

Glorius, F.
Organisch-Chemisches Institut,
Westfälische Wilhelms-Universität Münster,
Corrensstraße 40, 48149 Münster, Germany
(e-mail: glorius@uni-muenster.de)

Gröger, H.
Department of Chemistry and Pharmacy,
University of Erlangen-Nuremberg,
Henkestr. 42, 91054 Erlangen, Germany
(e-mail: harald.groeger@chemie.uni-erlangen.de)

Hirano, K.
Organisch-Chemisches Institut,
Westfälische Wilhlems-Universitat Münster,
Corrensstraße 40, 48149 Münster, Germany
(e-mail: khirano@uni-muenster.de)

Hüttl, M.R.M.
Institute of Organic Chemistry, RWTH Aachen University,
Landoltweg 1, 52074 Aachen, Germany

List, B.
Max-Planck-Institut für Kohlenforschung, Kaiser-Wilhelm-Platz 1,
45470 Mülheim an der Ruhr, Germany
(e-mail: list@mpi-muelheim.mpg.de)

Niemeier, O.
Institute of Organic Chemistry, RWTH Aachen University,
Landoltweg 1, 52074 Aachen, Germany

Pan, S.C.
Max-Planck-Institut für Kohlenforschung, Kaiser-Wilhelm-Platz 1,
45470 Mülheim an der Ruhr, Germany
(e-mail: mailto:subhas@mpi-muelheim.mpg.de)

Rueping, M
Institute of Organic Chemistry and Chemical Biology,
Johann Wolfgang Goethe-Universität Frankfurt am Main,
Max-von-Laue-Straße 7, 60438 Frankfurt am Main, Germany
(e-mail: M.rueping@chemie.uni-frankfurt.de)

Selig, P.
Lehrstuhl für Organische Chemie I, Technical University Munich,
Lichtenbergstr. 4, 85747 Garching, Germany

Sugiono, E.
Institute of Organic Chemistry and Chemical Biology,
Johann Wolfgang Goethe-Universität Frankfurt am Main,
Max-von-Laue-Straße 7, 60438 Frankfurt am Main, Germany

Zeitler, K.
Institut für Organische Chemie, Universität Regensburg,
Universitätsstr. 31, 93053 Regensburg, Germany

Ernst Schering Foundation Symposium Proceedings, Vol. 2, pp. 1–43
DOI 10.1007/2789_2008_084
© Springer-Verlag Berlin Heidelberg
Published Online: 30 April 2008

New Concepts for Organocatalysis

S.C. Pan, B. List

Max-Planck-Institut für Kohlenforschung, Kaiser-Wilhelm-Platz 1,
45470 Mülheim an der Ruhr, Germany
email: *list@mpi-muelheim.mpg.de*

1	Introduction: Organocatalysis	2
2	Enamine Catalysis	3
2.1	The Proline-Catalyzed Asymmetric Aldol Reaction: Scope, Mechanism and Consequences	5
2.2	Enamine Catalysis of Nucleophilic Addition Reactions	8
2.3	Enamine Catalysis of Nucleophilic Substitution Reactions	10
2.4	The Proline-Catalyzed Asymmetric Mannich Reactions	10
3	Brønsted Acid Catalysis	14
3.1	Catalytic Asymmetric Pictet–Spengler Reaction	15
3.2	Organocatalytic Asymmetric Reductive Amination	17
4	Iminium Catalysis	22
4.1	Organocatalytic Conjugate Reduction of α,β-Unsaturated Aldehydes	24
5	Asymmetric Counteranion Directed Catalysis	26
5.1	Asymmetric Counteranion-Directed Catalysis: Application to Iminium Catalysis	28
6	Conclusions	33
References		34

Abstract. Organocatalysis, catalysis with low-molecular weight catalysts in which a metal is not part of the catalytic principle or the reaction substrate, can be as efficient and selective as metal- or biocatalysis. Important discoveries in this area include novel Lewis base-catalyzed enantioselective processes and, more recently, simple Brønsted acid organocatalysts that rival the efficiency of traditional metal-based asymmetric Lewis acid-catalysts. Contributions to organocatalysis from our laboratories include several new and broadly useful concepts such as enamine catalysis and asymmetric counteranion-directed catalysis. Our laboratory has discovered the proline-catalyzed direct asymmetric intermolecular aldol reaction and introduced several other organocatalytic reactions.

1 Introduction: Organocatalysis

When chemists make chiral compounds—molecules that behave like object and mirror image, such as amino acids, sugars, drugs, or nucleic acids—they like to use asymmetric catalysis, in which a chiral catalyst selectively accelerates the reaction that leads to one mirror-image isomer, also called enantiomer. For decades, the generally accepted view has been that there are two classes of efficient asymmetric catalysts: enzymes and synthetic metal complexes (Nicolaou and Sorensen 1996). However, this view is currently being challenged, with purely organic catalysts emerging as a third class of powerful asymmetric catalysts (Fig. 1).

Most biological molecules are chiral and are synthesized in living cells by enzymes using asymmetric catalysis. Chemists also use enzymes or even whole cells to synthesize chiral compounds and for a long

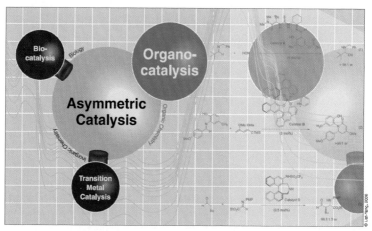

Fig. 1. The three pillars of asymmetric catalysis: biocatalyis, metal catalysis and organocatalysis

time, the perfect enantioselectivities often observed in enzymatic reactions were considered beyond reach for non-biological catalysts. Such biological catalysis is increasingly used on an industrial scale and is particularly favored for hydrolytic reactions. However, it became evident that high levels of enantioselectivity can also be achieved using synthetic metal complexes as catalysts. Transition metal catalysts are particularly useful for asymmetric hydrogenations, but may leave possibly toxic traces of heavy metals in the product.

In contrast, in organocatalysis, a purely organic and metal-free small molecule is used to catalyze a chemical reaction. In addition to enriching chemistry with another useful strategy for catalysis, this approach has some important advantages. Small organic molecule catalysts are generally stable and fairly easy to design and synthesize. They are often based on nontoxic compounds, such as sugars, peptides, or even amino acids, and can easily be linked to a solid support, making them useful for industrial applications. However, the property of organocatalysts most attractive to organic chemists may be the simple fact that they are organic molecules. The interest in this field has increased spectacularly in the last few years (Berkessel and Gröger 2005; List and Yang 2006).

Organocatalysts can be broadly classified as Lewis bases, Lewis acids, Brønsted bases, and Brønsted acids (for a review, see Seayad and List 2005). The corresponding (simplified) catalytic cycles are shown in Scheme 1. Accordingly, Lewis base catalysts (B:) initiate the catalytic cycle via nucleophilic addition to the substrate (S). The resulting complex undergoes a reaction and then releases the product (P) and the catalyst for further turnover. Lewis acid catalysts (A) activate nucleophilic substrates (S:) in a similar manner. Brønsted base and acid catalytic cycles are initiated via a (partial) deprotonation or protonation, respectively.

2   Enamine Catalysis

Enamine catalysis involves a catalytically generated enamine intermediate that is formed via deprotonation of an iminium ion and that reacts with various electrophiles or undergoes pericyclic reactions. The first example of asymmetric enamine catalysis is the Hajos–Parrish–Eder–

Sauer–Wiechert reaction (Eder et al. 1971; Hajos and Parrish 1974; for a review, see List 2002b; Scheme 2), an intramolecular aldol reaction catalyzed by proline. Despite its use in natural product and steroid synthesis, the scope of the Hajos–Parrish–Eder–Sauer–Wiechert reaction had not been explored, its mechanism was poorly understood, and its use was limited to a narrow context. Inspired by the development of elegant biocatalytic and transition metal complex-catalyzed direct asymmetric aldolizations (Barbas et al. 1997; Yamada et al. 1997), a revival of this chemistry was initiated with the discovery of the proline-catalyzed direct asymmetric intermolecular aldol reaction about 30 years later (List et al. 2000; also see Notz and List 2000). Since then, proline-catalyzed enantioselective intermolecular aldol reactions (Northrup and MacMillan 2002a; Bøgevig et al. 2002a; Chowdari et al. 2002; Córdova et al. 2002a,b; Sekiguchi et al. 2003; Bøgevig et al. 2003), Mannich reactions (List 2000; List et al. 2002; Hayashi et al. 2003a; Córdova et al. 2002c; Hayashi et al. 2003b; Córdova 2003; Notz et al. 2003; Enders et al. 2005; Ibrahem et al. 2004) and Michael additions (List et al. 2001; Enders and Seki 2002) have been developed (List 2004; Allemann et al. 2004).

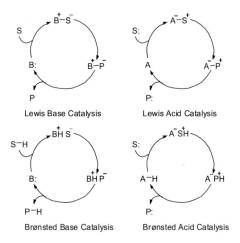

Scheme 1. Organocatalytic cycles

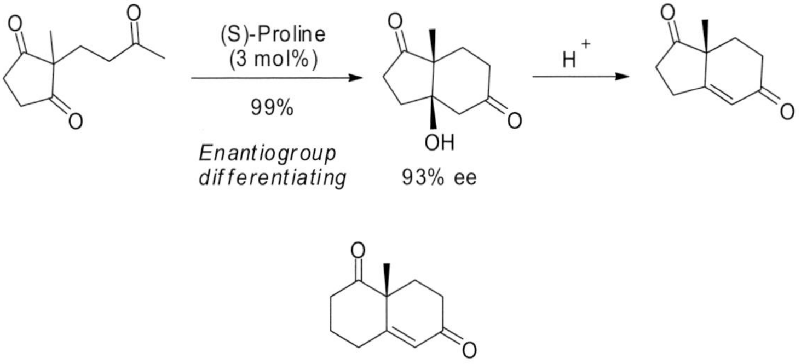

Scheme 2. The Hajos–Parrish–Eder–Sauer–Wiechert reaction

This concept has also been extended to highly enantioselective α-functionalizations of aldehydes and ketones such as aminations (List 2002a; Kumaragurubaran et al. 2002; Bøgevig et al. 2002b), hydroxylations (Brown et al. 2003a; Zhong 2003; Hayashi et al. 2003c, 2004; Bøgevig et al. 2004; Yamamoto and Momiyama 2005), alkylations (Vignola and List 2004), chlorination (Brochu et al. 2004; Halland et al. 2004), fluorination (Enders and Hüttl 2005; Marigo et al. 2005c; Steiner et al. 2005; Beeson and MacMillan 2005), bromination (Bertelsen et al. 2005), sulfenylation (Marigo et al. 2005a) and an intramolecular Michael reaction (Hechavarria Fonseca and List 2004) using proline, as well as other chiral secondary amines and chiral imidazolidinones as the catalysts.

2.1 The Proline-Catalyzed Asymmetric Aldol Reaction: Scope, Mechanism and Consequences

In addition to catalyzing the well-known Hajos–Parrish–Eder–Sauer–Wiechert reaction (Scheme 3; Eq. 1), we found in early 2000 that proline also catalyzes intermolecular aldolizations (e.g. Eq. 2). Thereafter, our reaction has been extended to other substrate combinations (aldehyde to aldehyde, aldehyde to ketone, and ketone to ketone; Eqs. 3–5) and to enolexo-aldolizations (Eq. 6; Northrup and MacMillan 2002a;

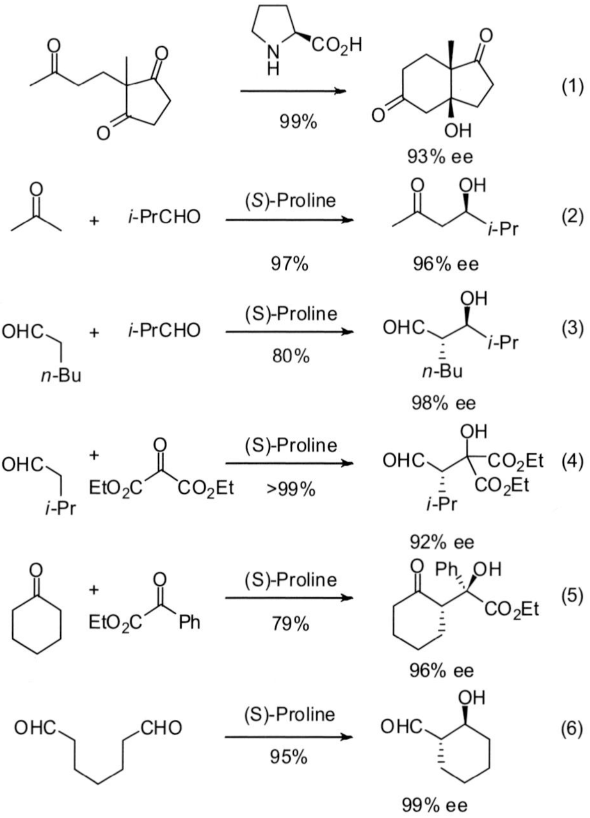

Scheme 3. Proline-catalyzed asymmetric aldol reactions

Bøgevig et al. 2002a; Pidathala et al. 2003; Tokuda et al. 2005). Proline seems to be a fairly general, efficient, and enantioselective catalyst of the aldol reaction and the substrate scope is still increasing continuously.

Both experimental and theoretical studies have contributed significantly to the elucidation of the reaction mechanism. We found that in contrast with earlier proposals (Agami et al. 1984, 1986, 1988; Puchot et al. 1986; Agami and Puchot 1986), proline-catalyzed aldol reactions

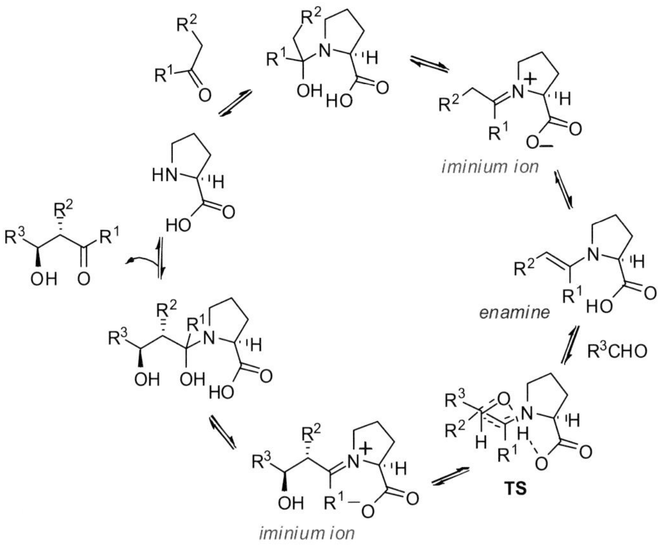

Scheme 4. The proposed mechanism and transition state of proline-catalyzed aldolizations

do not show any nonlinear effects in the asymmetric catalysis (Hoang et al. 2003; Klussmann et al. 2006). These lessons, as well as isotope incorporation studies, provided experimental support for our previously proposed single proline enamine mechanism and for Houk's similar density functional theory (DFT)-model of the transition state of the intramolecular aldol reaction (List et al. 2004; Bahmanyar and Houk 2001a,b; Clemente and Houk 2004; Cheong and Houk 2004; Allemann et al. 2004; Bahmanyar et al. 2003). On the basis of these results we proposed the mechanism shown in Scheme 4. Key intermediates are the iminium ion and the enamine. Iminium ion formation effectively lowers the lowest unoccupied molecular orbital (LUMO) energy of the system. As a result, both nucleophilic additions and α-deprotonation become more facile. Deprotonation leads to the generation of the enamine, which is the actual nucleophilic carbanion equivalent. Its reaction with the aldehyde then provides, via transition state **TS** and hydrolysis, the enantiomerically enriched aldol product.

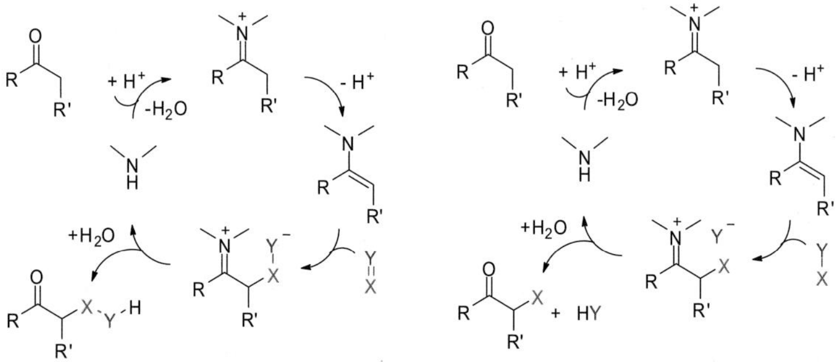

Scheme 5. Enamine catalysis of nucleophilic addition- and substitution reactions (*arrows* may be considered equilibria)

For us, the intriguing prospect arose that the catalytic principle of the proline-catalyzed aldol reaction may be far more general than originally thought. We reasoned that simple chiral amines including proline should be able to catalytically generate chiral enamines as carbanion equivalents, which then may undergo reactions with various electrophiles. We termed this catalytic principle enamine catalysis (Scheme 5; List 2001). Accordingly, the enamine, which is generated from the carbonyl compound via iminium ion formation can react with an electrophile X = Y (or X–Y) via nucleophilic addition (or substitution) to give an α-modified iminium ion and upon hydrolysis the α-modified carbonyl product (and HY).

Enamine catalysis has developed dramatically in the last few years and it turns out that its scope not only exceeds our most optimistic expectations but also that of the traditional stoichiometric enamine chemistry of Stork and others.

2.2 Enamine Catalysis of Nucleophilic Addition Reactions

Enamine catalysis using proline or related catalysts has now been applied to both intermolecular and intramolecular nucleophilic addition reactions with a variety of electrophiles. In addition to carbonyl compounds (C = O), these include imines (C = N) in Mannich reactions (List 2000; List et al. 2002; Hayashi et al. 2003a; Córdova et al. 2002c;

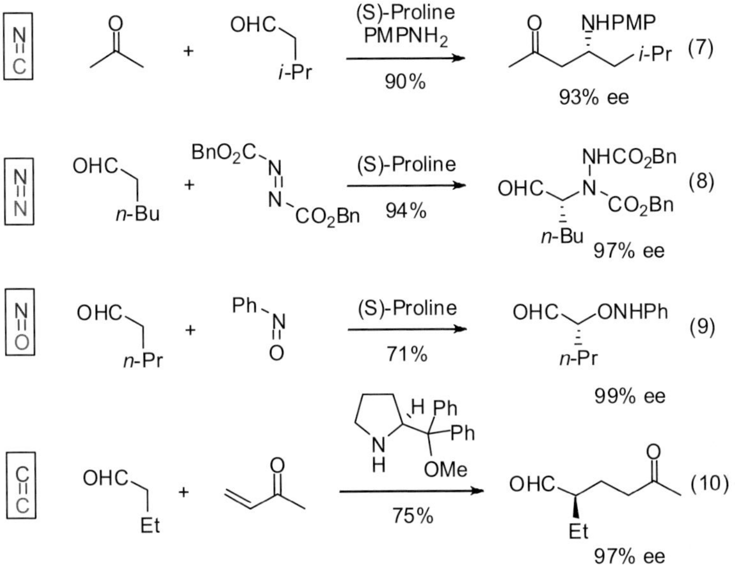

Scheme 6. Enamine catalysis of nucleophilic addition reactions

Hayashi et al. 2003b; Córdova 2003; Notz et al. 2003; Enders et al. 2005; Ibrahem et al. 2004), azodicarboxylates (N = N) (List 2002a; Kumaragurubaran et al. 2002; Bøgevig et al. 2002b), nitrosobenzene (O = N) (Brown et al. 2003a; Zhong 2003; Hayashi et al. 2003c; Hayashi et al. 2004; Bøgevig et al. 2004; Yamamoto and Momiyama 2005), and Michael acceptors (C = C) (Hechavarria Fonseca and List 2004; List et al. 2001; Halland et al. 2002; Peelen et al. 2005; Chi and Gellman 2005; Betancort and Barbas 2001; Alexakis and Andrey 2002; Wang et al. 2005; see Scheme 6; Eqs. 7–10 for selected examples).

Enamine catalysis often delivers valuable chiral compounds such as alcohols, amines, aldehydes, and ketones. Many of these are normally not accessible using established reactions based on transition metal catalysts or on preformed enolates or enamines, illustrating the complimentary nature of organocatalysis and metallocatalysis.

2.3 Enamine Catalysis of Nucleophilic Substitution Reactions

The first example of an asymmetric enamine catalytic nucleophilic substitution was a reaction that may have been considered impossible only a few years ago. We found that proline and certain derivatives such as α-methyl proline efficiently catalyze the asymmetric α-alkylation of aldehydes (Vignola and List 2004). Catalytic α-alkylation reactions of substrates other than glycine derivatives have been rare and that of aldehydes has been completely unknown before. In our process we could cyclize 6-halo aldehydes to give cyclopentane carbaldehydes in excellent yields and *ee*s (Scheme 7; Eq. 11). Other important and remarkably useful enamine catalytic nucleophilic substitution reactions have been developed subsequently and include enantioselective α-chlorinations (Brochu et al. 2004; Halland et al. 2004), α-fluorinations (Enders and Hüttl 2005; Marigo et al. 2005c; Steiner et al. 2005; Beeson and MacMillan 2005), α-brominations (Bertelsen et al. 2005), α-iodinations, and α-sulfenylations (Marigo et al. 2005a; Eqs. 12–16).

Once again, most of these reactions have never been realized before using preformed enamines or any other methodology but lead to highly valuable products of potential industrial relevance.

2.4 The Proline-Catalyzed Asymmetric Mannich Reactions

The catalytic asymmetric Mannich reaction is arguably the most useful approach to chiral β-amino carbonyl compounds. In the year 2000, we discovered a proline-catalyzed version of this powerful reaction (List 2000). Originally, ketones, aldehydes, and an aniline as the amine component were used in a catalytic asymmetric three-component reaction (Scheme 8; Eq. 17). After our report, proline-catalyzed Mannich reactions with aldehydes as the donor were also developed (Córdova et al. 2002c; Hayashi et al. 2003b; Eqs. 18,19). Despite its frequent use, both in an academic as well as an industrial context, the main limitation of the proline-catalyzed Mannich reaction has been the requirement to use anilines as the amine component. Although optically enriched *p*-anisidylamines are of potential utility in asymmetric synthesis, facile and efficient removal of the *N*-protecting group to yield the unfunctionalized amine is required. Generally, the removal of the most commonly

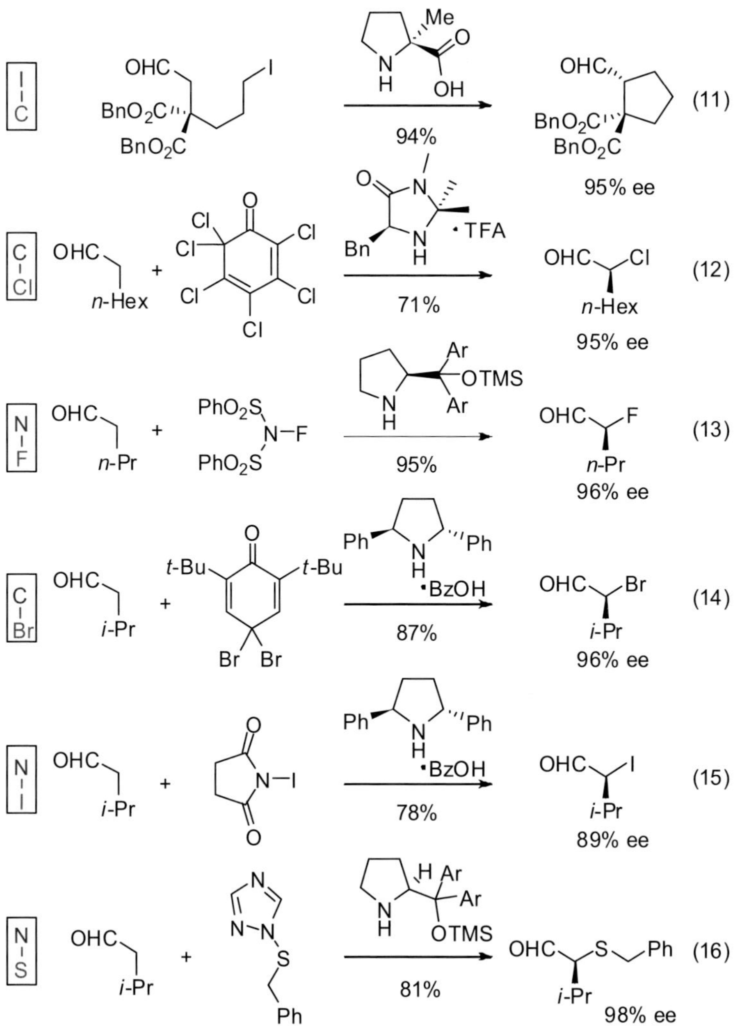

Scheme 7. Enamine catalysis of nucleophilic substitution reactions

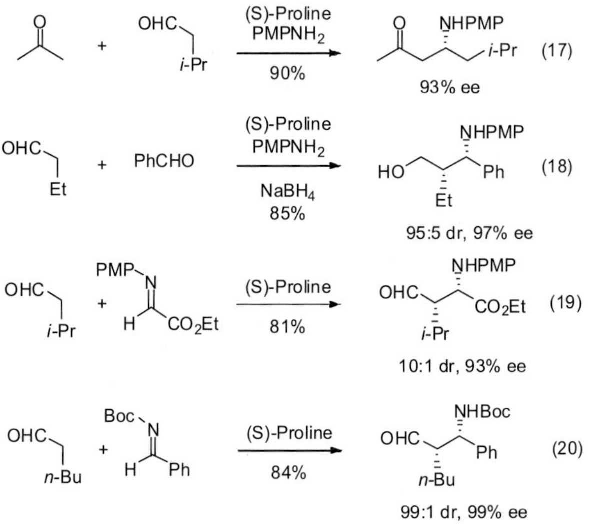

Scheme 8. Proline-catalyzed asymmetric Mannich reactions

used *p*-methoxyphenyl (PMP) group from nitrogen requires rather drastic oxidative conditions involving harmful reagents, such as ceric ammonium nitrate, that are not compatible with all substrates. We have now identified reaction conditions that allow for the use of simple preformed aromatic *N*-Boc-imines in proline-catalyzed Mannich reactions (Eq. 20). Remarkably, the reaction provides chiral β-amino aldehydes and ketones as stable, crystalline compounds in generally high diastereo- and enantioselectivities without the requirement for chromatographic purification [(Yang et al. 2007a,b); during our studies, Enders et al. also reported two examples of proline-catalyzed Mannich reaction between ketones and Boc-imines (Enders and Vrettou 2006; Enders et al. 2006)].

A typical experimental procedure is illustrated in Fig. 2. Mixing the 2-naphthaldehyde-derived *N*-Boc imine with isovaleraldehyde in the presence of (S)-proline (20 mol%) in acetonitrile at 0 °C resulted in an initially homogenous reaction mixture (Fig. 2a). After complete consumption of the starting material (10 h), a large amount of the de-

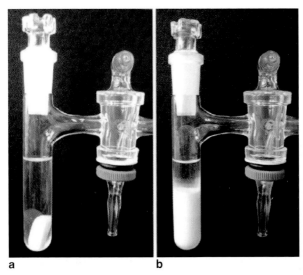

Fig. 2. The reaction of isovaleraldehyde with 2-naphthyl N-Boc-imine in the presence of (S)-proline (20 mol%) in CH_3CN. **a** Homogenous reaction mixture after mixing all components. **b** Reaction mixture after completion of the reaction (10 h)

sired product had precipitated and could easily be collected by filtration (Fig. 2b).

The N-Boc-imine-derived Mannich products can readily be converted into the corresponding α,β-branched-β-amino acids ($\beta^{2,3}$-amino acids). For example, oxidation of the product **1** to the carboxylic acid followed by acid-mediated deprotection provided the amino acid salt **2** without

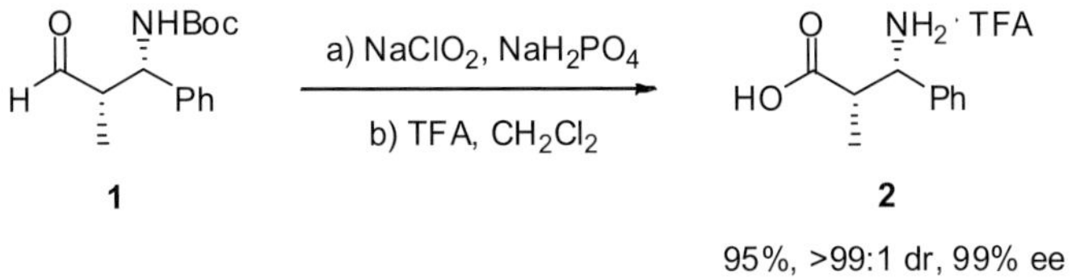

95%, >99:1 dr, 99% ee

Scheme 9. Conversion of the Mannich product **1** to β-amino acid **2**

loss of stereochemical integrity (Scheme 9, TFA, trifluoroacetic acid). Measuring NMR spectra and optical rotation of the corresponding HCl salt allowed us to confirm the expected absolute and relative configuration of the product.

3 Brønsted Acid Catalysis

In the proline-based enamine catalysis, proline actually plays a dual role. The amino-group of proline acts as Lewis base, whereas the carboxylic group acts as a Brønsted acid (Scheme 10).

The potential of using relatively strong chiral organic Brønsted acids as catalysts (Specific Brønsted acid catalysis) has been essentially ignored over the last decades. Achiral acids such as *p*-TsOH have been used as catalysts for a variety of reactions since a long time, but applications in asymmetric catalysis have been extremely rare. Only very recently, Akiyama et al. (2004, 2005a); also see Akiyama 2004; Akiyama et al. 2005b) and Terada et al. (Uraguchi and Terada 2004; Uraguchi

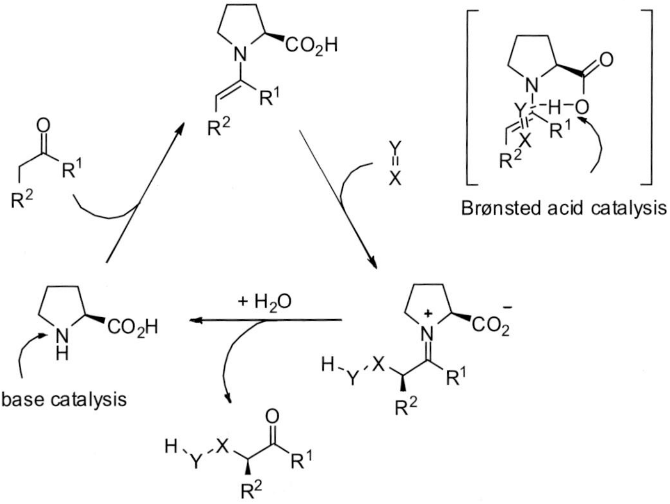

Scheme 10. Proline: a bifunctional catalyst

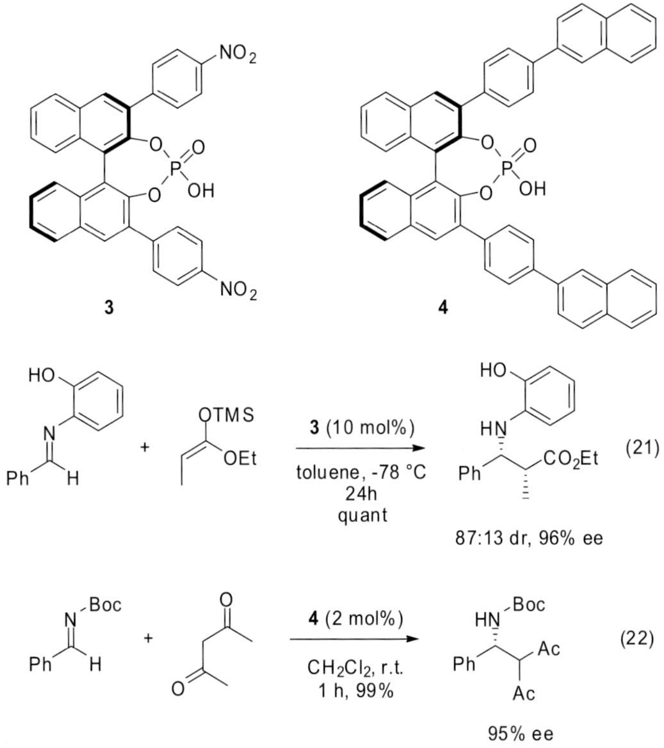

Scheme 11. Phosphoric acid catalysis pioneered by Akiyama and Terada

et al. 2004; also see Uraguchi et al. 2005; Terada et al. 2005) demonstrated, in pioneering studies, that relatively strong chiral binaphthol-derived phosphoric acids are efficient and highly enantioselective catalysts for addition reactions to aldimines (Scheme 11).

3.1 Catalytic Asymmetric Pictet–Spengler Reaction

The Pictet–Spengler reaction (Pictet and Spengler 1911; Tatsui 1928) is an important acid-catalyzed transformation frequently used in the laboratory as well as by various organisms for the synthesis of tetrahydro-βcarbolines and tetrahydroisoquinolines from carbonyl compounds and phenyl ethylamines or tryptamines, respectively (Eqs. 23 and 24).

$$\text{PhCH}_2\text{CH}_2\text{NH}_2 \xrightarrow[-\text{H}_2\text{O}]{\text{RCHO}} \text{tetrahydroisoquinoline} \quad (23)$$

$$\text{tryptamine} \xrightarrow[-\text{H}_2\text{O}]{\text{RCHO}} \text{tetrahydro-}\beta\text{-carboline} \quad (24)$$

Very recently, Taylor and Jacobsen (2004) reported the first truly catalytic approach by developing an elegant organocatalytic acyl-Pictet–Spengler reaction. The direct Pictet–Spengler reaction of aldehydes with aryl ethylamines, however, has been an illusive target for small molecule catalysis. As the reactions developed by Akiyama and Terada are assumed to involve chiral, hydrogen-bond assisted iminium–phosphate ion pairs we reasoned that the approach might be applicable to the Pictet–Spengler reaction, which also proceeds via iminium ion intermediates.

In line with observations by Jacobsen et al. attempts for acid catalysis of the Pictet–Spengler reaction of simple standard substrates such as unsubstituted tryptamines and phenylethyl amines proved unfruitful. For example, treating typtamine itself (**5**) with propionaldehyde and trifluoroacetic acid (TFA) gave only compound **6**, resulting from homo-aldol condensation and imine formation (Scheme 12; Eq. 25). We reasoned that one solution for this problem might be the use of more reactive substrates that are predisposed for cyclization. Specifically, we hypothesized that easily accessible geminally disubstituted tryptamines such as **7a** (Hagen et al. 1989) are promising substrates both for electronic reasons and by possibly favoring cyclization by virtue of a Thorpe–Ingold effect. Indeed, treatment of **7a** with TFA cleanly provided the desired product **8a** in high yield (Eq. 26). Encouraged by this result we set up a study towards identifying a chiral organic Brønsted acid as asymmetric catalyst. After screening several substituted chiral phosphoric acids we found catalyst **9** to give the highest enantioselectivity in the Pictet–Spengler reaction of tryptamine **7b** in the presence of Na_2SO_4 (Eq. 27;

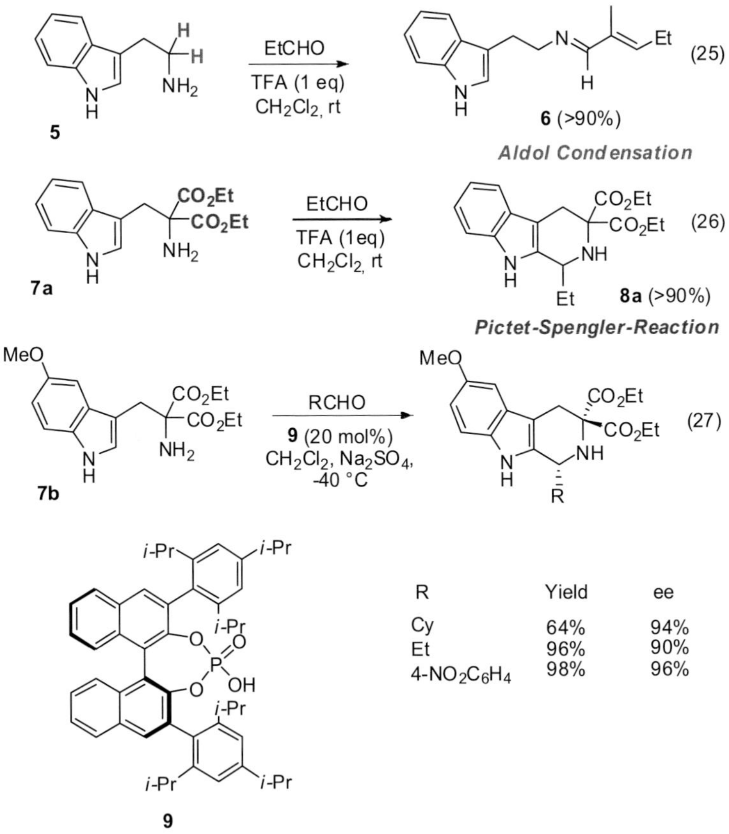

R	Yield	ee
Cy	64%	94%
Et	96%	90%
4-$NO_2C_6H_4$	98%	96%

Scheme 12. Brønsted acid-catalyzed Pictet–Spengler reaction

Seayad et al. 2006). Remarkably, the reaction tolerates a variety of different aldehydes with good results.

3.2 Organocatalytic Asymmetric Reductive Amination

The catalytic hydrogenation of alkenes, ketones, and imines is arguably one of the most important transformations in chemistry. Powerful asymmetric versions have been realized that require metal catalysts or the use of a stoichiometric amount of metal hydrides (Eq. 28; Noyori 2002; Knowles 2002).

$$R^1\underset{X=CR_2, O, NR}{\overset{X}{\underset{}{\|}}}R^2 \xrightarrow[\text{catalyst}]{+2\,[H]\atop\text{chiral}} R^1\overset{X\text{-}H}{\underset{}{\underset{*}{\|}}}R^2 \qquad (28)$$

Although effective and industrially relevant catalytic asymmetric hydrogenations and transfer hydrogenations of olefins and ketones have been developed, the corresponding imine reductions, although potentially highly useful for the synthesis of enantiomerically pure amines, are less advanced (for reviews, see: Taratov and Börner 2005; Ohkuma et al. 2000; Ohkuma and Noyori 2004; Nishiyama and Itoh 2000). Living organisms employ organic dihydropyridine cofactors such as nicotinamide adenine dinucleotide (NADH) in combination with enzyme catalysts for the reduction of imines (Sanwal and Zink 1961; Kula and Wandrey 1987).

reduced nicotinamide adenine dinucleotide (NADH)

Inspired by the recent observation that imines are reduced with Hantzsch esters in the presence of achiral Lewis or Brønsted acid catalysts (Itoh et al. 2004), we envisioned a catalytic cycle for the reductive amination of ketones which is initiated by protonation of the in situ generated ketimine **10** from a chiral Brønsted acid catalyst (Scheme 13). The resulting iminium ion pair, which may be stabilized by hydrogen bonding, is chiral and its reaction with the Hantzsch dihydropyridine **11** could give an enantiomerically enriched amine **12** and pyridine **13**.

After screening different phosphoric acid catalysts, catalyst **9** was found to be the best catalyst for this reaction. After optimizing the re-

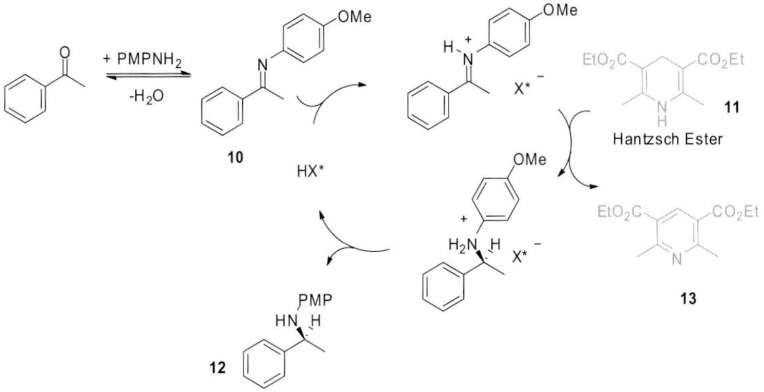

Scheme 13. Chiral Brønsted acid-catalyzed reductive amination

action conditions it was found that only 1 mol% of the catalyst is sufficient to give 93% *ee* of the product with an excellent yield of 96% (Scheme 14; Eq. 29; Hoffmann et al. 2005). A similar study by Rueping's research group using Akiyama's phosphoric acid catalyst **14** appeared during the preparation of our paper (Eq. 30; Rueping et al. 2005). MacMillan and co-workers also developed a reductive amination of different ketones using catalyst **15** (Eq. 31; Storer et al. 2006).

The previous examples are selected asymmetric reductive aminations of ketones to give chiral, α-branched amines (Eq. 32); however, the corresponding reactions of aldehydes are unknown. We reasoned that such a process might be realized if enolizable, α-branched aldehydes are used. Their asymmetric reductive amination should give β-branched amines via an enantiomer-differentiating kinetic resolution (Eq. 33).

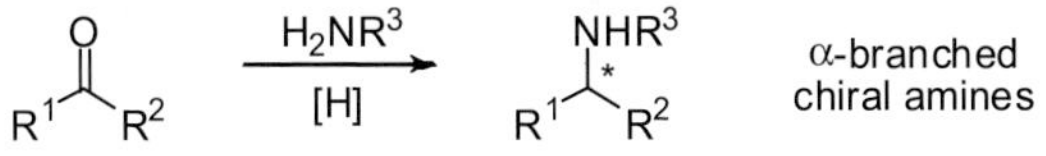

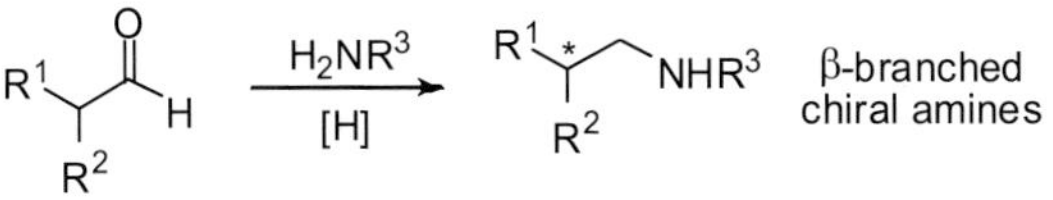

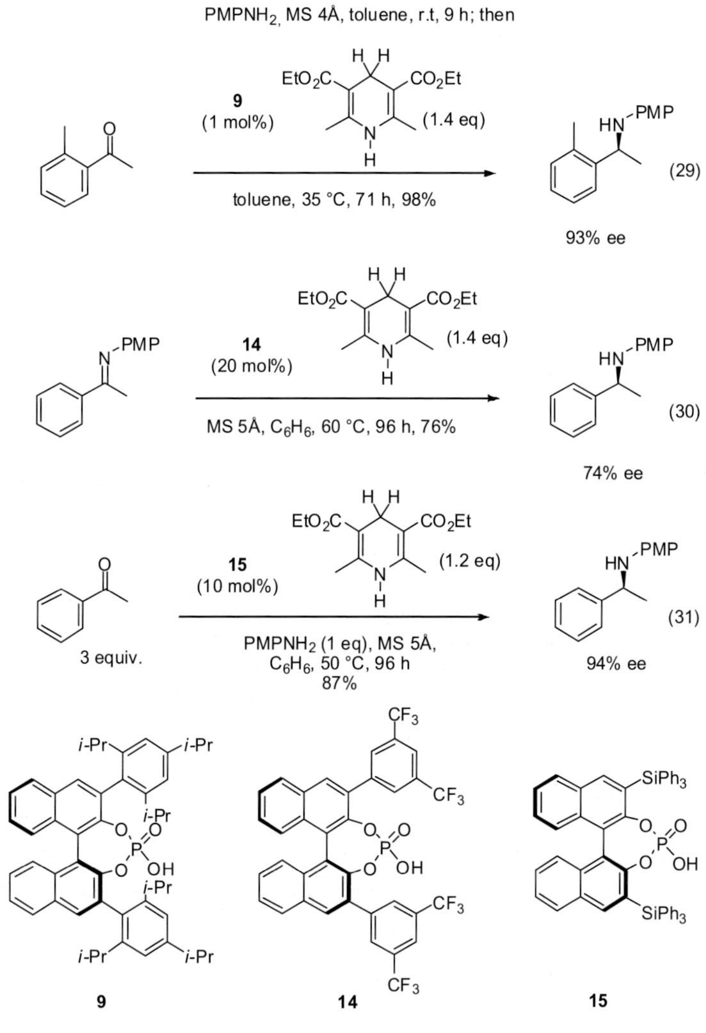

Scheme 14. Organocatalytic asymmetric reductive amination of ketones

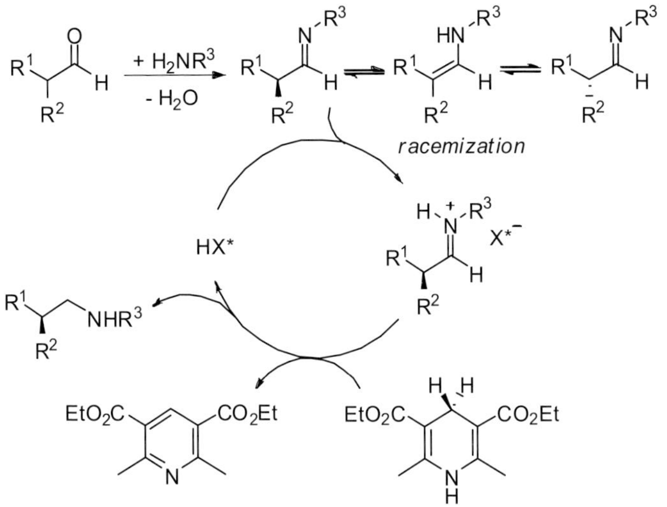

Scheme 15. Catalytic asymmetric reductive amination of aldehydes

At the onset of this study, we hypothesized that under our reductive amination conditions an α-branched aldehyde substrate would undergo a fast racemization in the presence of the amine and acid catalyst via an imine/enamine tautomerization. The reductive amination of one of the two imine enantiomers would then have to be faster than that of the other, resulting in an enantiomerically enriched product via a dynamic kinetic resolution (Scheme 15; for reviews, see: Noyori et al. 1995; Ward 1995; Caddick and Jenkins 1996; Stecher and Faber 1997; Huerta et al. 2001; Perllissier 2003).

Indeed, when we studied various phosphoric acid catalysts for the reductive amination of hydratopicaldehyde (**16**) with *p*-anisidine (PMPNH$_2$) in the presence of Hantzsch ester **11** to give amine **17**, the observed enantioselectivities and conversions are consistent with a facile in situ racemization of the substrate and a resulting dynamic kinetic resolution (Scheme 16). TRIP (**9**) once again turned out to be the most effective and enantioselective catalyst for this transformation and provided the chiral amine products with different α-branched aldehydes and amines in high enantioselectivities (Hoffmann et al. 2006).

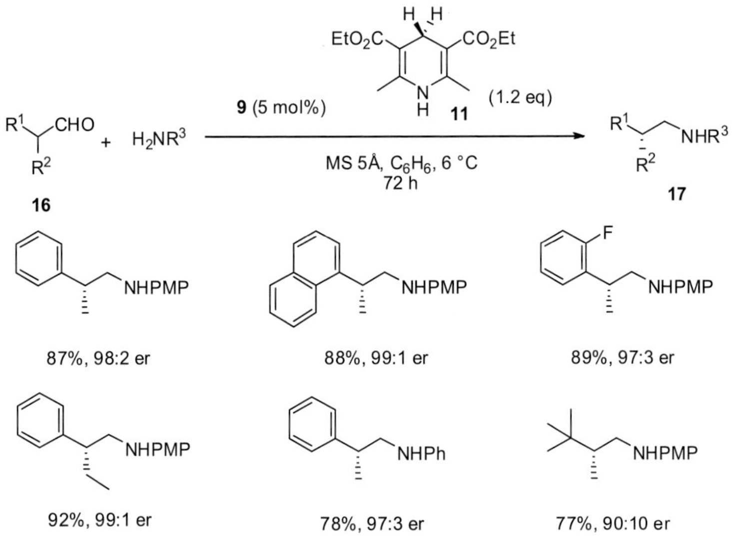

Scheme 16. Catalytic asymmetric reductive amination of aldehydes using catalyst **9**

We later developed an analogous enantioselective hydrogenation of aldehydes to the corresponding β-branched alcohols using [RuCl$_2$(xylyl-BINAP) (DPEN or DACH)] as the catalyst (Li and List 2007; for an independent study, see Xie et al. 2007).

4 Iminium Catalysis

The in situ generation of an iminium ion from a carbonyl compound lowers the LUMO energy of the system. Iminium catalysis is comparable to Brønsted- or Lewis acid activation of carbonyl compounds. The LUMO energy is lowered, the α-CH acidity increases, and nucleophilic additions including conjugate additions as well as pericyclic reactions are facilitated (Eq. 34).

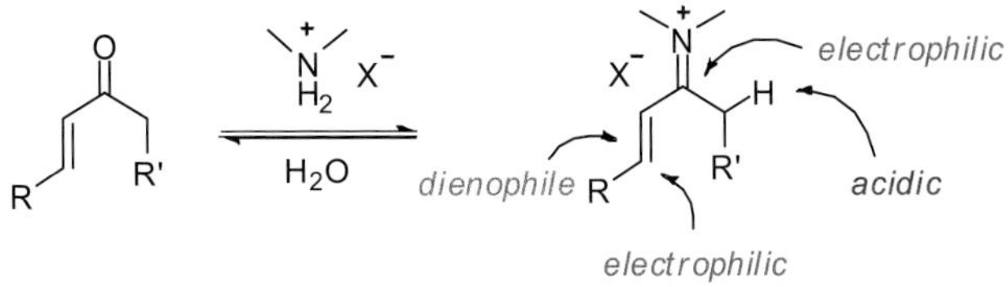

The first highly enantioselective examples of this catalysis strategy were reported by MacMillan et al. in 2000 (Ahrendt et al. 2000; also see: Wilson et al. 2005; Northrup and MacMillan 2002b), shortly after our first report on the proline-catalyzed intermolecular aldol reaction had appeared. The MacMillan group has quickly established that Diels–

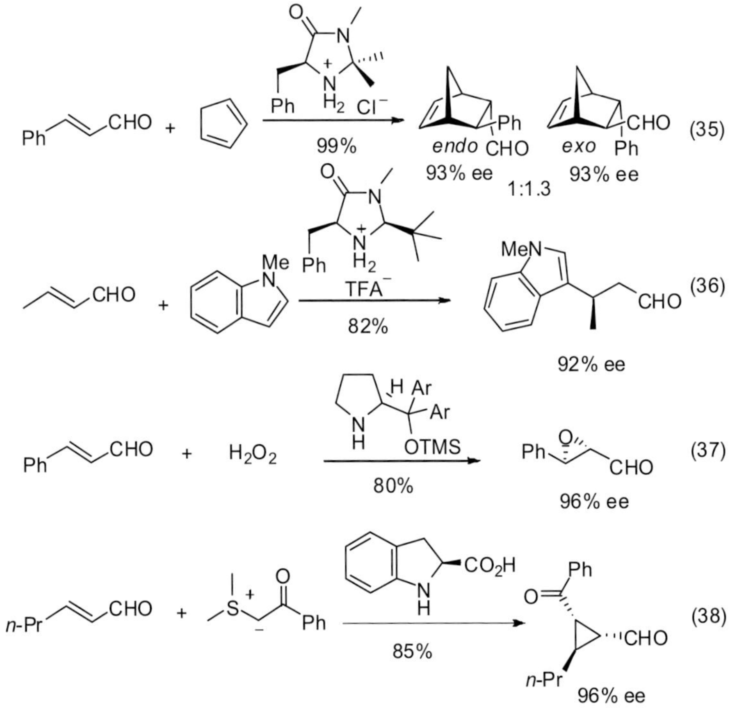

Scheme 17. Iminium catalytic asymmetric transformations

Alder reactions, 1,3-dipolar cycloadditions (Jen et al. 2000), and conjugate additions of electron rich aromatic and heteroaromatic compounds can be catalyzed using chiral amino acid derived imidazolidinones as catalysts (Scheme 17; Eqs. 35–38; Paras and MacMillan 2001, 2002; Austin and MacMillan 2002; Brown et al. 2003b). In addition, highly enantioselective epoxidations (Marigo et al. 2005b) and cyclopropanations (Kunz and MacMillan 2005) have recently been developed.

4.1 Organocatalytic Conjugate Reduction of α,β-Unsaturated Aldehydes

In 2001, we reasoned that this catalysis strategy might be applicable to the conjugate reduction of α,β-unsaturated carbonyl compounds if a suitable hydride donor could be identified (Scheme 18). Hantzsch ester **11** seemed to be particularly promising since its reaction with preformed α,β-unsaturated iminium ions had already been established (Makino et al. 1977; Baba et al. 1980).

This process was published in 2004 and constitutes the first metal-free organocatalytic transfer hydrogenation of α,β-unsaturated aldehy-

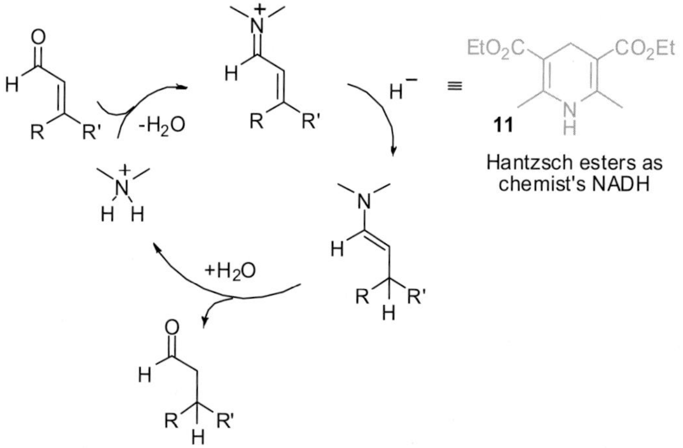

Scheme 18. Iminium catalytic transfer hydrogenation of α,β-unsaturated aldehydes

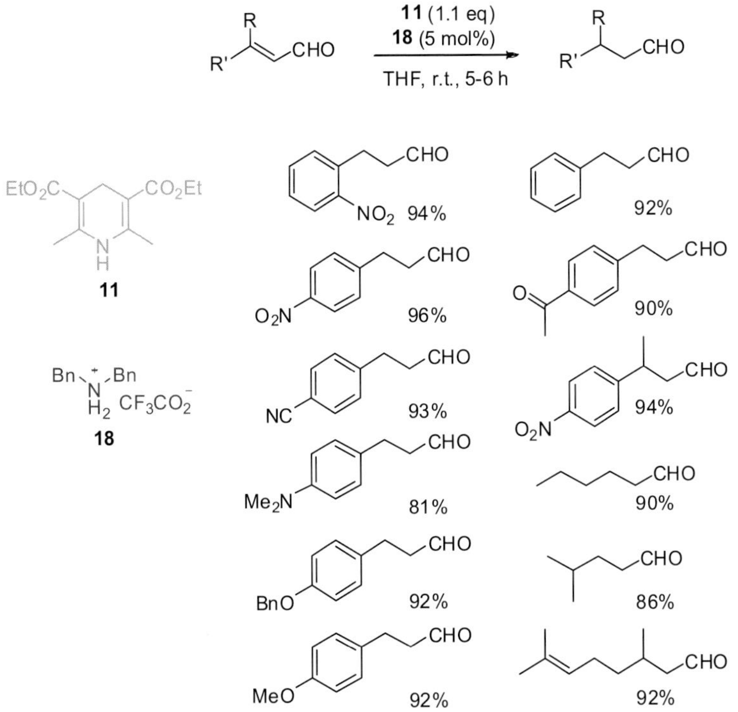

New Concepts for Organocatalysis

Scheme 19. Organocatalytic transfer hydrogenation of different α,β-unsaturated aldehydes

des (Yang et al. 2004). Dibenzylammonium trifluoroacetate (**18**), was found to be an efficient catalyst for this reaction. The reduction worked extremely well with a diverse set of unsaturated aldehydes, including substituted aromatic and aliphatic ones and the yields exceed 90% in almost all cases (Scheme 19). A variety of functional groups that are sensitive to standard hydrogenation condition (nitro, nitrile, benzyloxy, and alkene functional groups) were tolerated in the process.

An example of an asymmetric catalytic version was also presented in our first publication. This protocol was subsequently optimized and we developed a highly enantioselective variant using the trichloroacetate salt of MacMillan's second generation imidazolidinone (**19**) as the

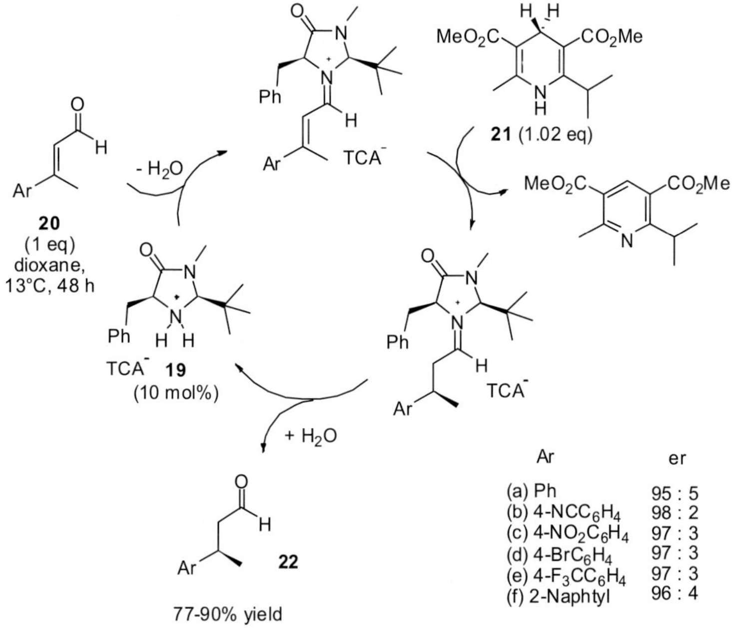

Scheme 20. Organocatalytic asymmetric transfer hydrogenation of different α,β-unsaturated aldehydes

catalyst (Yang et al. 2005; for an independent study on the same reaction, see Ouelett et al. 2005). We found that upon treating aromatic, trisubstituted α,β-unsaturated aldehydes **20** with a slight excess of dihydropyridine **21** and a catalytic amount of **19** at 13 °C in dioxane, the corresponding saturated aldehydes **22** were obtained in high yields and enantioselectivities (Scheme 20).

5 Asymmetric Counteranion Directed Catalysis

Most chemical reactions proceed via charged intermediates or transition states. In asymmetric Brønsted acid catalysis the substrate is protonated by the catalyst and a chiral H-bond-assisted ion pair is generated. We reasoned that, in principle, any reaction that proceeds via cationic in-

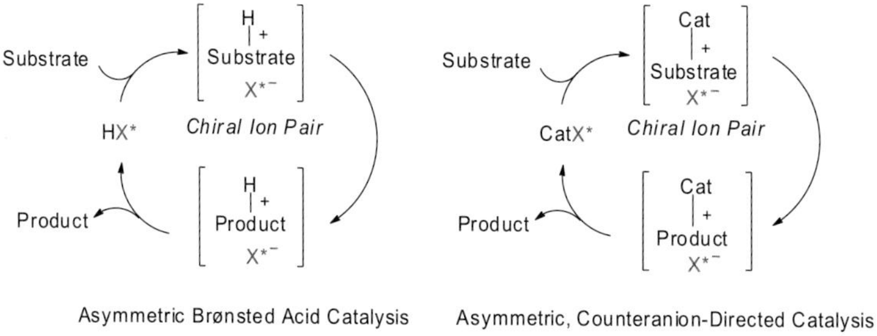

New Concepts for Organocatalysis 27

Scheme 21. Asymmetric counteranion-directed catalysis

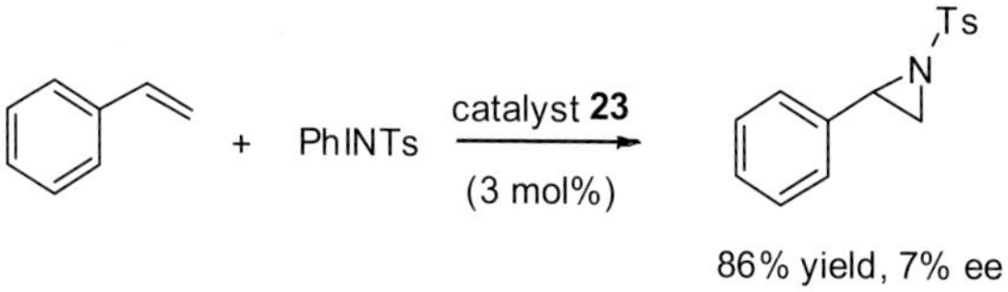

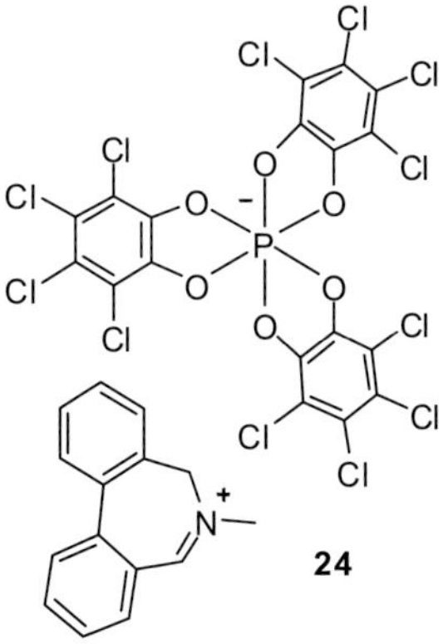

(39)

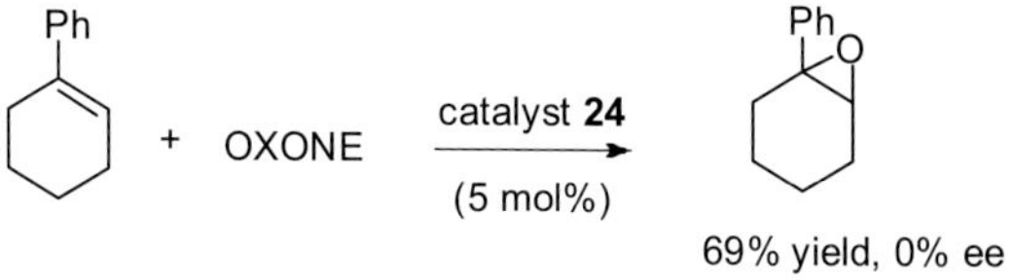

86% yield, 7% ee

(40)

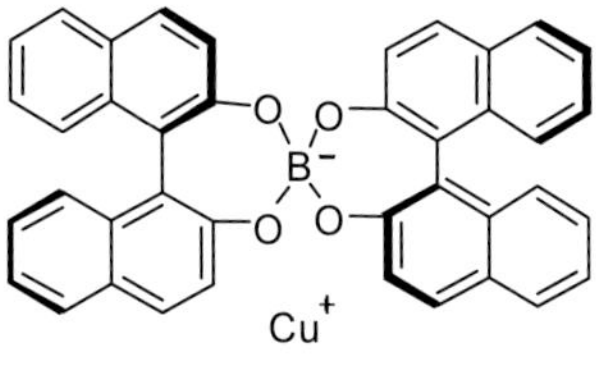

69% yield, 0% ee

Scheme 22. Previous attempts for asymmetric counteranion-directed catalysis

termediates can be conducted highly enantioselectively if a chiral counteranion is introduced into the catalyst, as a result of the generation of a chiral ion pair. We termed this new strategy as asymmetric counteranion-directed catalysis (ACDC) (Scheme 21).

Although efficient asymmetric catalytic transformations involving anionic intermediates with chiral, cationic catalysts have been realized (for several reviews, see Houk and List 2004), analogous versions of inverted polarity with reasonable enantioselectivity, despite attempts, have been illusive (Scheme 22; Eqs. 39–40; for a review, see Lacour and Hebbe-Viton 2003; see also: Llewellyn and Arndtsen 2005; Dorta et al. 2004; Carter et al. 2003).

5.1 Asymmetric Counteranion-Directed Catalysis: Application to Iminium Catalysis

In iminium catalysis, both we and the group of MacMillan had observed a strong counteranion effect on the yield and enantioselectivity of the reactions. Inspired by recent use of chiral phosphoric acid derivatives

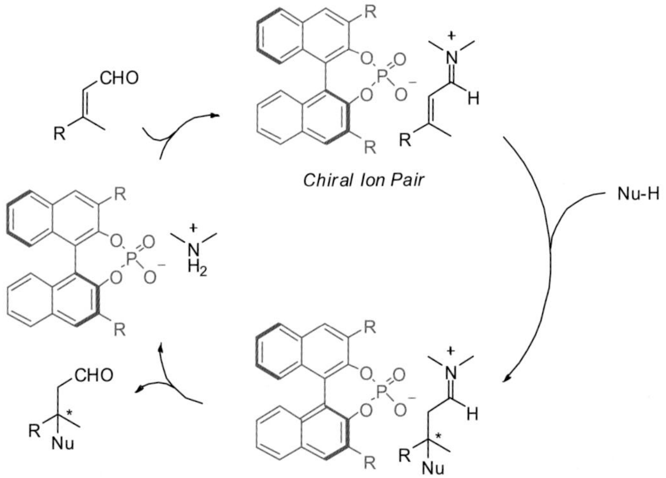

Scheme 23. Asymmetric counteranion-directed catalysis: application to iminium catalysis

New Concepts for Organocatalysis

as asymmetric catalysts, we hypothesized that catalytic salts of achiral amines and chiral phosphoric acids could induce asymmetry in these processes (Scheme 23).

We thought to start with the metal-free biomimetic transfer hydrogenation of α,β-unsaturated aldehydes as a model reaction which has been earlier discovered in our laboratory and independently in that of MacMillan et al. (Scheme 24). We have prepared a large number of ammonium salts as crystalline solids by mixing different primary and secondary amines with a chiral phosphoric acid. In particular, the ammonium salts of sterically hindered chiral phosphoric acids could catalyze

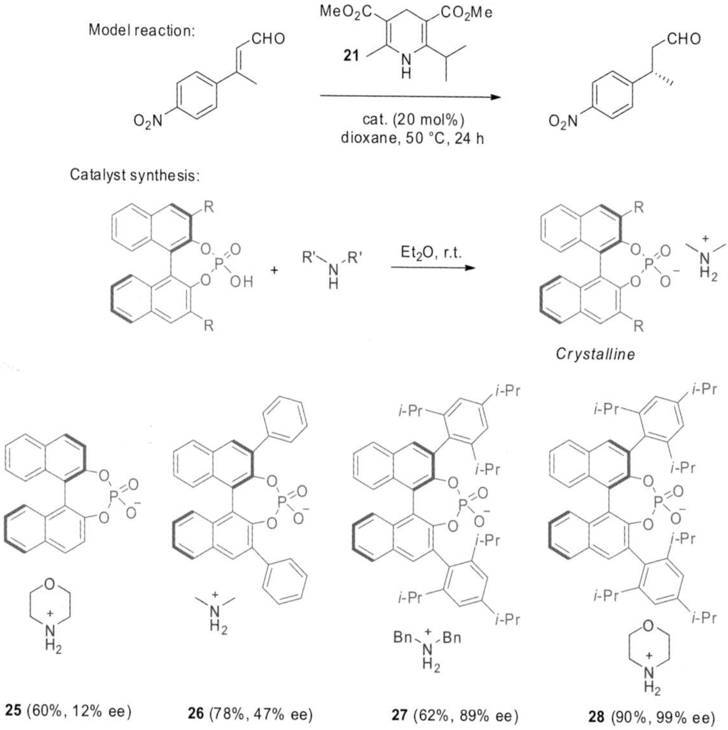

Scheme 24. Asymmetric counteranion-directed catalysis: catalyst synthesis and screening

the reaction with significant enantiomeric excess (*ee*) values (Scheme 24). After a thorough screening of various amines we identified morpholine salt **28** as a highly enantioselective catalyst (Mayer and List 2006).

Treating aromatic, trisubstituted α,β-unsaturated aldehydes **20** with a slight excess of dihydropyridine **21** and a catalytic amount of salt **28** at 50 °C in dioxane for 24 h, the corresponding saturated aldehydes **22** were obtained in high yields and in enantioselectivities of 96–99% *ee* (Scheme 25).

Significantly, the previously developed chiral amine based catalysts that we and MacMillan and co-workers have studied have not been of use for sterically nonhindered aliphatic substrates. For example, citral (**29**), of which the hydrogenation product citronellal (**30**) is an intermediate in the industrial synthesis of menthol and used as a perfume ingredient, could not readily be used (Scheme 26, Eq. 41). We could

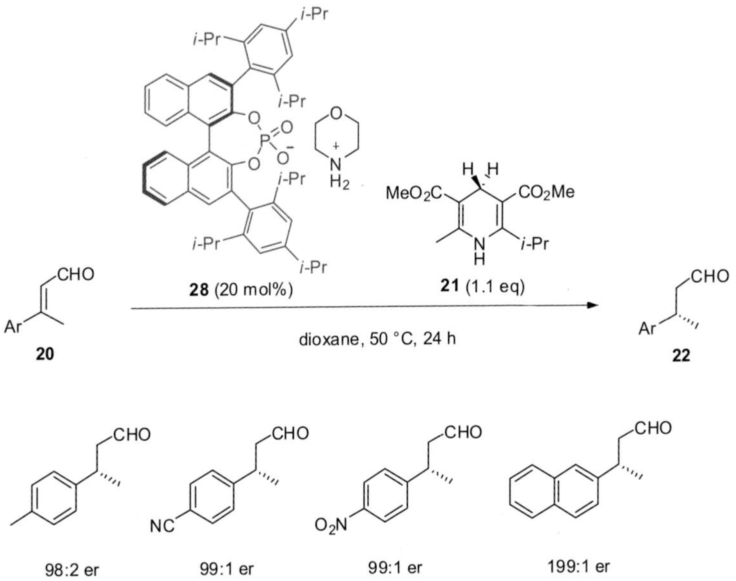

Scheme 25. Asymmetric counteranion-directed catalysis: transfer hydrogenation of enals

New Concepts for Organocatalysis

not achieve high enantioselectivity for this particular substrate with either our previous system (Yang et al. 2005) or with that of MacMillan and coworkers (Ouellet et al. 2005). However, with our novel chiral counteranion catalyst **28**, citral is converted into (R)-citronellal (**30**) with an e.r. value of 95:5. This is the highest enantioselectivity so far reported for a catalytic asymmetric (transfer) hydrogenation of citral (Rhone-Poulenc Industries 1979; Dang et al. 1982; Kortvelyessy 1985). Similarly, farnesal (**31**) gave (R)-dihydrofarnesal (**32**) in 77% yield and 96:4 e.r. (Scheme 26; Eq. 42).

We then thought to carry out conjugate reduction of α,β-unsaturated ketones. However, neither these ACDC-catalysts, nor the commonly

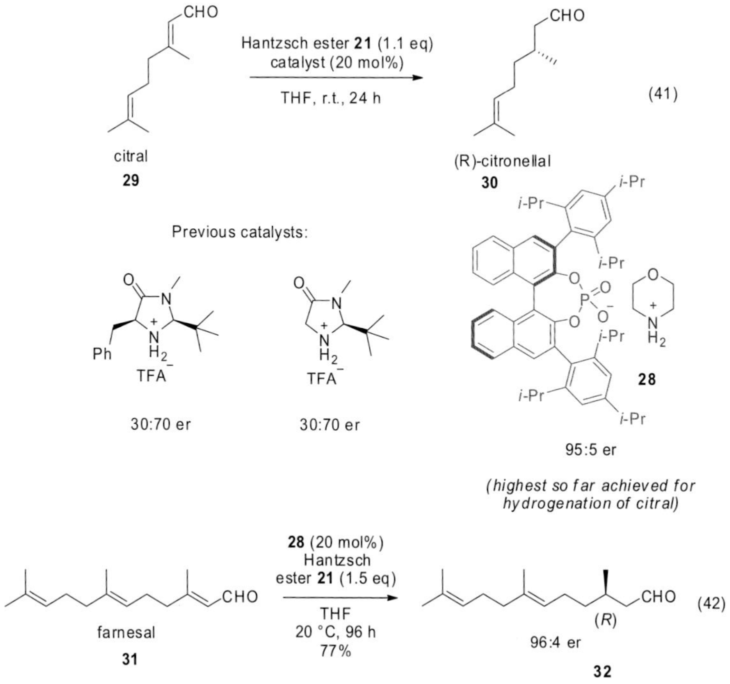

Scheme 26. Asymmetric counteranion-directed catalysis: transfer hydrogenation of citral and farnesal

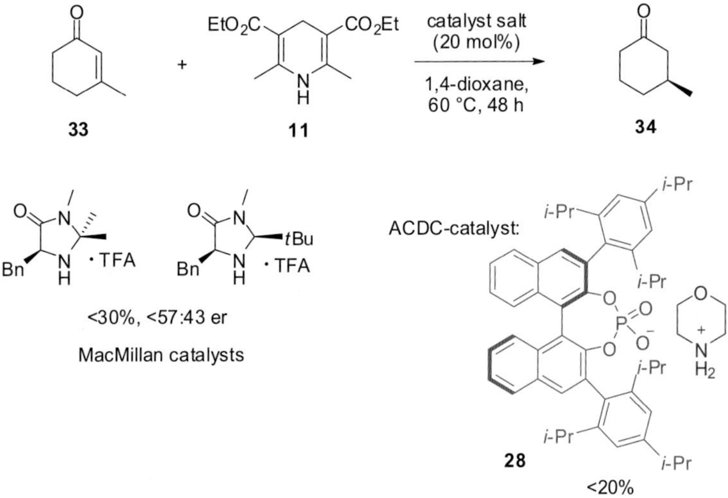

Scheme 27. Asymmetric counteranion-directed catalysis: transfer hydrogenation of enone

used chiral imidazolidinone catalysts gave satisfying yields or enantioselectivities in the conjugate reduction of 3-methyl cyclohexenone **33** to the product **34** (Scheme 27).

Hypothesizing that primary amine catalysts, due to their reduced steric requirements, might be suitable for the activation of ketones, we studied various salts of α-amino acid esters. (For pioneering use of primary amine salts in asymmetric iminium catalysis involving aldehyde substrates, see Ishihara and Nakano 2005; Sakakura et al. 2006; for the use of preformed imines of α,β-unsaturated aldehydes and amino acid esters in diastereoselective Michael additions, see Hashimot et al. 1977.) We have developed a new class of catalytic salts, in which both the cation and the anion are chiral. In particular, valine ester phosphate salt **35** proved to be an active catalyst for the transfer hydrogenation of a variety of α,β-unsaturated ketones **36** with commercially available Hantzsch ester **11** to give saturated ketones **37** in excellent enantioselectivities (Scheme 28; Martin and List 2006).

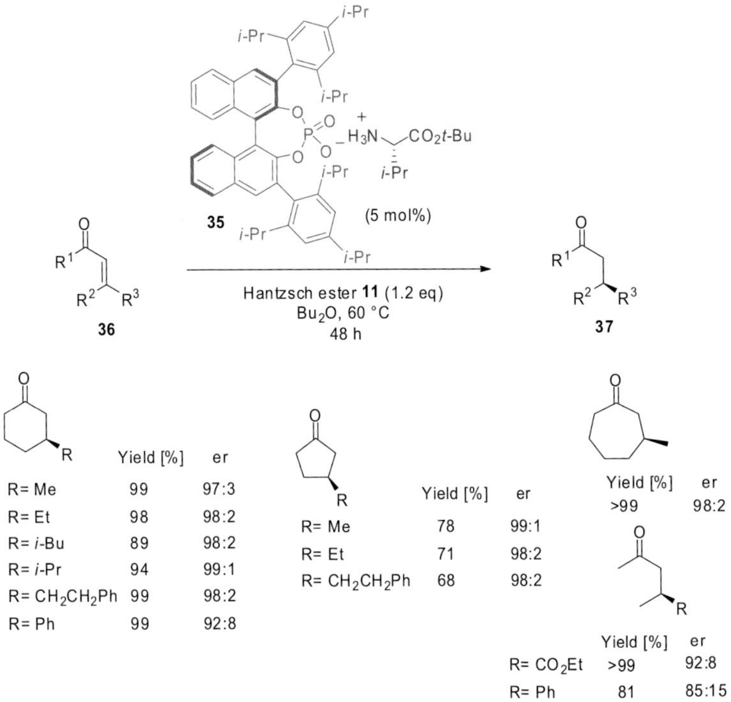

Scheme 28. Asymmetric counteranion-directed catalysis: transfer hydrogenation of different enones with catalyst **35**

Independently, MacMillan et al. developed an efficient catalyst system that is based on a chiral secondary amine (Tuttle et al. 2006).

6 Conclusions

Selected recent developments in the area of asymmetric organocatalysis in our laboratory have been briefly summarized. Enamine catalysis, Brønsted acid catalysis, and iminium catalysis turn out to be powerful new strategies for organic synthesis. Using Hantzsch ester as the hydride source, highly enantioselective transfer hydrogenantion reactions have been developed. We have also developed an additional new concept in asymmetric catalysis namely ACDC and successfully applied it to asymmetric iminium catalysis. Asymmetric induction presumably occurs in the cationic iminium ion transition state of the reaction by virtue of a sterochemical communication with the chiral phosphate counteranion, possibly via hydrogen bonding interaction. Our discovery may be of general applicability to other reactions that proceed via cationic intermediates. Despite its long roots, asymmetric organocatalysis is a relatively new and explosively growing field that, without doubt, will continue to yield amazing results for some time to come.

Acknowledgements. The present and past co-workers in my laboratory, whose names are given in the list of references, are highly acknowledged for their hard work, skill and enthusiasm. I thank the National Institute of Health for funding my work at Scripps. Generous support by the Max-Planck-Society and by Novartis (Young Investigator Award to BL) is gratefully acknowledged. I also thank the DFG (Priority Program Organocatalysis SPP1179), Degussa, Wacker, Merck, Saltigo, Sanofi-Aventis and BASF for general support and donating chemicals. I also thank Professor Kendall. N. Houk and Professor Walter Thiel for fruitful collaborations.

References

Agami C (1988) Mechanism of the proline-catalyzed enantioselective aldol reaction. Recent advances. Bull Soc Chim Fr 3:499–507

Agami C, Meynier F, Puchot C, Guilhem J, Pascard C (1984) New insights into the mechanism of the proline-catalyzed asymmetric Robinson cyclization; structure of two intermediates. Asymmetric dehydration. Tetrahedron 40:1031–1038

Agami C, Puchot C (1986) Kinetic analysis of dual catalysis by proline in an asymmetric intramolecular aldol reaction. J Mol Catal 38:341–343

Agami C, Puchot C, Sevestre H (1986) Is the mechanism of the proline-catalyzed enantioselective aldol reaction related to biochemical processes? Tetrahedron Lett 27:1501–1504

Ahrendt KA, Borths CJ, MacMillan DWC (2000) New strategies for organic catalysis: The first highly enantioselective organocatalytic Diels–Alder reaction. J Am Chem Soc 122:4243–4244

Akiyama T (2004) Preparation of chiral Bronsted catalysts in asym. synthesis and asym. Mannich, aza-Diels–Alder reaction, hydrophosphorylation therewith. PCT Int Appl WO 200409675, 2004-11-11

Akiyama T, Itoh J, Yokota K, Fuchibe K (2004) Enantioselective Mannich-type reaction catalyzed by a chiral Brønsted acid. Angew Chem Int Ed Engl 43:1566–1568

Akiyama T, Morita H, Itoh J, Fuchibe K (2005a) Chiral Brønsted acid catalyzed enantioselective hydrophosphonylation of imines: asymmetric synthesis of alpha-amino phosphonates. Org Lett 7:2583–2585

Akiyama T, Saitoh Y, Morita H, Fuchibe K (2005b) Enantioselective Mannich-type reaction catalyzed by a chiral Bronsted acid derived from TADDOL. Adv Synth Catal 347:1523–1526

Alexakis A, Andrey O (2002) Diamine-catalyzed asymmetric Michael additions of aldehydes and ketones to nitrostyrene. Org Lett 4:3611–3614

Allemann C, Gordillo R, Clemente FR, Cheong PH, Houk KN (2004) Theory of asymmetric organocatalysis of Aldol and related reactions: rationalizations and predictions. Acc Chem Res 37:558–569

Austin JF, MacMillan DWC (2002) Enantioselective organocatalytic indole alkylations. Design of a new and highly effective chiral amine for iminium catalysis. J Am Chem Soc 124:1172–1173

Baba N, Makino T, Oda J, Inouye Y (1980) Asymmetric reduction of alpha, beta-unsaturated iminium salts with 1,4-dehydronicotinamide sugar pyranosides. Can J Chem 58:387–392

Bahmanyar S, Houk KN (2001a) The origin of stereoselectivity in proline-catalyzed intramolecular aldol reactions. J Am Chem Soc 123:12911–12912

Bahmanyar S, Houk KN (2001b) Transition states of amine-catalyzed aldol reactions involving enamine intermediates: theoretical studies of mechanism, reactivity, and stereoselectivity. J Am Chem Soc 123:11273–11283

Bahmanyar S, Houk KN, Martin HJ, List B (2003) Quantum mechanical predictions of the stereoselectivities of proline-catalyzed asymmetric intermolecular aldol reactions. J Am Chem Soc 125:2475–2479

Barbas CF 3rd, Heine A, Zhong G, Hoffmann T, Gramatikova S, Bjoernstedt R, List B, Anderson J, Stura EA, Wilson I, Lerner RA (1997) Immune versus natural selection: antibody aldolases with enzymic rates but broader scope. Science 278:2085–2092

Beeson TD, MacMillan DWC (2005) Enantioselective organocatalytic alpha-fluorination of aldehydes. J Am Chem Soc 127:8826–8828

Berkessel A, Gröger H (2005) Asymmetric organocatalysis. Wiley-VCH, Weinheim

Bertelsen S, Halland N, Bachmann S, Marigo M, Braunton A, Jørgensen KA (2005) Organocatalytic asymmetric alpha-bromination of aldehydes and ketones. Chem Commun (Camb) 14:4821–4823

Betancort JM, Barbas CF 3rd (2001) Catalytic direct asymmetric Michael reactions: taming naked aldehyde donors. Org Lett 3:3737–3740

Bøgevig A, Juhl K, Kumaragurubaran N, Zhuang W, Jørgensen KA (2002b) Direct organo-catalytic asymmetric α-amination of aldehydes—a simple approach to optically active α-amino aldehydes, α-amino alcohols, and α-amino acids. Angew Chem Int Ed Engl 41:1790–1793

Bøgevig A, Kumaragurubaran N, Jørgensen KA (2002a) Direct catalytic asymmetric aldol reactions of aldehydes. Chem Commun (Camb) Mar 21:620–621

Bøgevig A, Poulsen TB, Zhuang W, Jørgensen KA (2003) Formation of optically active functionalized β-hydroxy nitrones using a proline catalyzed aldol reaction of aldehydes with carbonyl compounds and hydroxylamines. Synlett 2003:1915–1918

Bøgevig A, Sundéen H, Córdova A (2004) Direct catalytic enantioselective alpha-aminoxylation of ketones: a stereoselective synthesis of alpha-hydroxy and alpha,alpha'-dihydroxy ketones. Angew Chem Int Ed Engl 43:1109–1112

Brochu MP, Brown SP, MacMillan DWC (2004) Direct and enantioselective organocatalytic alpha-chlorination of aldehydes. J Am Chem Soc 126:4108–4109

Brown SP, Brochu MP, Sinz CJ, MacMillan DWC (2003a) The direct and enantioselective organocatalytic alpha-oxidation of aldehydes. J Am Chem Soc 125:10808–10809

Brown SP, Goodwin NC, MacMillan DWC (2003b) The first enantioselective organocatalytic Mukaiyama-Michael reaction: a direct method for the synthesis of enantioenriched gamma-butenolide architecture. J Am Chem Soc 125:1192–1194

Caddick S, Jenkins K (1996) Dynamic resolutions in asymmetric synthesis. Chem Soc Rev 25:447–456

Carter C, Fletcher S, Nelson A (2003) Towards phase-transfer catalysts with a chiral anion: inducing asymmetry in the reactions of cations. Tetrahedron Asymmetry 14:1995–2004

Cheong PH, Houk KN (2004) Origins of selectivities in proline-catalyzed alpha-aminoxylations. J Am Chem Soc 126:13912–13913

Chi Y, Gellman SH (2005) Diphenylprolinol methyl ether: a highly enantioselective catalyst for Michael addition of aldehydes to simple enones. Org Lett 7:4253–4256

Chowdari NS, Ramachary DB, Córdova A, Barbas CF 3rd (2002) Proline-catalyzed asymmetric assembly reactions: enzyme-like assembly of carbohydrates and polyketides from three aldehyde substrates. Tetrahedron Lett 43:9591–9595

Clemente FR, Houk KN (2004) Computational evidence for the enamine mechanism of intramolecular aldol reactions catalyzed by proline. Angew Chem Int Ed Engl 43:5766–5768

Córdova A (2003) One-pot organocatalytic direct asymmetric synthesis of γ-amino alcohol derivatives. Synlett 2004:1651–1654

Córdova A, Notz W, Barbas CF 3rd (2002a) Proline-catalyzed one-step asymmetric synthesis of 5-hydroxy-(2E)-hexenal from acetaldehyde. J Org Chem 67:301–303

Córdova A, Notz W, Barbas CF 3rd (2002b) Direct organocatalytic aldol reactions in buffered aqueous media. Chem Commun (Camb) Dec 21:3024–3025

Córdova A, Watanabe S, Tanaka F, Notz W, Barbas CF 3rd (2002c) A highly enantioselective route to either enantiomer of both alpha- and beta-amino acid derivatives. J Am Chem Soc 124:1866–1867

Dang TP, Aviron-Violet P, Colleuille Y, Varagnat J (1982) Catalysis of the homogeneous-phase hydrogenation of α,β-unsaturated aldehydes. Application to the asymmetric synthesis of citronellal. J Mol Catal 16:51–59

Dorta R, Shimon L, Milstein D (2004) Rhodium complexes with chiral counterions: achiral catalysts in chiral matrices. J Organomet Chem 689:751–758

Eder U, Sauer G, Wiechert R (1971) New type of asymmetric cyclization to optically active steroid CD partial structures. Angew Chem Int Ed Engl 10:496–497

Enders D, Grondal C, Vrettou M, Raabe G (2005) Asymmetric synthesis of selectively protected amino sugars and derivatives by a direct organocatalytic Mannich reaction. Angew Chem Int Ed Engl 44:4079–4083

Enders D, Grondal C, Vrettou M (2006) Efficient entry to amino sugars and derivatives via asymmetric organocatalytic Mannich reactions. Synthesis 2006:3597–3604

Enders D, Hüttl MRM (2005) Direct organocatalytic α-fluorination of aldehydes and ketones. Synlett 2005:991–993

Enders D, Seki A (2002) Proline-catalyzed enantioselective Michael additions of ketones to nitrostyrene. Synlett 2002:26–28

Enders D, Vrettou M (2006) Asymmetric synthesis of (+)-polyoxamic acid via an efficient organocatalytic Mannich reaction as the key step. Synthesis 13:2155–2158

Hagen TG, Narayanan K, Names J, Cook JM (1989) DDQ oxidations in the indole area. Synthesis of 4-alkoxy-beta-carbolines including the natural products crenatine and 1-methoxycanthin-6-one. J Org Chem 54:2170–2178

Hajos ZG, Parrish DR (1974) Asymmetric synthesis of bicyclic intermediates of natural product chemistry. J Org Chem 39:1615–1621

Halland N, Braunton A, Bachmann S, Marigo M, Jørgensen KA (2004) Direct organocatalytic asymmetric alpha-chlorination of aldehydes. J Am Chem Soc 126:4790–4791

Halland N, Hazell RG, Jørgensen KA (2002) Organocatalytic asymmetric conjugate addition of nitroalkanes to alpha,beta-unsaturated enones using novel imidazoline catalysts. J Org Chem 67:8331–8338

Hashimot S, Komeshima N, Yamada S, Koga K (1977) Asymmetric Michael reaction via chiral α,β-unsaturated aldimines. Tetrahedron Lett 18:2907–2908

Hayashi Y, Tsuboi W, Ashimine I, Urushima T, Shoji M, Sakai K (2003b) The direct and enantioselective, one-pot, three-component, cross-mannich reaction of aldehydes. Angew Chem Int Ed Engl 42:3677–3680

Hayashi Y, Tsuboi W, Shoji M, Suzuki N (2003a) Application of high pressure induced by water-freezing to the direct catalytic asymmetric three-component List-Barbas-Mannich reaction. J Am Chem Soc 125:11208–11209

Hayashi Y, Yamaguchi J, Hibino K, Shoji M (2003c) Direct proline catalyzed asymmetric α-aminooxylation of aldehydes. Tetrahedron Lett 44:8293–8296

Hayashi Y, Yamaguchi J, Sumiya T, Shoji M (2004) Direct proline-catalyzed asymmetric alpha-aminoxylation of ketones. Angew Chem Int Ed Engl 43:1112–1115

Hechavarria Fonseca MT, List B (2004) Catalytic asymmetric intramolecular Michael reaction of aldehydes. Angew Chem Int Ed Engl 43:3958–3960

Hoang L, Bahmanyar S, Houk KN, List B (2003) Kinetic and stereochemical evidence for the involvement of only one proline molecule in the transition states of proline-catalyzed intra- and intermolecular aldol reactions. J Am Chem Soc 125:16–17

Hoffmann S, Nicoletti M, List B (2006) Catalytic asymmetric reductive amination of aldehydes via dynamic kinetic resolution. J Am Chem Soc 128:13074–13075

Hoffmann S, Seayad AM, List B (2005) A powerful Brønsted acid catalyst for the organocatalytic asymmetric transfer hydrogenation of imines. Angew Chem Int Ed Engl 44:7424–7427

Houk KN, List B (2004) Guest editorial: asymmetric organocatalysis. Acc Chem Res 37:487

Huerta FF, Minidis ABE, Bäckvall JE (2001) Racemisation in asymmetric synthesis. Dynamic kinetic resolution and related processes in enzyme and metal catalysis. Chem Soc Rev 30:321–331

Ibrahem I, Casas J, Córdova A (2004) Direct catalytic enantioselective alpha-aminomethylation of ketones. Angew Chem Int Ed Engl 43:6528–6531

Ishihara K, Nakano K (2005) Design of an organocatalyst for the enantioselective Diels–Alder reaction with alpha-acyloxyacroleins. J Am Chem Soc 127:10504–10505

Itoh T, Nagata K, Miyazaki M, Ishikawa H, Kurihara A, Ohsawa A (2004) A selective reductive amination of aldehydes by the use of Hantzsch dihydropyridines as reductant. Tetrahedron 60:6649–6655

Jen WS, Wiener JJM, MacMillan DWC (2000) New strategies for organic catalysis: The first enantioselective organocatalytic 1,3-dipolar cycloaddition. J Am Chem Soc 122:9874–9875

Klussmann M, Iwamura H, Mathew SP, Wells DH Jr, Pandya U, Armstrong A, Blackmond DG (2006) Thermodynamic control of asymmetric amplification in amino acid catalysis. Nature 441:621–623

Knowles WS (2002) Asymmetric hydrogenations (Nobel lecture). Angew Chem Int Ed Engl 41:1998–2007

Kortvelyessy G (1985) Preparation of derivatives of citronellal. Acta Chim Hung 119:347–354

Kula MR, Wandrey C (1988) Continuous enzymic transformation in an enzyme-membrane reactor with simultaneous NADH regeneration. Methods Enzymol 136:9–21

Kumaragurubaran N, Juhl K, Zhuang W, Bøgevig A, Jørgensen KA (2002) Direct L-proline-catalyzed asymmetric alpha-amination of ketones. J Am Chem Soc 124:6254–6255

Kunz RK, MacMillan DWC (2005) Enantioselective organocatalytic cyclopropanations. The identification of a new class of iminium catalyst based upon directed electrostatic activation. J Am Chem Soc 127:3240–3241

Lacour J, Hebbe-Viton V (2003) Recent developments in chiral anion mediated asymmetric chemistry. Chem Soc Rev 32:373–382

Li X, List B (2007) Catalytic asymmetric hydrogenation of aldehydes. Chem Commun 17:1739–1741

List B (2000) The direct catalytic asymmetric three-component Mannich reaction. J Am Chem Soc 122:9336–9337

List B (2001) Asymmetric aminocatalysis. Synlett 2001:1675–1686

List B (2002a) Direct catalytic asymmetric alpha-amination of aldehydes. J Am Chem Soc 124:5656–5657

List B (2002b) Proline-catalyzed asymmetric reactions. Tetrahedron 58:5573–5590

List B (2004) Enamine catalysis is a powerful strategy for the catalytic generation and use of carbanion equivalents. Acc Chem Res 37:548–557

List B, Hoang L, Martin HJ (2004) New mechanistic studies on the proline-catalyzed aldol reaction. Proc Natl Acad Sci U S A 101:5839–5842

List B, Lerner RA, Barbas CF 3rd (2000) Proline-catalyzed direct asymmetric aldol reactions. J Am Chem Soc 122:2395–2396

List B, Pojarliev P, Martin HJ (2001) Efficient proline-catalyzed Michael additions of unmodified ketones to nitro olefins. Org Lett 3:2423–2425

List B, Pojarliev P, Biller WT, Martin HJ (2002) The proline-catalyzed direct asymmetric three-component Mannich reaction: scope, optimization, and application to the highly enantioselective synthesis of 1,2-amino alcohols. J Am Chem Soc 124:827–833

List B, Yang JW (2006) Chemistry. The organic approach to asymmetric catalysis. Science 313:1584–1586

Llewellyn DB, Arndtsen BA (2005) Synthesis of a library of chiral α-amino acid-based borate counteranions and their application to copper catalyzed olefin cyclopropanation. Tetrahedron Asymmetry 16:1789–1799

Makino T, Baba N, Oda J, Inouye Y (1977) Asymmetric reduction of alpha, beta-unsaturated iminuum salt with N-glucopyranosyl-1,4-dihydronicotinamides. Chem Ind 1977:277–278

Marigo M, Fielenbach D, Braunton A, Kjaersgaard A, Jørgensen KA (2005c) Enantioselective formation of stereogenic carbon-fluorine centers by a simple catalytic method. Angew Chem Int Ed Engl 44:3703–3706

Marigo M, Franzen J, Poulsen TB, Zhuang W, Jørgensen KA (2005b) Asymmetric organocatalytic epoxidation of alpha,beta-unsaturated aldehydes with hydrogen peroxide. J Am Chem Soc 127:6964–6965

Marigo M, Wabnitz TC, Fielenbach D, Jørgensen KA (2005a) Enantioselective organocatalyzed alpha sulfenylation of aldehydes. Angew Chem Int Ed Engl 44:794–797

Martin NJ, List B (2006) Highly enantioselective transfer hydrogenation of alpha,beta-unsaturated ketones. J Am Chem Soc 128:13368–13369

Mayer S, List B (2006) Asymmetric counteranion-directed catalysis. Angew Chem Int Ed Engl 45:4193–4195

Nicolaou KC, Sorensen EJ (1996) Classics in total synthesis. Wiley-VCH, Weinheim, p 344

Nishiyama H, Itoh K (2000) Asymmetric hydrosilylation and related reactions. In: Ojima I (ed) Catalytic asymmetric synthesis, 2nd edn. Wiley-VCH, New York, p 111–144

Northrup AB, MacMillan DWC (2002a) The first direct and enantioselective cross-aldol reaction of aldehydes. J Am Chem Soc 124:6798–6799

Northrup AB, MacMillan DWC (2002b) First general enantioselective catalytic Diels–Alder reaction with simple alpha,beta-unsaturated ketones. J Am Chem Soc 124:2458–2460

Notz W, List B (2000) Catalytic asymmetric synthesis of anti-1,2-diols. J Am Chem Soc 122:7386–7387

Notz W, Tanaka F, Watanabe S, Chaudari NS, Turner JM, Thayumanavan R, Barbas CF 3rd (2003) The direct organocatalytic asymmetric mannich reaction: unmodified aldehydes as nucleophiles. J Org Chem 68:9624–9634

Noyori R (2002) Asymmetric catalysis: science and opportunities (Nobel lecture). Angew Chem Int Ed Engl 41:2008–2022

Noyori R, Tokunaga M, Kitamura M (1995) Stereoselective organic synthesis via dynamic kinetic resolution. Bull Chem Soc Jpn 68:36–55

Ohkuma T, Kitamura M, Noyori R (2000) Asymmetric hydrogenation. In: Ojima I (ed) Catalytic asymmetric synthesis, 2nd edn. Wiley-VCH, New York, p 1–110

Ohkuma T, Noyori R (2004) Hydrogenation of imino groups. In: Jacobsen EN, Pfaltz A, Yamamoto H (eds) Comprehensive asymmetric catalysis, suppl 1. Springer, New York, p 43

Ouellet SG, Tuttle JB, MacMillan DWC (2005) Enantioselective organocatalytic hydride reduction. J Am Chem Soc 127:32–33

Paras NA, MacMillan DWC (2001) New strategies in organic catalysis: the first enantioselective organocatalytic Friedel-Crafts alkylation. J Am Chem Soc 123:4370–4371

Paras NA, MacMillan DWC (2002) The enantioselective organocatalytic 1,4-addition of electron-rich benzenes to alpha,beta-unsaturated aldehydes. J Am Chem Soc 124:7894–7895

Peelen TJ, Chi Y, Gellman SH (2005) Enantioselective organocatalytic Michael additions of aldehydes to enones with imidazolidinones: cocatalyst effects and evidence for an enamine intermediate. J Am Chem Soc 127:11598–11599

Perllissier H (2003) Dynamic kinetic resolution. Tetrahedron 59:8291–8327

Pictet A, Spengler T (1911) Formation of isoquinoline derivatives by the action of methylal on phenylethylamine, phenylalanine and tyrosine. Ber Dtsch Chem Ges 44:2030–2036

Pidathala C, Hoang L, Vignola N, List B (2003) Direct catalytic asymmetric enolexo aldolizations. Angew Chem Int Ed Engl 42:2785–2788

Puchot C, Samuel O, Dunach E, Zhao S, Agami C, Kagan HB (1986) Nonlinear effects in asymmetric synthesis. Examples in asymmetric oxidations and aldolization reactions. J Am Chem Soc 108:2353–2357

Rhone-Poulenc Industries (1979) Optical active citronellal. (Rhone-Poulenc Industries, France). Patent JP 78-80630, 1979-02-03

Rueping M, Sugiono E, Azap C, Theissmann T, Bolte M (2005) Enantioselective Brønsted acid catalyzed transfer hydrogenation: organocatalytic reduction of imines. Org Lett 7:3781–3783

Sakakura A, Suzuki K, Nakano K, Ishihara K (2006) Chiral 1,1′-binaphthyl-2,2′-diammonium salt catalysts for the enantioselective Diels–Alder reaction with α-acyloxyacroleins. Org Lett 8:2229–2232

Sanwal BD, Zink MW (1961) L-Leucine dehydrogenase of Bacillus cereus. Arch Biochem Biophys 94:430–435

Seayad J, List B (2005) Asymmetric organocatalysis. Org Biomol Chem 3:719–724

Seayad J, Seayad AM, List B (2006) Catalytic asymmetric Pictet-Spengler reaction. J Am Chem Soc 128:1086–1087

Sekiguchi Y, Sasaoka A, Shimomoto A, Fujioka S, Kotsuki H (2003) High-pressure-promoted asymmetric aldol reactions of ketones with aldehydes catalyzed by L-proline. Synlett 2003:1655–1658

Stecher H, Faber K (1997) Biocatalytic deracemization techniques: dynamic resolutions and stereoinversions. Synthesis 1997:1–16

Steiner DD, Mase N, Barbas III CF (2005) Direct asymmetric alpha-fluorination of aldehydes. Angew Chem Int Ed Engl 44:3706–3719

Storer RI, Carrera DE, Ni Y, MacMillan DWC (2006) Enantioselective organocatalytic reductive amination. J Am Chem Soc 128:84–86

Taratov VI, Börner A (2005) Approaching highly enantioselective reductive amination. Synlett 2005:203–211

Tatsui G (1928) Synthesis of carboline derivatives. J Pharm Soc Jpn 48:453–459

Taylor MS, Jacobsen EN (2004) Highly enantioselective catalytic acyl-pictet-spengler reactions. J Am Chem Soc 126:10558–10559

Terada M, Uraguchi D, Sorimachi K, Shimizu H (2005) Process for production of optically active amines by stereoselective nucleophilic addition reaction of imines with C nucleophiles using chiral phosphoric acid derivative. PCT Int Appl WO 2005070875 2005-08-04

Tokuda O, Kano T, G Gao W, Ikemoto T, Maruoka K (2005) A practical synthesis of (S)-2-cyclohexyl-2-phenylglycolic acid via organocatalytic asymmetric construction of a tetrasubstituted carbon center. Org Lett 7:5103–5105

Tuttle JB, Ouellet SG, MacMillan DWC (2006) Organocatalytic transfer hydrogenation of cyclic enones. J Am Chem Soc 128:12662–12663

Uraguchi D, Sorimachi K, Terada M (2004) Organocatalytic asymmetric aza-Friedel-Crafts alkylation of furan. J Am Chem Soc 126:11804–11805

Uraguchi D, Sorimachi K, Terada M (2005) Organocatalytic asymmetric direct alkylation of alpha-diazoester via C-H bond cleavage. J Am Chem Soc 127:9360–9361

Uraguchi D, Terada M (2004) Chiral Brønsted acid-catalyzed direct Mannich reactions via electrophilic activation. J Am Chem Soc 126:5356–5357

Vignola N, List B (2004) Catalytic asymmetric intramolecular alpha-alkylation of aldehydes. J Am Chem Soc 126:450–451

Wang W, Wang J, Li H (2005) Direct, highly enantioselective pyrrolidine sulfonamide catalyzed Michael addition of aldehydes to nitrostyrenes. Angew Chem Int Ed Engl 44:1369–1371

Ward RS (1995) Dynamic kinetic resolution. Tetrahedron Asymmetry 6:1475–1490

Wilson RM, Jen WS, MacMillan DWC (2005) Enantioselective organocatalytic intramolecular Diels–Alder reactions. The asymmetric synthesis of solanapyrone D. J Am Chem Soc 127:11616–11617

Xie JH, Zhou ZT, Kong WL, Zhou QL (2007) Ru-catalyzed asymmetric hydrogenation of racemic aldehydes via dynamic kinetic resolution: efficient synthesis of optically active primary alcohols. J Am Chem Soc 129:1868–1869

Yamada YM, Yoshikawa N, Sasai H, Shibasaki M (1997) Direct catalytic asymmetric aldol reactions of aldehydes with unmodified ketones. Angew Chem Int Ed Engl 36:1871–1873

Yamamoto H, Momiyama N (2005) Rich chemistry of nitroso compounds. Chem Commun (Camb) Jul 28:3514–3525

Yang JW, Hechavarria Fonseca MT, List B (2004) A metal-free transfer hydrogenation: organocatalytic conjugate reduction of alpha,beta-unsaturated aldehydes. Angew Chem Int Ed Engl 43:6660–6662

Yang JW, Hechavarria Fonseca MT, Vignola N, List B (2005) Metal-free, organocatalytic asymmetric transfer hydrogenation of alpha,beta-unsaturated aldehydes. Angew Chem Int Ed Engl 44:108–110

Yang JW, Stadler M, List B (2007a) Proline-catalyzed mannich reaction of aldehydes with *N*-Boc-imines. Angew Chem Int Ed Engl 46:609–611

Yang JW, Stadler M, List B (2007b) Practical Proline-catalyzed asymmetric Mannich reaction of aldehydes with *N*-Boc-imines. Nat Protoc 2:1937–1942

Zhong G (2003) A facile and rapid route to highly enantiopure 1,2-diols by novel catalytic asymmetric alpha-aminoxylation of aldehydes. Angew Chem Int Ed Engl 42:4247–4250

Ernst Schering Foundation Symposium Proceedings, Vol. 2, pp. 45–124
DOI 10.1007/2789_2007_069
© Springer-Verlag Berlin Heidelberg
Published Online: 14 February 2008

Biomimetic Organocatalytic C–C-Bond Formations

D. Enders(✉), M.R.M. Hüttl, O. Niemeier

Institute of Organic Chemistry, RWTH Aachen University, Landoltweg 1, 52074 Aachen, Germany
email: *Enders@rwth-aachen.de*

1	Introduction	46
2	Organocatalytic Approach Using Proline and Its Derivatives	47
2.1	Asymmetric Biomimetic C–C-Bond Formations with a DHAP-Equivalent	47
2.2	Proline-Catalyzed Asymmetric Synthesis of Ulosonic Acid Precursors	69
2.3	First Direct Organocatalytic α-Fluorination of Aldehydes and Ketones	73
2.4	Asymmetric Organocatalytic Multicomponent Domino Reactions	75
3	C–C-Bond Formations Employing N-Heterocyclic Carbene Catalysis	82
3.1	Enzymes as Archetypes	82
3.2	The Benzoin Condensation	83
3.3	The Stetter Reaction	104
4	Conclusion	107
References		108

Abstract. Mother Nature utilizes simple precursors to build up complex organic molecules efficiently. One important example is the C_3 building block dihydroxyacetone phosphate, which is used in various enzyme-catalyzed reactions. Following this biosynthetic strategy the DHAP equivalent 'dioxanone' can be used in organocatalytic reactions to synthesize sugars, aminosugars, carbasugars, polyoxamic acids and sphingosines. In this respect, organocatalytic domino reactions can also be seen as biomimetic as they are reminiscent of tandem reactions that may occur during biosyntheses of complex natural products. In nature, the coenzyme thiamin (vitamin B_1), a natural thiazolium salt, is used in biochemical nucleophilic acylations ('Umpolung'). The catalytic active species is a nucleophilic carbene. Mimicking this approach, organocatalytic carbene catalysis has emerged to an exceptionally fruitful research area, which is used in asymmetric C–C bond formations.

1 Introduction

The development and application of biomimetic strategies plays an outstanding role in the synthesis of complex organic molecules from simple precursors. In recent years these strategies have been used successfully in organic synthesis, but the development of synthetic equivalents of the naturally occurring building blocks is an ongoing challenge in asymmetric biomimetic synthesis. Our group has already developed some stereoselective methods using dihydroxyacetone phosphate (DHAP) or phosphoenol pyruvate (PEP) equivalents in combination with the SAMP/RAMP hydrazone methodology. With the exploration of organocatalysis as a powerful tool of asymmetric catalysis, new biomimetic protocols come within reach, where small chiral organic molecules act as catalytic active species mimicking the activation pathways of enzymes. These organocatalysts are metal free, usually non-toxic, readily available and often very robust. A significant advantage of many organocatalysts is the capability of promoting reactions via different activation modes. In this chapter we report our recent results in the development of methods for biomimetic organocatalytic C–C-bond formations and their applications in the asymmetric synthesis of natural products and bioactive compounds in general.

2 Organocatalytic Approach Using Proline and Its Derivatives

2.1 Asymmetric Biomimetic C–C-Bond Formations with a DHAP-Equivalent

DHAP (**A**) is used in nature as a C_3-methylene component for the efficient and stereoselective formation of carbohydrates (Scheme 1). In this pathway DHAP reacts with D-glyceraldehyde-3-phosphate (**1**) via an enzyme-catalyzed aldol reaction to form **2**. After dephosphonylation the aldol adduct **3** is then transformed to various carbohydrates and derivatives (Calvin 1962).

To mimic the biosynthesis of carbohydrates and other valuable compounds possessing similar structural motifs we used 2,2-dimethyl-1,3-dioxan-5-one (**4**, dioxanone) (Enders and Bockstiegel 1989; Enders et al. 2002a), which can function as a protected dihydroxy acetone-equivalent (DHA). To use the dioxanone methodology in asymmetric biomimetic syntheses, two concepts have been developed. First, the SAMP/RAMP hydrazone methodology was used to form a chiral hydrazone which, after formation of the aza-enolate, can be trapped with a great variety of electrophiles. In addition, the stoichiometric approach was extended by using a bulky silyl group in the α-position of the dioxanone (**4**) as a traceless directing group in order to obtain very high asymmetric induction in problematic cases. Both methods were successfully applied in C–C-bond forming reactions and in natural product syntheses (Fig. 1) (for a review see: Enders et al. 2005a). Secondly, an organocatalytic approach was established that opened up a new entry to carbohydrates and amino sugars by using proline-catalyzed aldol and Mannich reactions (Fig. 1).

2.1.1 Direct Organocatalytic De Novo Carbohydrate Synthesis

Inspired by the biosynthesis of carbohydrates we envisaged a direct de novo synthesis of carbohydrate derivatives by using the DHA-equivalent **4** in a C_3+C_n-strategy. As can be seen from the retrosynthetic analysis, the desired building blocks **5** should be prepared from the dioxanone (**4**) and an aldehyde component **6** in an organocatalytic aldol reaction

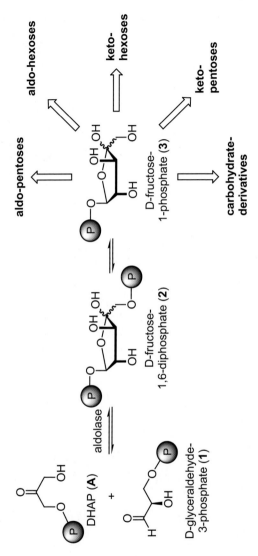

Scheme 1. Nature's pathway to carbohydrates employing DHAP (A)

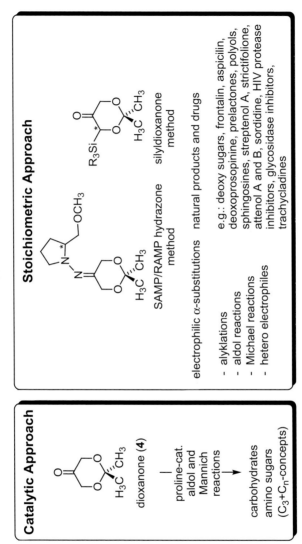

Fig. 1. The dioxonanone methodology in asymmetric synthesis

(Scheme 2). Based on the work of List et al. the proper catalyst for this reaction should be proline (List et al. 2000; Notz and List 2000).

For our first example we chose 2-methylpropanal as a model system for the aldol reaction with dioxanone and optimized the reaction conditions in terms of chemical yield, enantiomeric excess, and *anti/syn* ratio. The best reaction conditions so far call for (*S*)-proline as the catalyst, dimethylformamide (DMF) as the solvent, and a temperature of 2°C. The *anti* aldol product **7** was obtained diastereoselectively with an excellent yield of 97%, an *anti/syn* ratio of >98:2, and a high enantiomeric excess of 94% *ee* (Enders and Grondal 2005). Subsequently we were also able to show that the aldol reaction of **4** with the α-branched aldehydes proceeds with good to very good yields, excellent *anti/syn* ratios, and enantiomeric excesses in all cases (Scheme 3). When a linear aldehyde was used, the aldol product **7** was isolated in only moderate yield (40%), but still excellent stereoselectivity (*anti/syn* >98:2, 97% *ee*).

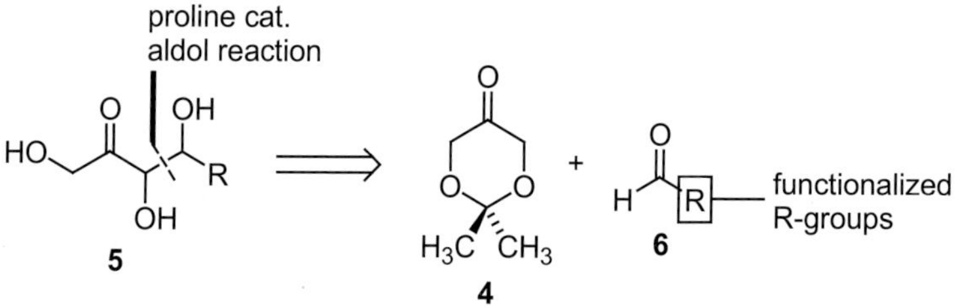

Scheme 2. Retrosynthetic analysis of the aldol adducts **5**

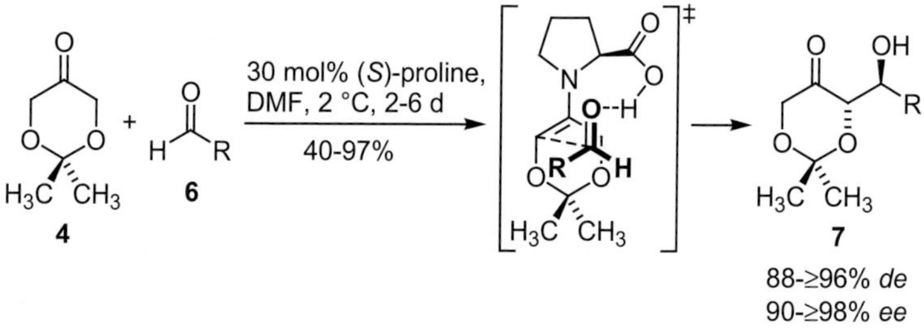

Scheme 3. Proline-catalyzed aldol reaction of **4**

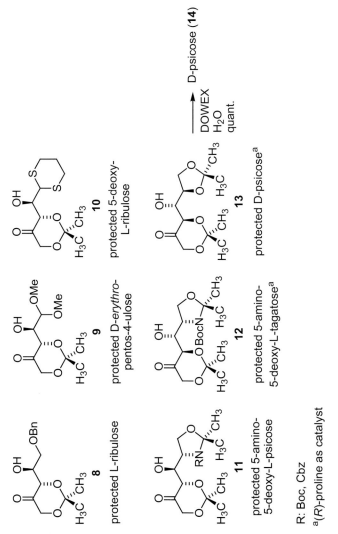

Scheme 4. Several protected sugars and amino sugars **8–13** available by the C_3+C_n strategy

The lower yield may be explained by the fact that linear aldehydes also undergo self-aldol condensation, which is in direct competition with the crossed-aldol reaction. Aromatic aldehydes as the carbonyl component led to reduced diastereoselectivity. For example, the (*S*)-proline-catalyzed aldol reaction of **4** with *ortho*-chlorobenzaldehyde proceeded with a good yield of 73%, but with an *anti*/*syn* ratio of only 4:1 and enantiomeric excesses of 86% *ee* (*anti*) and 70% *ee* (*syn*).

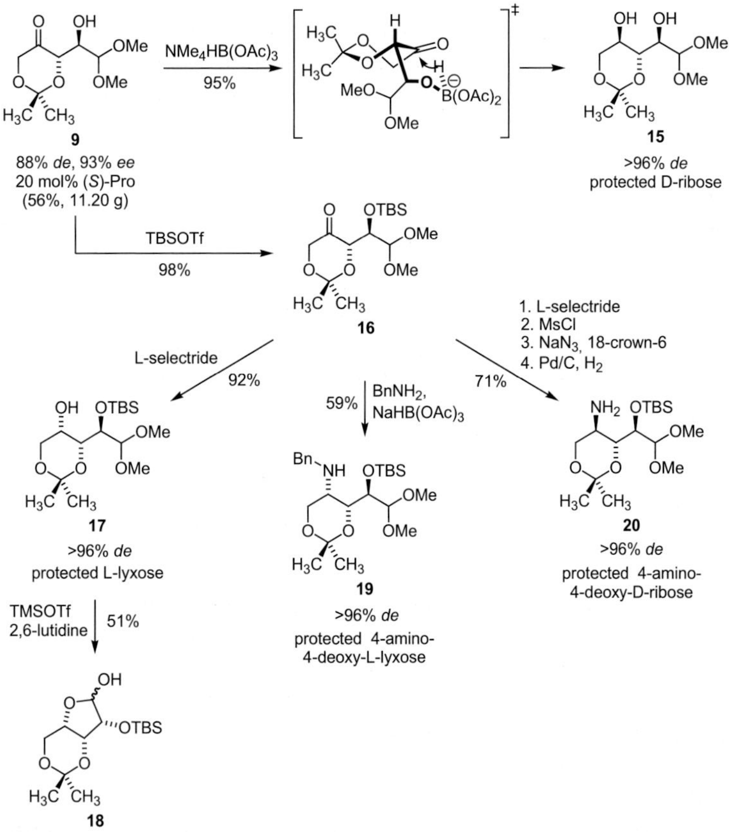

Scheme 5. Inversion strategy and further functionalizations for the diversity oriented synthesis of carbohydrate derivatives

Our biomimetic C_3+C_n concept allows the synthesis of selectively and partly orthogonal double protected sugars and amino sugars in one step. For example L-ribulose (**8**), D-*erythro*-pentos-4-ulose (**9**), 5-deoxy-L-ribulose (**10**), 5-amino-5-deoxy-L-psicose (**11**), 5-amino-5-deoxy-L-tagatose (**12**) and D-psicose (**13**) were prepared in this way (Enders and Grondal 2005; Enders and Grondal 2006). The double acetonide protected D-psicose **13** was quantitatively deprotected with an acidic ion-exchange resin (Dowex W50X2-200) to give the parent D-psicose (**14**, Scheme 4).

The stereoselective reduction of the ketone function of **9** leads to a direct entry to selectively protected aldopentoses ('inversion strategy') (Borysenko et al. 1989), which greatly expand the potential of this new protocol (Scheme 5). Following Evans' protocol the tetramethylammonium triacetoxyborohydride-mediated reduction provides the *syn*-diol **15** constituting a protected D-ribose (95%, >96% *de*). The *anti*-selective reduction to **17** was obtained after silyl protection of the free hydroxyl group of **9** to the OTBS-ether **16** using L-selectride. The aldopentose **18** was then accessible via chemoselective acetal cleavage followed by in situ cyclization (47% over two steps, >96% *de*).

Besides reduction, other transformations were performed, for example, reductive amination, nucleophilic 1,2-addtion, deoxygenation or olefination/reduction and thionation (Enders and Grondal 2006; Grondal 2006).

2.1.2 Direct Organocatalytic Entry to Sphingoids

Sphingoids are long-chain amino-diol and -triol bases that form the backbone and characteristic structural unit of sphingolipids, which are important membrane constituents and play vital roles in cell regulation as well as signal transduction (see selected reviews: (Kolter and Sandhoff 1999; Brodesser et al. 2003; Kolter 2004; Liao et al. 2005)). Furthermore, glycosphingolipids show important biological activities, e.g., antitumor, antiviral, antifungal or cytotoxic properties (Naroti et al. 1994; Kamitakahara et al. 1998; Kobayashi et al. 1998; Li et al. 1995). Phytosphingosines, one of the major classes of sphingoids, have been isolated and identified either separately or as parts of sphingolipids found in plants, marine organisms, fungi, yeasts and even mammalian tissues

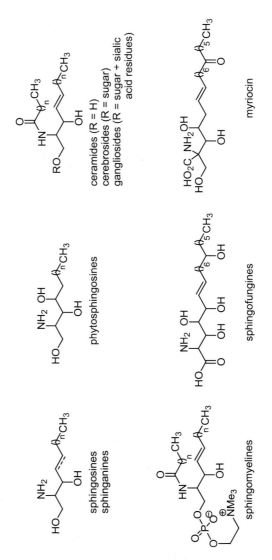

Fig. 2. Representative sphingolipids and analogues

(Carter et al. 1954; Kawano et al. 1988; Li et al. 1984; Oda 1952; Thorpe and Sweeley 1967; Karlsson et al. 1968; Barenholz and Gatt 1967; Takamatsu et al. 1992; Okabe et al. 1968; Wertz et al. 1985; Vance and Sweeley 1967). Due to the physiological importance of these compounds a large number of syntheses have been reported, which usually involve many steps and extensive protecting group strategies. A number of representative sphingolipids and analogues are depicted in Fig. 2.

Our group previously established an asymmetric stoichiometric approach to build up several sphingosines (Enders et al. 1995a) and sphinganines (Enders et al. 1995a; Enders and Müller-Hüwen 2004), which we recently extended by a direct and flexible organocatalytic approach to sphingoids demonstrated by the efficient asymmetric synthesis of D-*arabino*- and L-*ribo*-phytosphingosine **21** and **22** (Fig. 3).

Our retrosynthetic analysis of the desired sphingoids relies on the previously developed diastereo- and enantioselective (*S*)-proline-catalyzed aldol reaction of the readily available dioxanone (**4**). In a second step, the amino group should be installed by reductive amination (Scheme 6) (Enders et al. 2006a).

After extensive optimization of the reaction conditions regarding yield as well as diastereo- and enantioselectivity, we were able to obtain the aldol product **26** with 60% yield and excellent diastereo- and enantiomeric excesses (>99% *de*, 95% *ee*). Thus, the simple (*S*)-proline-catalyzed aldol reaction of **4** with pentadecanal directly delivered gram-amounts of the selectively acetonide protected ketotriol precursor **26** of the core unit of phytosphingosines in excellent stereoisomeric purity (Scheme 7).

In order to create stereoselectively the *syn*- and the *anti*-1,3-aminoalcohol function of the stereotriad, we first envisaged a diastereoselective

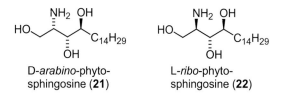

Fig. 3. Structures of the phytosphingosines **21** and **22**

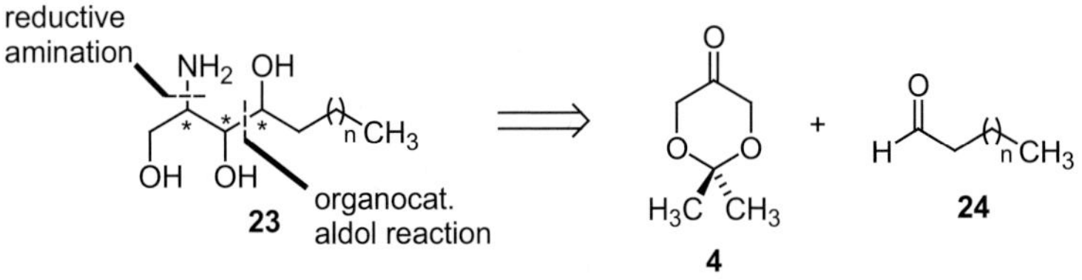

Scheme 6. Retrosynthetic analysis of the phytosphingosine structure **23**

reductive amination of **26**. Initially, we investigated this reductive amination of **26** with BnNH$_2$ and NaHB(OAc)$_3$ in the presence of acetic acid, but unfortunately we obtained only a 1:1-epimeric mixture of the corresponding 1,3-aminoalcohol in 72% yield. Therefore, we attempted the reductive amination with the corresponding *O*TBS-protected aldol derivative **27**, which can be easily obtained in excellent yield (95%) using TBSOTf and 2,6-lutidine (Enders and Grondal 2006). The *anti*-1,3-aminoalcohol **28** was isolated in almost quantitative yield (94%) and virtually complete diastereoselectivity (*de*>99%, Scheme 7). Thus, our six-step organocatalytic protocol affords via orthogonal and selectively protected intermediates D-*arabino*-phytosphingosine (**21**) in 49% overall yield and of high diastereo- and enantiomeric purity. Needless to say, the corresponding enantiomer can be obtained using (*R*)-proline instead of (*S*)-proline as the organocatalyst. Because the direct and stereoselective reductive amination of **26** or **27** to afford the corresponding *syn*-1,3-aminoalcohol was not possible, we decided to synthesize the *syn*-isomer via a substitution reaction by inversion of the stereogenic centre (Enders and Müller-Hüwen 2004). Therefore, **27** was first transformed to the corresponding *anti*-1,3-diol **30** by a highly diastereoselective reduction with L-selectride (Scheme 8). The newly generated secondary alcohol **30** was then converted into the mesylate (91%) and subsequently into azide (80%). The substitution of the mesylate by NaN$_3$ in the presence of a crown ether (18-c-6) proceeded with complete inversion of the stereogenic centre (>99:1, determined by gas chromatography). Subsequent reduction of the azide with lithium aluminium hydride and acidic cleavage of the two protecting groups afforded the L-*ribo*-phytospingosine (**22**) in 41% overall yield (Scheme 8).

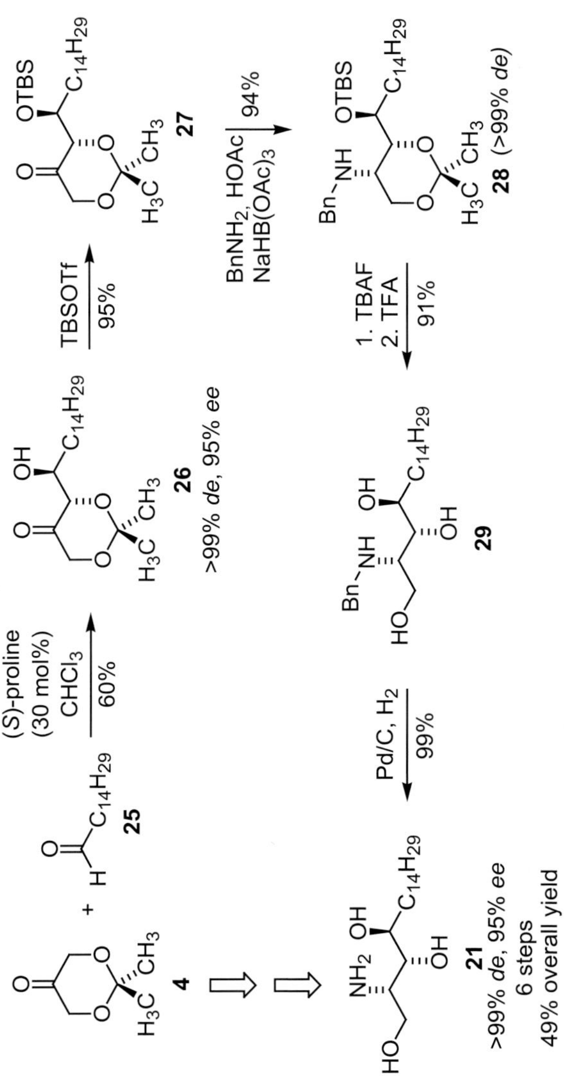

Scheme 7. Six-step asymmetric synthesis of the D-*arabino*-phytospingosine (**21**)

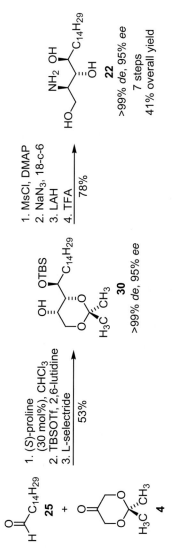

Scheme 8. Seven-step synthesis of the l-*ribo*-phytospingosine (**22**)

2.1.3 Direct Organocatalytic Entry to Carbasugars

Carbasugars (Sollogoub and Sinay 2006; Suami and Ogawa 1990), also known as pseudosugars (McCasland et al. 1966), are characterized by the replacement of the ring oxygen of monosaccharides by a methylene group (for reviews see: Suami 1987; Suami 1990; Ogawa 1988) (Fig. 4). Not only saturated carbasugars are known, but also unsaturated ones bearing a ring double bond. Interestingly, they are often recognized by enzymes instead of the original sugar. Because of the lack of the acetal moiety, such carbasugars are stable towards hydrolysis (Berecibar et al. 1999). Furthermore, they often show interesting biological properties, for instance, they are glycosidase inhibitors, antibiotics, antivirals or plant growing inhibitors (Musser 1992; Witczak 1997; Dwek 1996). Typical examples of naturally occurring carbasugars are streptol (Isogai et al. 1987), valienamine (Horii et al. 1971), validamine (Kameda and Horii 1972; Kameda et al. 1984), cyclophellitol (Atsumi et al. 1990a,b), (+)-MK7607 (Yoshikawa et al. 1994) or the family of gabosines (Bach et al. 1993). (+)-MK7607 has effective herbicidal activity and is the 4-epimer of streptol, a plant-growth inhibitor. They are two representative examples of eight possible diastereomers of the class of the unsaturated 5a-carbasugars characterized by an exocyclic hydroxymethyl moiety (Fig. 4).

Altogether, four diastereomers are already known, three of them are naturally occurring and the fourth one has been synthesized in racemic form. Most interestingly, all of these compounds are bioactive, but unfortunately direct and flexible approaches to synthesize different stereoisomers or derivatives have not yet been reported. (For 5a-carbasugar syntheses, see: Ogawa and Tsunoda 1992; Chupak et al. 1998; Lubineau and Billault 1998; Rassu et al. 2000; Mehta and Lakshminath 2000; Song et al. 2001; Holstein Wagner and Lundt 2001; Ishikawa et al. 2003). We therefore developed a modular strategy for the synthesis of carbasugars, which was demonstrated by an efficient and straightforward synthesis of 1-*epi*-(+)-MK7607 (**31**) (Grondal and Enders 2006). The retrosynthetic analysis for **31** is depicted in Scheme 9 and involves the construction of the cyclohexene core via a ring-closing metathesis. The second disconnection is a (*R*)-proline-catalyzed aldol reaction be-

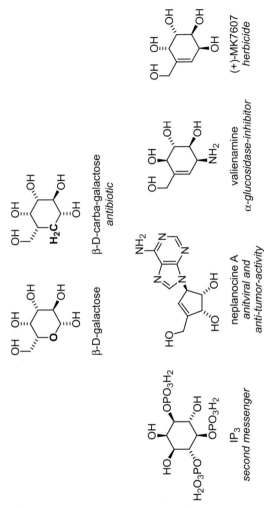

Fig. 4. Structures of the representative carbasugars

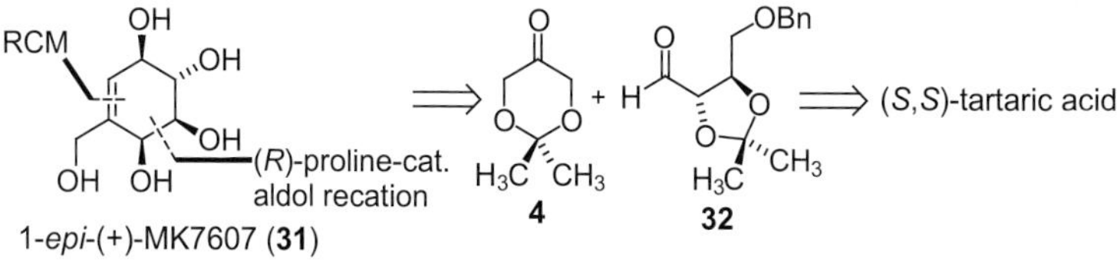

Scheme 9. Retrosynthetic analysis of 1-*epi*-(+)-MK7607 (**31**)

tween dioxanone (**4**) and the aldehyde **32**, easily available from (*S*,*S*)-tartaric acid in four steps (Mukaiyama et al. 1990).

The first step of the total synthesis of **31** is the (*R*)-proline-catalyzed aldol reaction between **4** and **32**, which gave the aldol adduct **33** with a good yield (69%) and nearly perfect stereocontrol ($\geq$96% *de*, >99% *ee*, Scheme 10). The same results were observed when the reaction was carried out on a 40-mmol scale yielding 5.22 g of **33** without a decrease of selectivity. The free hydroxyl group of **33** was quantitatively protected as MOM-ether. After hydrogenolytic debenzylation the aldehyde-ketone was obtained after Dess-Martin oxidation followed by a double Wittig reaction to provide the bisolefine **34** in 41% yield over 4 steps (Scheme 10).

34 was then converted into the protected bis-acetonide **35** via ring-closing metathesis employing Grubbs' second-generation catalyst. To our delight, the desired cyclohexene **35** was smoothly formed with 90% yield after 5 h in refluxing dichloromethane, although it represents a pentafunctionalized cyclohexene and is the part of a tricycle. The relative configuration of **35** was determined by ^{1}H-NMR spectroscopy and NOE measurements and is in agreement with the relative configuration of the aldol product **33**. Finally, the treatment of **35** with the acidic ion-exchange resin DOWEX in methanol at 70°C led to the complete removal of both acetonide groups and the MOM-ether in one operation to liberate the desired carbasugar 1-*epi*-(+)-MK7607. The seven-step synthesis provides **31** in 23% overall yield (Grondal and Enders 2006).

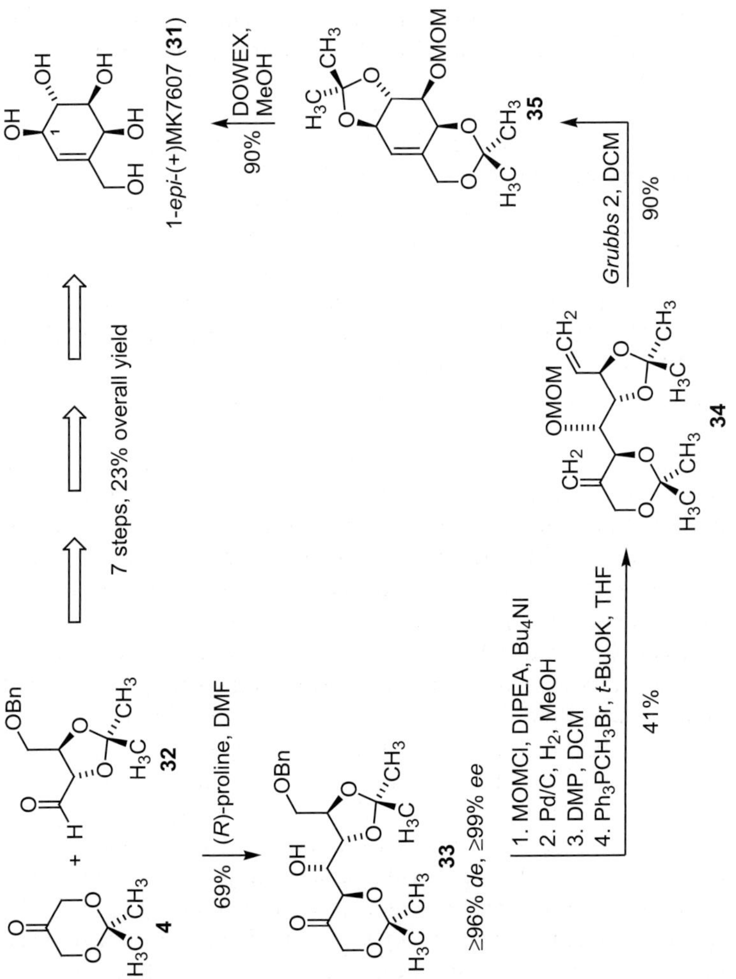

Scheme 10. Asymmetric synthesis of the carbasugar 1-*epi*-(+)-MK7607 (**31**)

2.1.4 Asymmetric Synthesis of Selectively Protected Amino Sugars and Derivatives via Direct Organocatalytic Mannich Reaction

In the classic Mannich reaction the corresponding β-aminocarbonyl compounds are formed from formaldehyde, an amine, and an enolizable carbonyl component (Mannich and Krösche 1912). These so-called Mannich bases have found broad applications as synthetic building blocks (Arend et al. 1998; Kobayashi and Ishitani 1999), most importantly in the preparation of natural products and biologically active compounds (Traxler et al. 1995; Dimmock et al. 1993; Kleemann and Engel 1982). The main disadvantage of the classic Mannich reaction has been the lack of stereocontrol and the formation of by-products. As a result, the development of more selective and particularly diastereo- and enantioselective protocols for this important C–C bond-forming reaction has been of substantial interest. In 1985 our research group, in cooperation with Steglich et al., disclosed for the first time a procedure for a stereoselective Mannich reaction, by which enamines together with acyliminoacetates could be transformed into diastereo- and enantiomerically pure α-amino-γ-keto esters (Kober et al. 1985). Later on we developed a first practical procedure for the regio- and enantioselective α-aminomethylation of ketones with the assistance of a directing silyl group at the α-position to the carbonyl group (Enders et al. 1996d, 2000, 2002b; Enders and Oberbörsch 2002). Interest in catalytic asymmetric variants of the Mannich reaction has grown considerably in recent years. In particular, the application of metal-free catalysts is highly desirable in accomplishing diastereo- and enantioselective Mannich reactions. Special notice should be taken of the proline-catalyzed three-component Mannich reaction developed by List et al. (List 2000; List et al. 2002). In this sophisticated organocatalytic method, enolizable aldehydes and ketones are treated with in situ generated imines to afford the corresponding Mannich products with good-to-excellent stereoselectivities.

Based on our organocatalytic C_3+C_n concept for the direct synthesis of carbohydrates, we envisaged the successful development of a diastereo- and enantioselective Mannich variant that paves the way to selectively protected amino sugars and their derivatives. These amino sugars are a class of carbohydrates in which one or more hydroxyl functions are replaced by amino groups. They are found as parts of glycoproteins, glycolipids, aminoglycosides, and many biologically active secondary metabolites containing free, methylated or acetylated amino groups (Wong CH 2003; Weymouth-Wilson 1997). The substitution of a hydroxy function by an amino group may alter the properties of the sugar significantly, for example, its hydrogen bonding properties, solubility, and charge. Consequently, amino sugars play important physiological roles in many glycoconjugates and are of interest for the development of new drugs (Wong 2003).

Initially, the (*S*)-proline-catalyzed three-component Mannich reaction of **6** with dioxanone (**4**) and *para*-anisidine (**36**), as the amine component, was achieved (Enders et al. 2005b). Thus, in the presence of 30 mol% (*S*)-proline in DMF at 2°C we obtained the Mannich product **39** in 91% yield and excellent stereoselectivities (>99% *de*, 98% *ee*, see Scheme 11). After recrystallization from heptane/2-propanol (9:1) **39** could be obtained in practically diastereo- and enantiomerically pure form. Analogous conditions employed for the α-branched aldehydes also led to very good yields and selectivities. In the case of linear aldehydes such as **41** the results obtained under the above reaction conditions were not as good. Following extensive optimization of the reaction conditions **41** was obtained in 77% yield and with improved stereoselectivities (88% *de*, 96% *ee*) using **38** as a catalyst, acetonitrile as a solvent, and five equivalents of water. In all cases the *syn* configuration of the Mannich products was observed, which was confirmed both by *nuclear Overhauser effect* (NOE) measurements and an X-ray crystal structural analysis. The result is consistent with the transition state proposed by List et al.

Furthermore, several derivatizations of the Mannich products were possible, for example, via diastereoselective reduction of the ketone function or by direct reductive amination, as illustrated for **44** and **45** (Scheme 12) (Enders et al. 2005b, 2006b). Thus, the reduction of **39** with L-selectride proceeded with high stereocontrol to yield the all-*syn*-configured β-amino alcohol **45**, which in its protected form belongs to the class of the biologically very important 2-amino-2-deoxy sugars (Enders et al. 2005b). Alternatively, the *anti*-aminoalcohol **44** was available by $Me_4NHB(OAc)_3$-mediated reduction. The direct reductive amination was carried out using $NaHB(OAc)_3$, $BnNH_2$ and acetic acid.

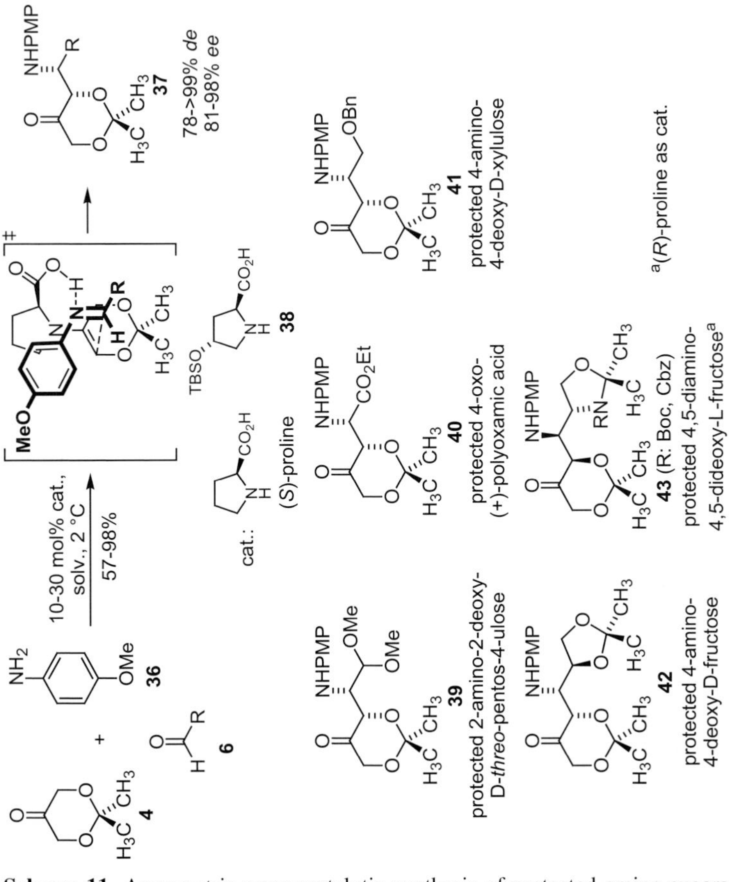

Scheme 11. Asymmetric organocatalytic synthesis of protected amino sugars **39–43**

This protocol led to the diamine **46**, which also represents a protected amino sugar (2,4-diamino-2,4-dideoxy-L-xylose). Interestingly, the direct reductive amination of **47** was followed by in situ cyclization to afford the lactam **48** as the major diastereomer.

Our efficient asymmetric catalytic approach provides a viable alternative to the conventional, relatively elaborate, and less flexible methods

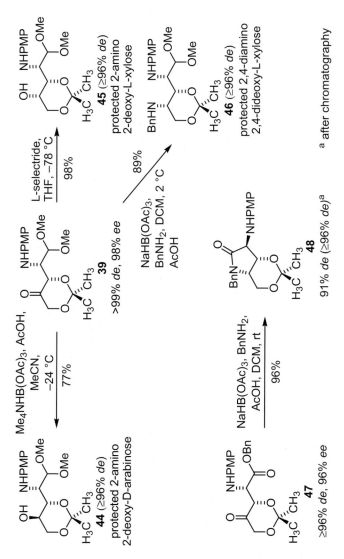

Scheme 12. Further extension of the direct organocatalytic Mannich reaction

for the synthesis of amino sugars. Our protocol facilitates the synthesis of different amino pentoses and hexoses with high diversity in only one to two steps.

2.1.5 Asymmetric Organocatalytic Synthesis of (+)-Polyoxamic Acid

The polyoxins are a family of important agricultural pest control agents isolated from *Streptomyces cacaoi* var. *asoensis* (Isono et al. 1969). They act by inhibiting the synthesis of chitin, which constitutes an important component of the fungal cell wall structure. They have also been found to be of potential therapeutic value against the human fungal pathogen *Candida albicans* (Shenbagamurthi et al. 1983). A common motif of several members of the polyoxin family is 5-*O*-carbamoylpolyoxamic acid from which (+)-polyoxamic acid (**49**) is derived upon mild hydrolysis (Fig. 5). In the past three decades, numerous syntheses of **49** have been reported in the literature (Casiraghi et al. 1995), most of which utilize the existing chiral pool, although stereoselective approaches have also been published (Enders and Vrettou 2006).

Based on our previously developed organocatalytic Mannich methodology (Enders et al. 2005b) we planned a synthetic strategy towards (+)-polyoxamic acid (**49**) starting from a suitable Mannich base, which would utilize a diastereoselective reduction of the ketone functionality to the corresponding secondary alcohol and an oxidation step to con-

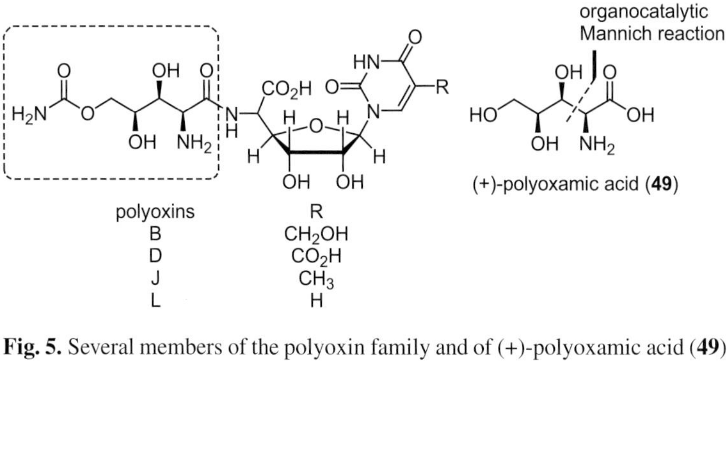

Fig. 5. Several members of the polyoxin family and of (+)-polyoxamic acid (**49**)

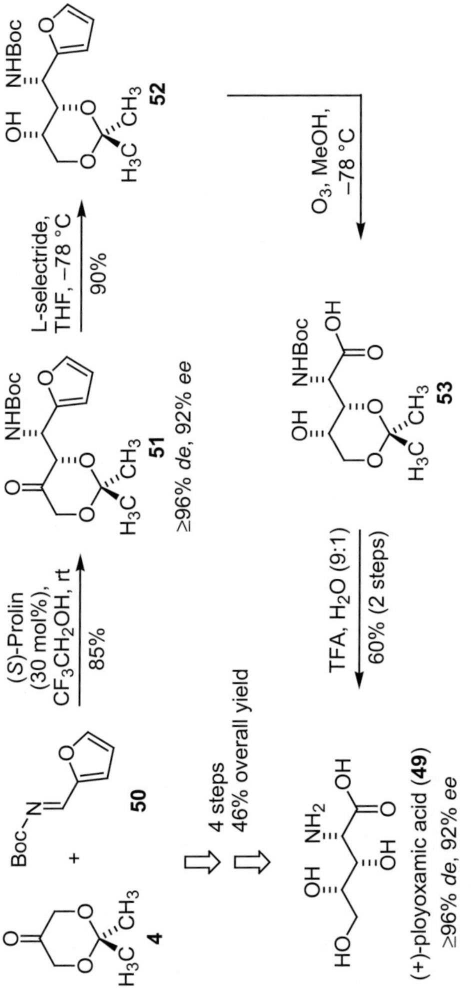

Scheme 13. Synthesis of (+)-polyoxamic acid (**49**) starting from Mannich base **51**

Biomimetic Organocatalytic C–C-Bond Formations 69

vert a furan moiety into the corresponding carboxylic acid group (Enders and Vrettou 2006). Thus, starting from **4** and the known imine **50** (Wenzel and Jacobsen 2002) we were able to access the corresponding Mannich base **51** in 85% yield and with excellent selectivity (>96% *de* as judged by ^{1}H NMR, 92% *ee* as determined by HPLC). Following optimization of the reaction, (*S*)-proline was found to be the optimum catalyst at room temperature and trifluoroethanol the most suitable solvent. The desired *syn*-configuration of the Mannich base was determined by NOE experiments.

Diastereoselective reduction of the ketone **51** using L-selectride afforded the desired *syn*-isomer **52** in 90% yield as a single diastereoisomer as judged by ^{1}H-NMR and NOE experiments. With the intermediate amino alcohol **52** at hand, the corresponding carboxylic acid **53** was obtained by ozonolysis at −78°C in MeOH. Subsequent deprotection with aqueous trifluoroacetic acid afforded the free (+)-polyoxamic acid (**49**) in a good yield of 60% over two steps. Thus, it was possible to synthesize **49** in four steps from the known imine **50** in 46% overall yield as a single diastereoisomer, as judged by ^{1}H-NMR, and with 92% *ee* (Scheme 13) (Enders and Vrettou 2006).

2.2 Proline-Catalyzed Asymmetric Synthesis of Ulosonic Acid Precursors

The forthcoming advent of a post-antibiotic era has driven much research towards developing new tools to fight the emergence of devastating diseases and has prompted much effort to identify the biological functions of carbohydrates in physiological processes (Varki 1993; Dwek 1996; Sears and Wong 1998). Naturally occurring 2-keto-3-deoxy-nonulosonic acids such as Neu5Ac (**54**) and KDN (**55**), generally known as sialic acids, have been significantly implicated in the pathogenesis of microorganisms and various disease states (Unger 1981; Schauer 1982; Schauer R 1982; Alexander and Rietschel 1999; Angata and Varki 2002; Kiefel and von Itzstein 2002; Varki 1992; Troy 1992; and references therein). Likewise, pivotal biological roles are constantly ascribed to widely diffuse higher 3-deoxy-2-ulosonic acids. For example, 3-deoxy-D-manno-2-octulosonic acid (KDO, **56**), present in the outer membrane lipopolysaccharide (LPS) of Gram-negative bacteria, is essential for their replication. The 7-phosphate of the 3-deoxy-D-arabino-2-heptulosonic acid (DAH, **57**) is a key intermediate in the biosynthesis of aromatic amino acids via the shikimate pathway. The phosphorylated form of 2-keto-3-deoxy-D-glucosonic acid (D-KDG, **58**) is part of the Entner-Doudoroff pathway (Fig. 6).

Over recent years a number of useful chemical and enzymatic methodologies have been reported and implemented to develop efficient syntheses of sialic and ulosonic acids (Danishefsky et al. 1988; DeNinno 1991; von Itzstein and Kiefel 1997; Banwell et al. 1998; Voight et al. 2002; Silvestri et al. 2003; Sugai et al. 1993; Dondoni et al. 1994; Li and Wu 2002; Banaszek and Mlynarski 2005) as well as of certain analogues. However, in enzyme-catalyzed reactions the loss of stereocontrol regarding the substrate scope is often a problem and the total synthesis approaches have suffered from long reaction sequences due to protecting group manipulations. Consequently the need for short and practical synthetic routes remained a challenging endeavor of great interest. Previous efforts from our laboratories led to structurally modified deoxygenated ulosonic acids via metalated SAMP-/RAMP-hydrazones as efficient chiral equivalents of phosphoenol pyruvate (PEP) (Enders et al.

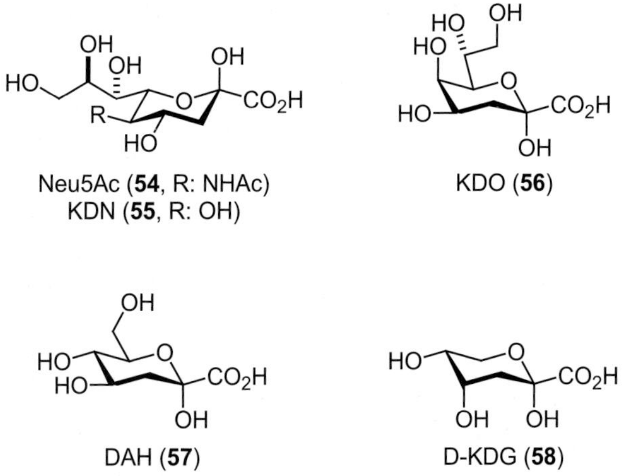

Neu5Ac (**54**, R: NHAc)
KDN (**55**, R: OH)

KDO (**56**)

DAH (**57**)

D-KDG (**58**)

Fig. 6. Sialic and ulosonic acids

1992; Enders et al. 1993a). This strategy resembled the natural biosynthetic pathway, whereby PEP undergoes C–C linkage to aldehydes by means of class I aldolase catalyzed reactions. Pursuing a biomimetic route, we investigated the asymmetric organocatalytic synthesis of sialic and ulosonic acids that led us to obtain a direct precursor of D-KDG (**58**) as well as advanced intermediates of KDO (**56**) and analogues (Enders and Gasperi 2007). In our biomimetic approach we chose the pyruvic aldehyde dimethyl acetal **59** as masked pyruvic acid in aldol reactions with various aldehydes **6** (Scheme 14). Since a number of amine-catalytic systems gave different results with respect to the substrates used, we initially tested the enantiopure pyrrolidine derivatives (30 mol%) in the reaction of **59** with 2-methyl propanal (**6**, R = *i*-Pr) as a model carbonyl component by performing the reaction in dimethyl-sulfoxide (DMSO) (Scheme 14).

The best results were observed in the reactions with (*S*)-proline and the tetrazole **61** affording the aldol product **60** in reasonable yield (51%–53%) and good enantioselectivity (73%–75% *ee*). As both catalysts produced comparable results, the optimization of the reaction was performed with the much less expensive and proteinogenic amino acid proline. While screening diverse solvents of varying polarity failed to improve either the yield or the enantiomeric excess, cooling the reaction mixture to 4°C and using an excess of **59** revealed that the aldol **60**

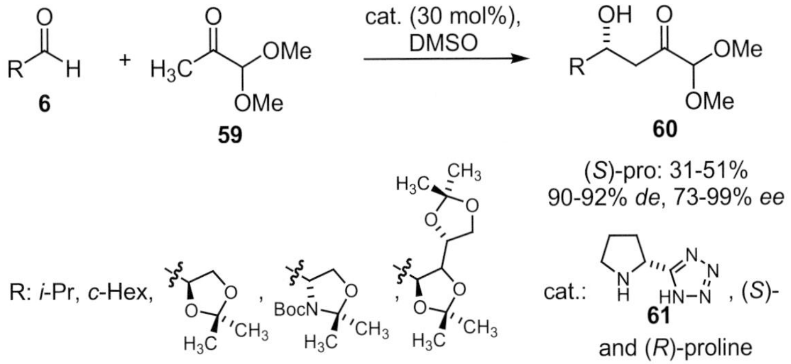

Scheme 14. Organocatalyzed aldol reactions of the pyruvic aldehyde dimethyl acetal **59** with aldehydes **6**

was obtained with a considerably higher enantioselectivity (93% *ee*). However, the Mannich-elimination side reaction could not be avoided, as well as the formation of the acetal self-aldolization product, which

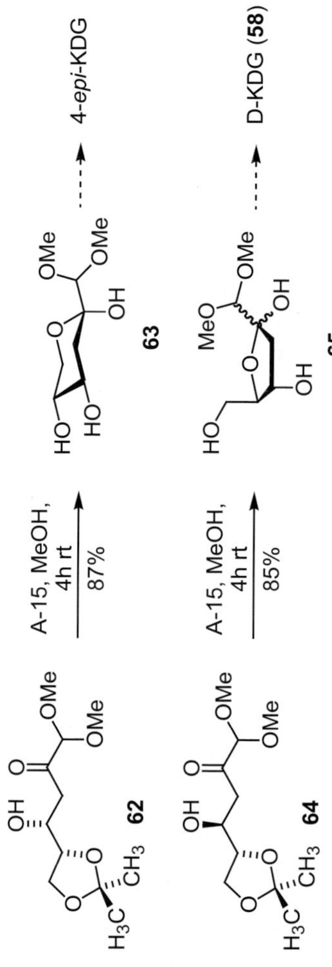

Scheme 15. Deprotection of the aldol products **62** and **64**

should lead to a decreasing efficiency of the catalyst towards the desired pathway. Afterwards, the developed conditions were evaluated using the α-branched aldehydes **6** as carbonyl components. In spite of the modest yields (31%–45%), in all cases the expected aldol product was obtained in very good diastereomeric excess (90%–92% *de*) with an increase in the reaction rate when aldehydes bearing a heteroatom in the β position were used. The given stereochemical assignments are based on the X-ray crystal structure analysis, which proved an (*R*)-configuration at the newly formed stereogenic centre and is in agreement with the transition state model for the proline-catalyzed aldol reaction. The epimeric aldol products **62** and **64** were easily deprotected with an acidic ionic-exchange resin (Amberlyst 15) to give the hemiketals **63** and **65** (Scheme 15).

Upon acidic treatment **62** afforded only the pyranose form **63** as a single anomer, which could be straightforwardly converted into the 4-*epi*-KDG. In contrast, the aldol **64** provided only the furanose ring **65** in both possible anomeric forms (α:β = 1:1), whose stereogenic centers C(4) and C(5) have the correct configurations as a direct precursor of 2-keto-3-deoxy-D-glucosonic acid (D-KDG, **58**). In conclusion, we have developed a direct entry to precursors of ulosonic and sialic acids by means of an organocatalytic approach closely resembling the natural pathway. Despite the modest yields, the proline catalyzed aldol reaction might easily be scaled up to yield gram amounts of key intermediates for the synthesis of biologically challenging molecules. The high stereoselectivity of our protocol as well as the broad range of proline organocatalysis makes it suitable to a wide scope of substrates.

2.3 First Direct Organocatalytic α-Fluorination of Aldehydes and Ketones

Due to the unique properties of fluorinated compounds there has been a growing interest in the selective introduction of one or more fluorine atoms into organic molecules (Shimizu and Hiyama 2005; Welch and Eswarakrishnan 1991; Soloshonok 1999; Chambers 2004). In particular, the resistance to many metabolic transformations, as well as the increased lipophilicity and stability of fluorine-containing compounds is of importance to the pharmaceutical and agrochemical industry

(Shimizu and Hiyama 2005; Welch and Eswarakrishnan 1991; Sološhonok 1999; Chambers 2004; for recent reviews on catalytic α-fluorination, see: Taylor et al. 1999; Muñiz 2003; Ma and Cahard 2004; Pihko 2006; Prakash and Beier 2006; Bobbio and Gouverneur 2006; Hamashima and Sodeoka 2006). In 1997 our group developed an asymmetric synthesis of α-fluoroketones using chiral α-silylketones (Enders and Potthoff 1997; Enders et al. 2001). Eight years later we expanded the entry to α-fluorocompounds by an organocatalytic approach (Enders and Hüttl 2005). Inspired by the organocatalytic enantioselective α-chlorination and α-bromination of aldehydes and ketones, the direct proline-catalyzed α-fluorination was investigated. Cyclohexanone (**69**) was used as the model substrate and Selectfluor (**68**) as the fluorinating agent (Scheme 16). Due to the low solubility of Selectfluor and (*S*)-proline in a range of common organic solvents, acetonitrile emerged as the reaction medium of choice giving the highest conversion to the α-fluorinated ketone **70**. Because it is well known that the addition of acids can promote enamine formation as well as increase the reactivity of Selectfluor, equal amounts of trifluoro acetic acid and (*S*)-proline were added to the reaction mixture. Furthermore, the solubility of proline in acetonitrile could be significantly increased by the addition of the acid. Accordingly, the monofluorinated cyclohexanone **70** could be obtained in a good yield of 73% after 1.5 days.

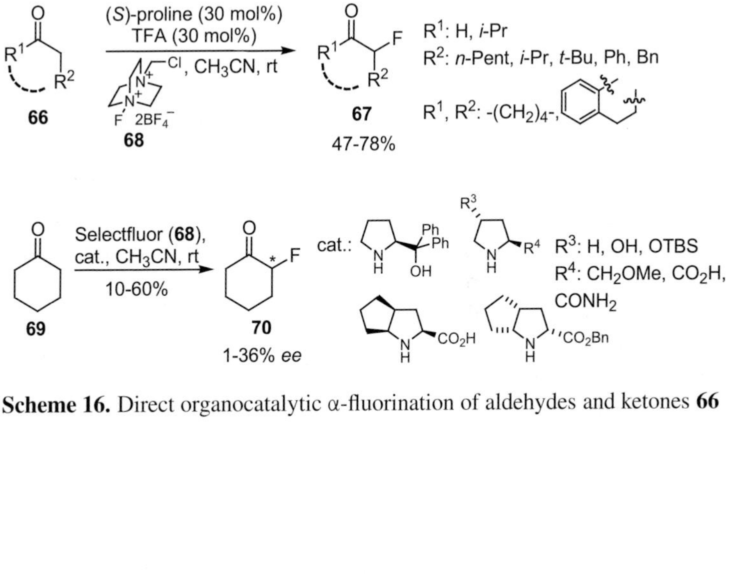

Scheme 16. Direct organocatalytic α-fluorination of aldehydes and ketones **66**

Using the procedure above, several aldehydes and ketones could be smoothly transformed to the fluorinated carbonyl compounds **67** (Scheme 16) (Enders and Hüttl 2005). It turned out that the aldehydes were highly reactive forming the corresponding α-fluoroaldehydes after 3.5 h in good yields and by performing the fluorination at 0°C side reactions, such as homo aldol reaction, were suppressed. As expected the ketones were less reactive, thus longer reaction times of up to 6 days were necessary to give satisfying yields.

The enantiomeric excess of the (*S*)-proline-catalyzed α-fluorination was generally found to be rather low. For example, the enantiomeric excess of **70** was determined to be 29%. Because of the importance of enantiopure organofluorine compounds we studied the efficiency of several organocatalysts with respect of the asymmetric α-fluorination of cyclohexanone as the model substrate and Selectfluor as the fluorinating agent (Scheme 16). Unfortunately, neither (*S*)-proline nor its derivatives were able to catalyze the α-fluorination with high stereoselectivity. The highest enantiomeric excess was observed with *trans*-4-hydroxy-(*S*)-proline (36% *ee*).

Although the highly enantioselective organocatalytic α-fluorination of aldehydes was reported in 2005 independently by three different research groups (Marigo et al. 2005a; Steiner et al. 2005; Beeson and MacMillan 2005) using NFSi as fluorinating agent and diarylprolinol silylethers or imidazolidinone derivatives, respectively, the efficient asymmetric amine-catalyzed α-fluorination of ketones still remains unsolved.

2.4 Asymmetric Organocatalytic Multicomponent Domino Reactions

Efficient and elegant syntheses of complex organic molecules with multiple stereogenic centers continue to be important in both academic and industrial laboratories (Nicolaou et al. 2003, 2006). In particular, catalytic asymmetric multicomponent domino reactions, used in the course total syntheses of natural products and synthetic building blocks, are highly desirable (Nicolaou et al. 2003, 2006; Tietze and Beifuss 1993; Tietze 1996; Tietze and Haunert 2000; Wasilke et al. 2005; Ramón and Yus 2005; Guo and Ma 2006; Pellissier 2006; Pellisier 2006; Tietze et al. 2006; Chapman and Frost 2007). These reactions avoid time-consuming and costly processes, including the purification of intermediates and steps involving the protection and deprotection of functional groups, and they are environmentally friendly and often proceed with excellent stereoselectivities (Tietze and Beifuss 1993; Tietze 1996; Tietze and Haunert 2000; Wasilke et al. 2005; Ramón and Yus 2005; Guo and Ma 2006; Pellissier 2006; Pellisier 2006; Tietze et al. 2006; Chapman and Frost 2007). Therefore, the design of new catalytic and stereoselective cascade reactions is a continuing challenge at the forefront of synthetic chemistry (for a review see Enders et al. 2007a). In addition, organocatalytic cascade reactions can be described as biomimetic, as they are reminiscent of tandem reactions that may occur during biosyntheses of complex natural products (AL Lehninger 1993; Katz 1997; Koshla 1997; Koshla et al. 1999; Mann 1999; Staunton and Weissmann 2001). In particular, secondary amines-capable of both enamine as well as iminium catalysis-can be used to design domino processes (Enders et al. 2007a). Recently, List (Yang et al. 2005), MacMillan (Huang et al. 2005) and Jørgensen (Marigo et al. 2005b) showed examples of this property by merging first iminium and then enamine activation. In contrast, we embarked on a reverse strategy using enamine activation of the first substrate to start a triple cascade. The retrosynthetic analysis that we have devised is depicted in Fig. 7.

The assembly of the cyclohexene derivatives **D** should be feasible via an organocatalyzed domino Michael/Michael/aldol condensation sequence starting from a linear aldehyde **A**, a nitroalkene **B** and an α,β-unsaturated aldehyde **C**.

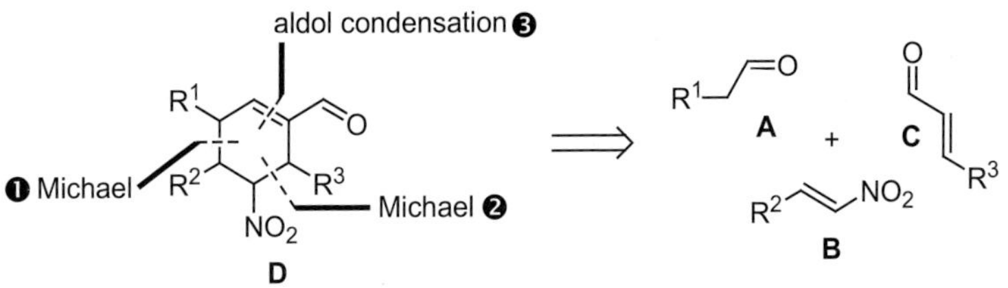

Fig. 7. Retrosynthetic analysis of the tetrasubstituted cyclohexene carbaldehydes D

This catalytic cascade was first realized using propanal, nitrostyrene and cinnamaldehyde in the presence of catalytic amounts of *O*TMS-protected diphenylprolinol ((*S*)-**71**, 20 mol%), which is capable of catalyzing each step of this triple cascade. In the first step, the catalyst (*S*)-**71** activates component **A** by enamine formation, which then selectively adds to the nitroalkene **B** in a Michael-type reaction (Hayashi et al. 2005). The following hydrolysis liberates the catalyst, which is now able to form the iminium ion of the α,β-unsaturated aldehyde **C** to accomplish in the second step the conjugate addition of the nitroalkane (Prieto et al. 2005). In the subsequent third step, a further enamine reactivity of the proposed intermediate leads to an intramolecular aldol condensation. Hydrolysis returns the catalyst for further cycles and releases the desired tetrasubstituted cyclohexene carbaldehyde **72** (Fig. 8) (Enders and Hüttl 2006).

After optimization of the reaction conditions, we carried out the domino reaction in toluene between 0°C and room temperature; the reaction was finished between 16 hours and 3 days. Nearly stoichiometric amounts of all three components were used (**A**:**B**:**C** =1.20:1.00:1.05). Complete enantiocontrol (97% to ≥99% *ee*, except in one case only 79% *ee*) was obtained.

The best results concerning yield and selectivity were obtained with 20 mol% of the TMS-ether (*S*)-**71**, which is easily available from the natural amino acid (*S*)-proline. Using the (*R*)-proline derived catalyst (*R*)-**71** affords the enantiomeric products *ent*-**72** (Scheme 17).

The fact that the residues R^1–R^3 of the precursors **A**, **B** and **C** can be broadly varied demonstrates the high flexibility of our approach. R^1 of component **A** can bear simple to demanding residues as well as valuable functional groups (Scheme 17). R^2 is limited to aromatic and heteroaromatic substituents, due to the lower reactivity of the aliphatic nitroalkenes. The residue R^3 of component **C** allows the broadest diversity. Aliphatic as well as aromatic moieties are tolerated. Furthermore, acrolein (R^3 = H) can be used, affording trisubstituted cyclohexene carbaldehydes. The best yields were obtained with aromatic substituents R^2 and R^3 (38%–60%). The replacement of R^3 by aliphatic residues led to lower yields (25% and 29%), whereas sterically demanding aldehydes **A** had less influence on the yield. In contrast, the variation of the residues had only a small impact on the diastereoselectivity (68:32-

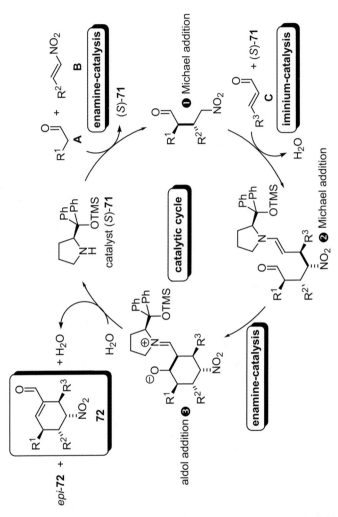

Fig. 8. Proposed catalytic cycle of the triple cascade

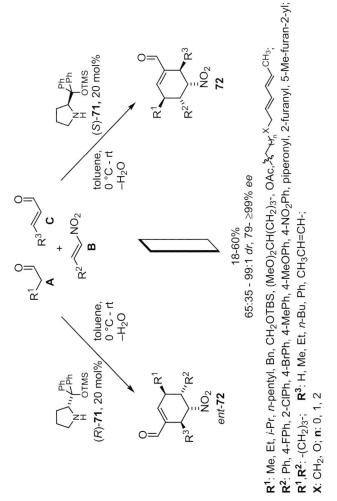

Scheme 17. Asymmetric, organocatalytic three-component multistep reaction cascade

99:01 *dr*), and we were pleased that in all cases nearly complete enantiocontrol (97% to ≥98% ee) was reached.

This cascade reaction generates four stereogenic centers, and theoretically could give rise to $2^4 = 16$ different stereoisomers. We note that in fact this asymmetric cascade forms enantioselectively just two diastereomers (α-nitro epimers), which are easy to separate by flash chromatography. The reason for the high stereoselectivity is the first Michael addition, which is known to proceed with high diastereo- and enantioselectivity (List et al. 2001; Enders and Seki 2002). Clearly this selectivity is kept or enhanced in the second step via a sterically favorable interaction between the iminium species and the nitroalkane (Prieto et al. 2005). The relative and absolute configurations of the cyclohexene carbaldehydes **72** were determined by ^{1}H-NMR (NOE) experiments and X-ray crystallography in several cases, which are in agreement with respective related organocatalytic conjugate additions.

In addition, we used this method to develop a one-pot procedure to construct polyfunctionalized tricyclic carbon frameworks, which contain up to eight stereogenic centers, with high stereocontrol (Scheme 18) (Enders et al. 2007b). Such functionalized decahydroacenaphthylene (*n* = 0) and decahydrophenalene (*n* = 1) carbon cores are typical structural features of diterpenoid natural products, such as the hainanolides and amphilectanes (Piers and Romero 1993; Li et al. 2004). The assembly of the condensed polycyclic structures **76** and **77** was accomplished by using the organocatalyzed domino Michael/Michael/aldol condensation sequence (Enders and Hüttl 2006) starting from the simple aldehyde **73**, which already bears a diene moiety, and the substrates **74** and **75**, followed by an intramolecular Diels-Alder reaction (IMDA).

As shown in Scheme 18, the domino reaction was carried out without any change in the reaction conditions. After consumption of the starting materials, the mixture was diluted with dichloromethane and cooled to −78°C, and an excess of dimethylaluminum chloride was added to ensure complete conversion of the intermediates. When the reaction was complete, the title compounds **76** or **77** were successfully isolated after separation of the minor isomers by flash chromatography. The one-pot synthesis was applicable to a range of different substrates, and through variation of the chain length of the residue R^1, the ring size can be adjusted to five- or six-membered rings. We observed very good yields

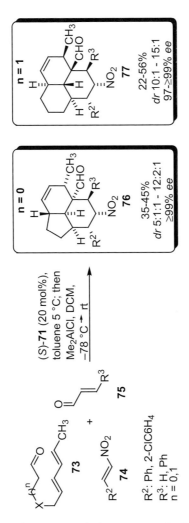

Scheme 18. Asymmetric, organocatalytic one-pot synthesis of **76** and **77**

for a four-step procedure (35%–56%). In this process, five new C–C bonds and seven or eight new stereocenters were generated with complete enantioselectivity ($\geq$99% *ee*). Interestingly, the reactions of structures that consisted of three six-membered rings **77** gave only two diastereomers with a ratio of 10:1 to 15:1, whereas reactions of the more strained structures **76** gave three diastereomers in ratios of 5:1:1 to 12:2:1. One of the diastereomers emanates from the triple cascade as an epimer in the position α to the nitro group, the other diastereomer is formed in the course of the IMDA reaction. The relative and absolute configuration of the complex structures was determined by X-ray analysis of the compound **76** and also by NOE measurements based on the known configuration of the intermediates. The relative and absolute configuration of compounds **76** and **77** supports the mechanism that we proposed earlier (Enders and Hüttl 2006), and also the selectivity for *trans*-fused *endo* configuration of the intramolecular cycloaddition promoted by a Lewis acid (Enders et al. 2007b).

A comparison of the relevant transition states of the intramolecular Diels-Alder reaction can explain the observed configuration of the products **76** and **77**. In the case of **76** ($n = 0$), the diene moiety is more likely to approach from underneath the enal face in an '*endo* manner' (the Alder rule) because of a steric interaction with the phenyl group. The other structures **77** that contain a longer side chain ($n = 1$) allow the approach from both beneath and above the enal face, but NOE analyses of the isolated products **77** revealed that the diene approaches the dienophile from the top. Thus, in both systems, the trans-fused endo-configuration is preferred because of steric interactions with the phenyl substituents and the nitro group (Enders et al. 2007b).

3 C–C-Bond Formations Employing N-Heterocyclic Carbene Catalysis

3.1 Enzymes as Archetypes

In Nature, realms of complex biochemical reactions are catalyzed by enzymes, many which lack metals in their active site. Among them, nucleophilic acylation reactions are catalyzed by transketolase enzymes in the presence of coenzyme thiamine (**78**, vitamin B_1), a natural thia-

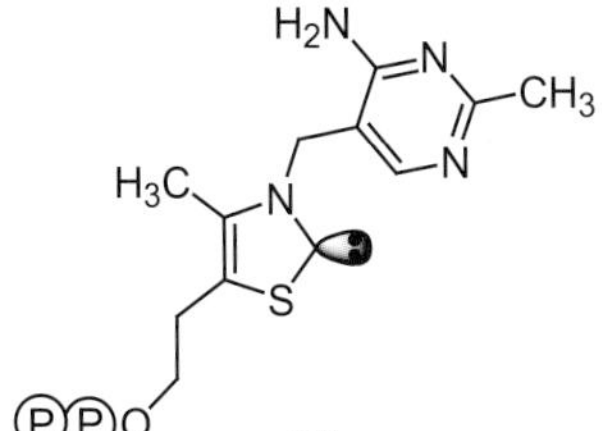

Biomimetic Organocatalytic C–C-Bond Formations 83

Fig. 9. The coenzyme thiamine (vitamin B_1)

zolium salt (Stryer 1995). *Inter alia*, this coenzyme is present in beaker's yeast and might serve as an example of highly selective chemical reactions that are accomplished in vivo. In 1951, Mizuhara et al. found that the catalytic active species of the coenzyme thiamine is a nucleophilic carbene (Mizuhara et al. 1951). The biochemistry of thiamine-dependent enzymes has been studied in detail resulting in broad applications as synthetic tools (Fig. 9) (Jordan 2003; Schoerken and Sprenger 1998; Sprenger and Pohl 1999).

An X-ray structure of a thiamine dependent transketolase enzyme was determined by Schneider et al. after isolation from *Saccharomyces cerevisiae* in the 1990s and is shown in Fig. 10 (Sundström et al. 1993; Nilsson et al. 1997). The thiamine cofactor is embedded in a narrow channel in the centre of the enzyme. From the complex surrounding of the heart of this enzyme it seems to be obvious that chemical reactions at the catalytically active site in this channel proceed inevitably with high selectivities.

3.2 The Benzoin Condensation

3.2.1 The Breslow Mechanism

The benzoin condensation catalyzed by N-heterocyclic carbenes has been investigated intensively. First investigations date back to 1832 when Wöhler and Liebig discovered the cyanide-catalyzed coupling of benzaldehyde to benzoin (Wöhler and Liebig 1832). In 1903 Lapworth postulated a mechanism for this reaction in which an intermediate carbanion is formed by hydrogen cyanide addition to benzaldehyde fol-

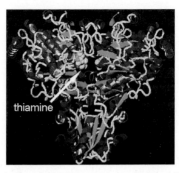

Fig. 10. Structure of the transketolase enzyme isolated from *Saccharomyces cerevisiae*

lowed by deprotonation (Lapworth 1903). Here, the former carbonyl carbon features an inverted, nucleophilic reactivity. This intermediate 'active aldehyde' exemplifies the 'Umpolung' concept of Seebach and coworkers (Seebach 1979). In 1943 Ukai et al. recognized that thiazolium salts could also be used as catalysts in the benzoin condensation (Ugai et al. 1943). Breslow et al. proposed in 1958 on the basis of the work of Lapworth a mechanistical model for the thiazolium salt-catalyzed benzoin condensation (Breslow 1958). In this mechanism, the catalytically active species is a thiazolin-2-ylidene **80**, a carbene compound, which is formed in situ by deprotonation of the thiazolium salt **79**. The postulated catalytic cycle is shown in Scheme 19.

Breslow assumed that the thiazolium salt **79** is deprotonated at its most acidic position to form the thiazolin-2-ylidene **80**. Nucleophilic attack of the carbonyl function of an aldehyde molecule **81** then generates the alkylthiazolium salt **82**. Deprotonation/reprotonation leads to the active aldehyde in form of the resonance stabilized enaminol-type Breslow intermediate **83**. This nucleophilic acylation reagent **83** (equivalent to a d^1-synthon in the terminology of Seebach et al.) reacts again with an electrophilic substrate such as the carbonyl group of a second aldehyde molecule. The intermediate **84** eliminates benzoin (**85**) and the original carbene catalyst **80** is regenerated.

Stetter et al. were the first to use thiazolium salts as catalysts for the preparation of acyloin and benzoin compounds on a preparative

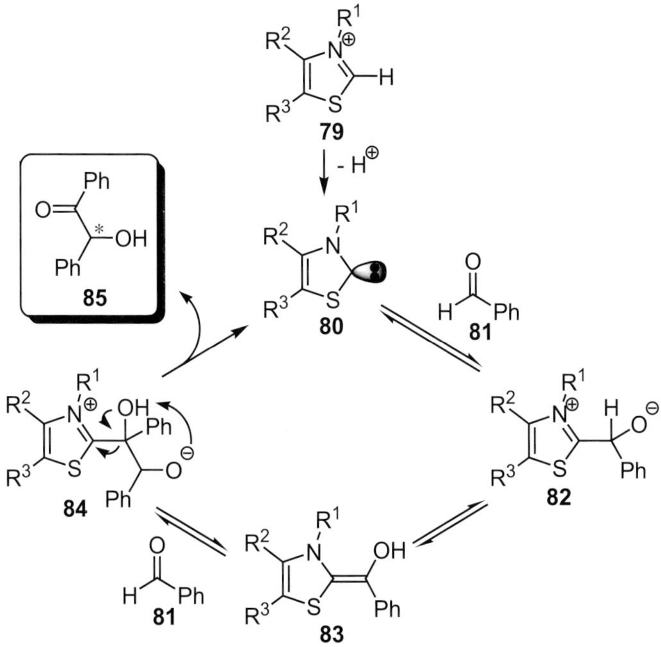
79
- H⊕
80
81
82
83
84
85
81
Ph
OH
Ph
H
Ph
OH
Ph

Scheme 19. Catalytic cycle of the benzoin condensation as proposed by Breslow

scale. They could utilize this synthetic concept for the preparation of numerous α-hydroxy ketones. Aliphatic aldehydes showed the best results with catalyst **86a**, whereas, for aromatic substrates the thiazolium salts **86b** or **86c** were better catalysts (Fig. 11) (Stetter et al. 1976; Stetter and Kuhlmann 1991).

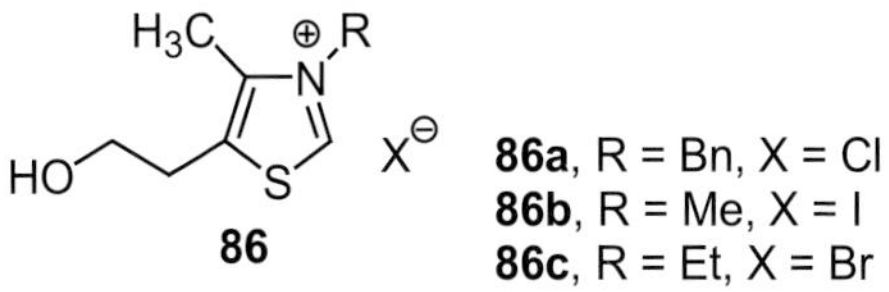
86
86a, R = Bn, X = Cl
86b, R = Me, X = I
86c, R = Et, X = Br

Fig. 11. Thiazolium salts utilized by Stetter et al.

3.2.2 Stable Carbenes

Side-by-side with the research on carbene catalyzed benzoin condensations (for reviews see: Enders and Balensiefer 2004; Zeitler 2005; Enders et al. 2007c), much effort has been devoted towards on the isolation of stable carbenes. The chemistry of nucleophilic carbenes was intensively studied by Wanzlick et al. in the 1960s (Wanzlick 1962; Wanzlick and Kleiner 1963). Until then, the divalent carbenes were considered as highly reactive intermediates. In retrospective, Wanzlick et al. had N-heterocyclic carbenes in their hands, but did not believe that they would be able to actually isolate them. Only the dimeric forms, the electron rich olefins, were reported. It was more than 25 years later that Bertrand et al. (Igau et al. 1988; Igau et al. 1989) and Arduengo et al. (Arduengo et al. 1991; Arduengo et al. 1992) presented carbene compounds that were stable at room temperature. The best known representatives, the imidazolin-2-ylidenes, for example **87**, and derivatives such as **88**, are often called today Wanzlick carbenes (Fig. 12).

However, the carbene character of Bertrand´s phosphinocarbene **89** was initially doubted, as its chemical reactivity appeared to resemble more a phosphaacetylene **90** (Dixon et al. 1991; Soleihavoup et al. 1992; Bourissou and Bertrand 1999), but recent studies seem to support its carbene character (Scheme 20) (Despagnet et al. 2002).

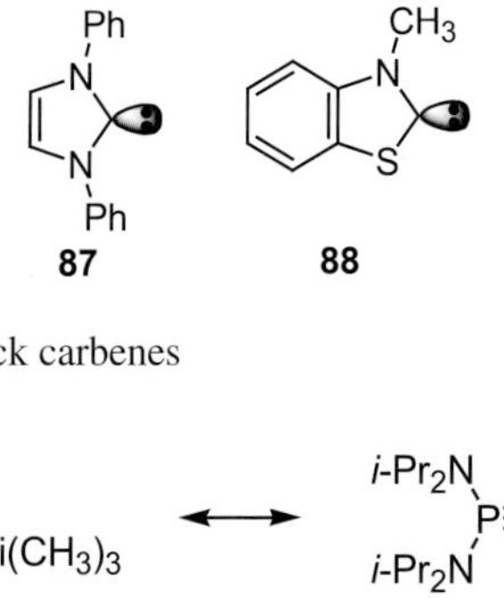

Fig. 12. Typical Wanzlick carbenes

Scheme 20. Phosphinocarbene reported by Bertrand et al.

Thus, the imidazolin-2-ylidene **87a** synthesized by Arduengo et al. in 1991 is now referred to as the first N-heterocyclic carbene to be isolated and characterized. Its preparation was achieved by deprotonation of the corresponding imidazolium salt **91**. Later, Arduengo et al. also synthesized stable carbenes with less bulky substituents. It should be noted that the structure of these carbenes is identical with the Wanzlick carbenes. Progress in laboratory methods were finally decisive for the successful isolation of these carbenes (Scheme 21).

Since Arduengo's initial reports, numerous stable N-heterocyclic carbenes of various types have been synthesized (for reviews on the history see: Arduengo and Kraftczyk 1998; see also Herrmann 2002; César et al. 2004; Kauer Zinn et al. 2004; Díez-González and Nolan 2005; Garrison and Youngs 2005; Tekavec and Louie 2007; for reviews see Herrmann and Koecher 1997; Arduengo 1999; Bourissou et al. 2000; Perry and Burgess 2003; Korotkikh et al. 2005; Hahn 2006). Applications in organocatalysis and metal-based catalysis have developed rapidly.

Inspired by this success and based on our work on the asymmetric Stetter reaction in the late 1980s, our research group together with that of Teles studied the triazole heterocycle as an alternative structure.

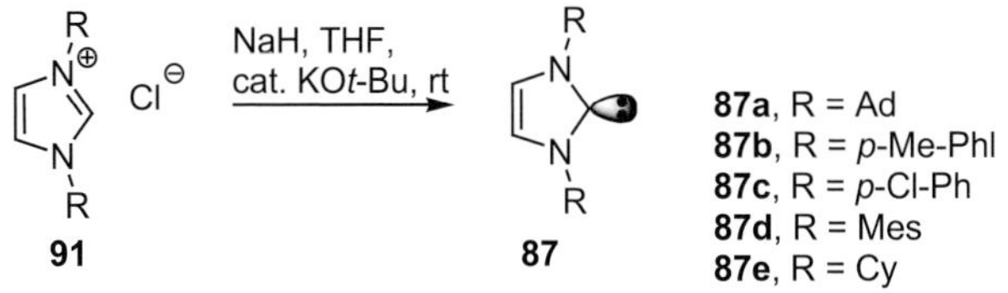

87a, R = Ad
87b, R = p-Me-Phl
87c, R = p-Cl-Ph
87d, R = Mes
87e, R = Cy

Scheme 21. Preparation of stable imidazolin-2-ylidenes by Arduengo et al.

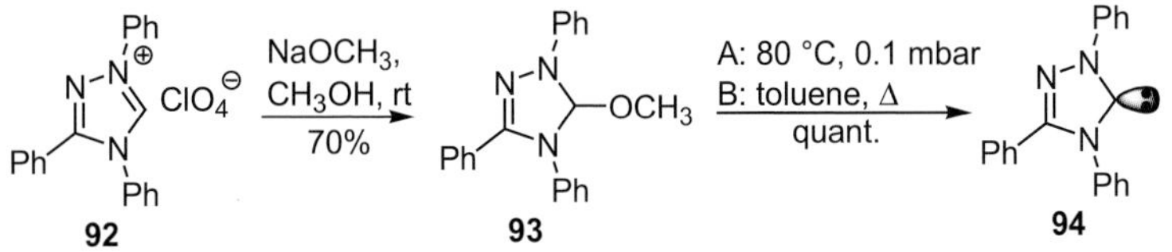

Scheme 22. Synthesis of the stable carbene **94** developed by Enders, Teles et al.

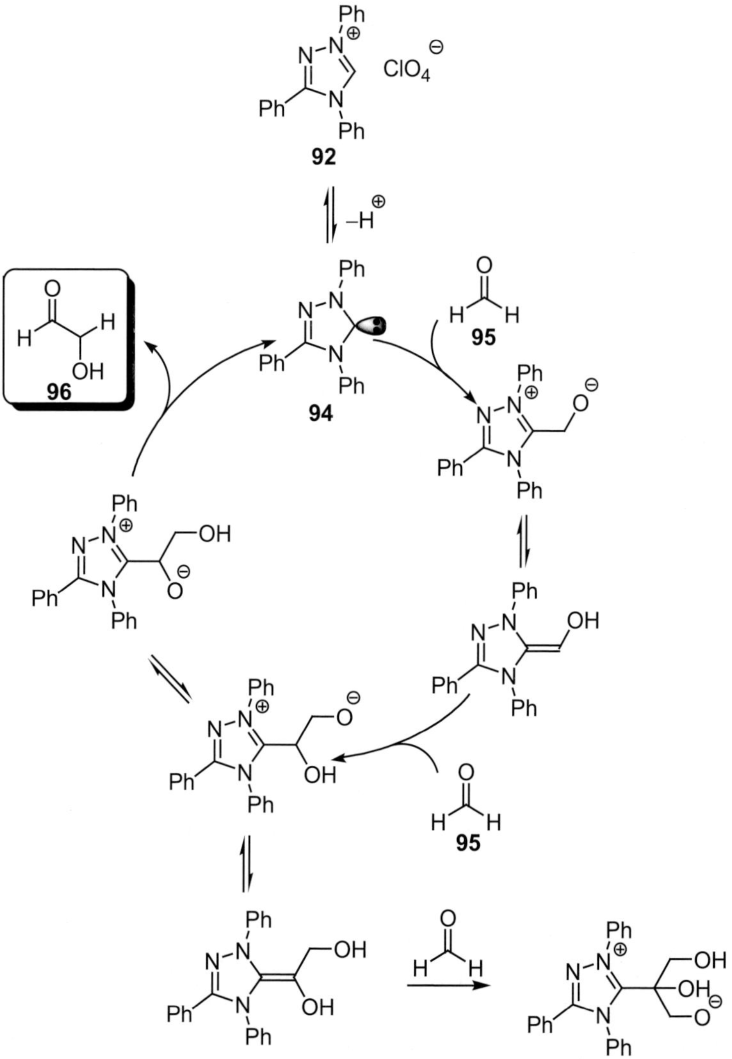

Scheme 23. Postulated mechanism of the formoin condensation reported by Teles et al.

In 1995 we synthesized the triazol-5-ylidene **94** starting from the triazole precursor **92** (Enders et al. 1995b). The crystalline carbene **94** was obtained from **92** by addition of methanolate to give the adduct **93** followed by α-elimination of methanol, either at 80°C under vacuum or in refluxing toluene. **94** proved to be stable at temperatures up to 150°C in absence of air and moisture (Enders et al. 1996a; Enders et al. 1997a; Enders et al. 1997b; Enders et al. 2003). Ab initio calculations on the parent heterocyclic system suggested that the π-donation from the adjacent nitrogen atoms lead to a significant transfer of electron density into the formally vacant p-orbital at the carbon atom (Raabe et al. 1996). **94** showed the typical behavior of a nucleophilic Wanzlick-type carbene and was the first carbene to be commercially available (Scheme 22).

On an industrial scale, methane, the main component of natural gas, can easily be converted to formaldehyde. An efficient catalytic condensation of formaldehyde to dihydroxyactone or glycolaldehyde would thus provide a route to C_2- and C_3-chemicals from methane. The triazolin-5-ylidene **94** turned out to be a powerful catalyst for the conversion of formaldehyde (**95**) to glycolaldehyde (**96**) in the 'formoin reaction' (Teles et al. 1996). This reactivity is a useful complement to the catalytic properties of thiazolium salts which mainly afford 1,3-dihydroxy acetone as product (Scheme 23) (Castells et al. 1980; Matsumoto and Inoue 1983; Matsumoto et al. 1984). As triazolium ylides are much more stable than thiazolium ylids, the elimination of glycolaldehyde occurs faster than the addition of the third formaldehyde molecule.

3.2.3 Asymmetric Intermolecular Benzoin Reactions

The promising results of triazolium salt catalysis inspired our research group to synthesize a variety of chiral triazolium salts for the asymmetric benzoin condensation (Enders et al. 1996b; Enders and Breuer 1999; Teles et al. 1999). Extensive investigations have shown that the enantiomeric excesses and catalytic activities are highly dependent on the substitution pattern of the triazolium system. The most active catalyst (*S*, *S*)-**97**, which is readily available from an intermediate of the industrial chloramphenicol synthesis, provided benzoin (**85′**) in its (*R*)-configuration with 75% *ee* and a good yield of 66%. Remarkably, only low catalyst loadings were necessary (1.25 mol%), which indicated that the activity increased by almost two orders of magnitude compared with the chiral thiazolium salts used before. The application of the catalyst system to other electron-rich aromatic aldehydes **6** gave the corresponding α-hydroxy-ketones **85** with enantiomeric excesses up to 86%. Electron-deficient aldehydes showed significantly lower inductions (Scheme 24).

Attempts to apply catalyst (*S*, *S*)-**97** to aliphatic aldehydes resulted in very low enantiomeric excesses and yields. Utilizing triazolium salt (*S*)-**98** yielded only low enantiomeric excesses up to 26% and only moderate yields (Scheme 25) (Breuer 1997).

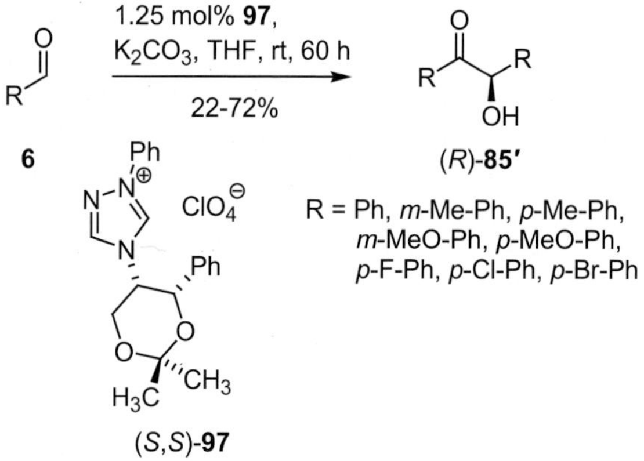
1.25 mol% 97,
K2CO3, THF, rt, 60 h
22-72%
6
(R)-85'
R = Ph, m-Me-Ph, p-Me-Ph,
m-MeO-Ph, p-MeO-Ph,
p-F-Ph, p-Cl-Ph, p-Br-Ph
(S,S)-97

Scheme 24. Asymmetric benzoin condensation

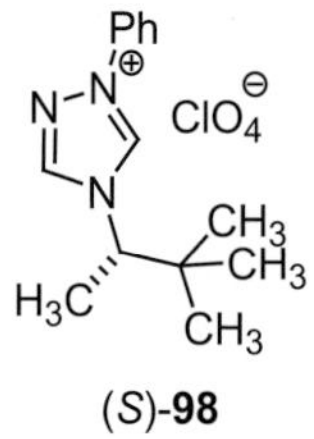
(S)-98

Scheme 25. Precatalyst **98** for benzoin reactions with aliphatic substrates

Further contributions to the research on the asymmetric benzoin condensation were made by Leeper et al. using novel chiral, bicyclic thiazolium salts, which led to enantiomeric excesses up to 21% and yields up to 50% (Knight and Leeper 1997). Another thiazolium catalyst containing a norbonane backbone gave benzoin in quantitative yields with an enantiomeric excess of 26% (Gerhards and Leeper 1997). In 1998, Leeper et al. reported novel chiral, bicyclic triazolium salts that produced aromatic acyloins with varying enantioselectivities (20%–83% ee) (Knight and Leeper 1998).

Inspired by these results, a new chiral, bicyclic triazolium salt was developed in our research group in 2002 on the basis of (*S*)-*tert*-leucin. Utilizing a modified three-step procedure of Knight and Leeper the oxazolidin-2-one **99** was methylated with Meerwein's reagent to form the corresponding imino ether **100**. This was treated with phenylhydrazine to yield the corresponding hydrazone **101** and cyclized with trimethyl orthoformate in methanol to give the triazolium salt **102** in 50% yield as a crystalline solid (Scheme 26).

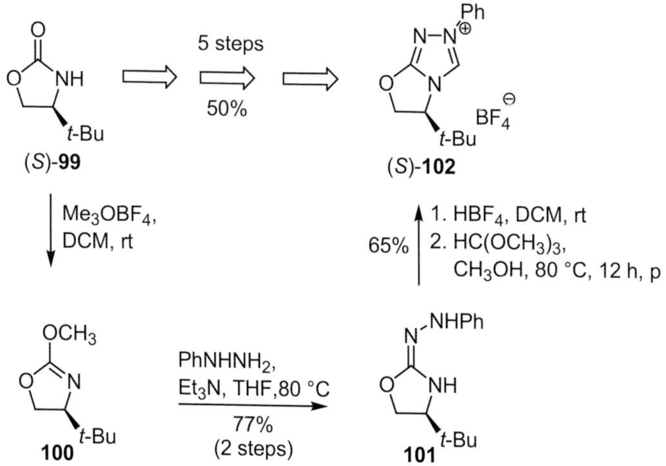

Scheme 26. Synthesis of the triazolium salt **102**

Applying (S)-**102** in the asymmetric benzoin condensation (S)-benzoin (**6**, R = Ph) was produced in very good enantioselectivity (90% ee, 83% yield) (Enders and Kallfass 2002). The condensation of numerous other aromatic aldehydes **6** yielded the corresponding α-hydroxy ketones **85** with excellent enantiomeric excesses up to 99%. As previously observed, electron-rich aldehydes gave higher asymmetric inductions than the electron-deficient ones. Lower reaction temperatures (0°C instead of room temperature) or lower amounts of catalyst caused decreased yields but slightly enhanced enantiomeric excesses (Scheme 27).

The transition state **103**, shown in Fig. 13, has been proposed to explain the observed absolute configuration. The *Si*-face of the assumed Breslow-type intermediate would be sterically shielded by the *tert*-butyl-group of the bicyclic catalyst. The second aldehyde molecule would then attack the hydroxy enamine from its *Re*-face leading to an (S)-configured product, which is actually observed in the experiments. In addition, the phenyl substituent of the enol moiety (via π-stacking) as well as the hydroxyl group (via intermolecular H-bridge activation of the aldehyde) may lead to a pre-organized transition state.

However, the (*E*/*Z*)-geometry of the Breslow intermediate has not yet been determined, but is of relevance for the pre-orientation of the second aldehyde molecule. As shown in transition state **103**, the corresponding (*E*)-isomer would probably favor a *Si-Si*-attack and therefore a (*R*)-configuration in the product. An unfavorable sterical interaction between the phenyl substituent of the enol moiety and the phenyl sub-

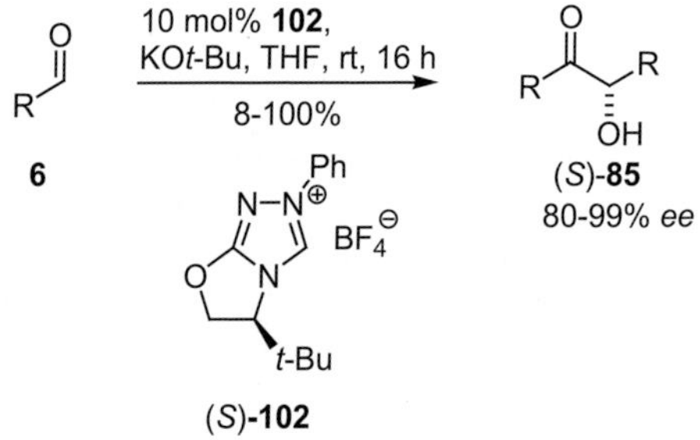

Scheme 27. Asymmetric benzoin condensation of aromatic aldhydes

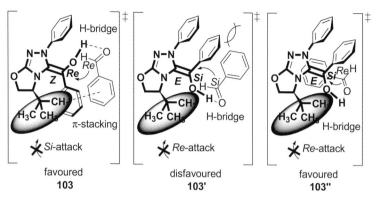

Fig. 13. Transition states for the asymmetric benzoin condensation proposed by Enders et al. and Houk et al.

stituent of the attacking aldehyde in **103'** probably disfavors this transition state. The most stable transition state has been determined to be **103"** after computational calculation by Houk et al. (Dudding and Houk 2004). In this intermediate no π-stacking occurs, but the substituent of the approaching aldehyde resides in an open pocket of the catalyst with minimum steric repulsion. Thus, the formation of (*S*)-benzoin would be favored again.

3.2.4 Asymmetric Intramolecular Benzoin Reactions

In contrast to the previously described intermolecular condensation reaction, intramolecular crossed benzoin reactions have been developed much less for a long time (Cookson and Lane 1976).

Synthetic thiazolium salts (**86c**), developed by Stetter and his co-workers and similar to thiamine itself (Stetter 1976), have been successfully used by Suzuki et al. for a diastereoselective intramolecular crossed aldehyde-ketone benzoin reaction in the course of an elegant synthesis of preanthraquinones (Hachisu et al. 2003). By using the highly functionalized isoxazole **104** as substrate, the tetracyclic α-hydroxy ketone **105** was obtained by base-promoted cyclocondensation in good yield. The high diastereoselectivity was induced by pre-existing stereocenters in the substrates (Scheme 28).

In order to develop a general method, we started investigations on simple aldehyde ketones **106** as substrates for the carbene catalyzed crossed intramolecular benzoin condensation. Independently both our group and Suzuki et al. reported that indeed various five- and six-membered cyclic acyloins **107/109/111** can be obtained as their racemic mixtures by using commercially available thiazolium salts **10** as precatalysts (Scheme 29) (Enders and Niemeier 2004; Suzuki et al. 2004; Niemeier 2006). The tetralone **107a** was obtained in significantly higher yield than that of tetralone **109a** with interchanged functionalities. This may be due to activating effect of the aromatic aldehyde in the precursor of **107a**. Substrates lacking the obviously favorable benzannulation in the resulting acyloin (**111**) showed only moderate yields as the competing intermolecular reaction was also observed. The quinone **115** was isolated starting from the corresponding dialdehyde. The α-hydroxy ketone **114** is only formed as an intermediate and undergoes air oxidation to the quinone **115** during work-up.

Unfortunately the bicyclic triazolium salt that had successfully been used in our research group for the enantioselective intermolecular benzoin condensation (Enders and Kallfass 2002) did not show any catalytic activity in the intramolecular reaction. We thus searched for alternative, easily accessible enantiopure polycyclic γ-lactams as precursors for the synthesis of novel triazolium salts (Enders et al. 2006c; for a related study see: Takikawa et al. 2006). The rigid polycyclic structure of the catalysts should allow high asymmetric inductions. A first tar-

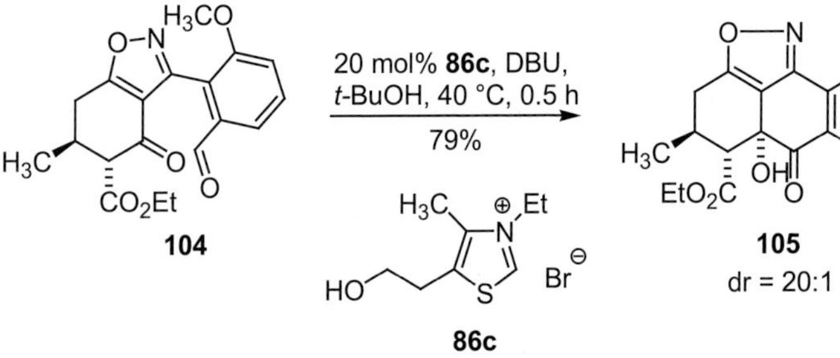

Scheme 28. Intramolecular crossed benzoin reaction by Suzuki et al.

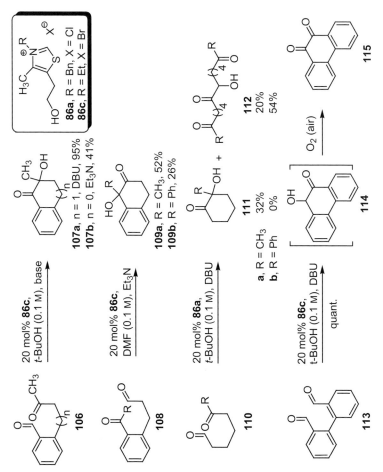

Scheme 29. Thiazolin-2-ylidene catalyzed intramolecular crossed benzoin reactions

get was found in the *cis*-bicyclic lactam **118**, which was obtained in a diastereo- and enantioselective manner in five steps starting from cyclopentanone (**116**) in 36% yield following a modified procedure developed by Omar and Frahm (Scheme 30) (Omar and Frahm 1989, 1990).

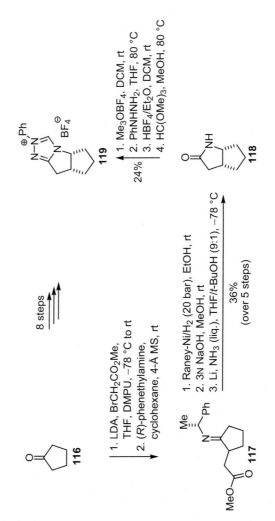

Scheme 30. Preparation of the tricyclic triazolium salt **119**

After γ-alkylation of **116** with methyl bromoacetate and subsequent condensation with (*R*)-phenethylamine, the resulting imine **117** was diastereoselectively hydrogenated, cyclized, and deprotected to give lactam **118**. The *cis*-tricyclic triazolium salt **119** was then obtained as a solid in 24% yield by a modified three-step procedure reported by Knight and Leeper (1998).

Two different concepts were chosen to increase the steric demand of the carbene catalysts and thus increase the asymmetric induction. A triazolium salt synthesis starting from L-pyroglutamic acid (**120**), which is among the cheapest chiral sources available, might generate an extremely flexible catalyst system. The *tert*-butyldimethylsilyl (TBS) and triisopropylsilyl (TIPS) protected lactams **122** could be synthesized in almost quantitative yields from L-pyroglutamic acid (**120**) by reduction to the hydroxymethyl-substituted lactam **121** and subsequent reaction with the silyl chlorides (Acevedo et al. 2001). The bulkiness at the stereogenic centre might be modified by a simple exchange of the protecting group. Conversion into the triazolium salts could be achieved by following an optimized one-pot procedure recently reported by Rovis and co-workers (Kerr et al. 2005). The lactams **122** were methylated with Meerwein's reagent to form the corresponding amidates. These were treated in situ with phenylhydrazine to yield the hydrazonium salts, which were directly cyclized with trimethyl orthoformate in methanol to give the triazolium salts **123** in moderate yields as crystalline solids (Scheme 31).

For the structural optimization of the tricyclic triazolium salt **119** the *cis*-tricyclic lactam **126** was chosen as the precursor for the synthesis of the tetracyclic triazolium salt **127**. The diastereo- and enantiopure γ-lactam **126** was synthesized following a procedure reported by Ennis et al. (Scheme 32) (Ennis et al. 1996; Nieman and Ennis 2000). α-Tetralone (**124**) was α-alkylated with ethyl bromoacetate and subsequently hydrolyzed to the corresponding carboxylic acid. Condensation with (*R*)-phenylglycinol yielded the lactam **125** as a single stereoisomer. Stereoselective reduction, dehydration of the alcohol, and acid-catalyzed enamine hydrolysis provided the *cis*-tricyclic lactam **126**. The one-pot procedure that had previously been successful in the synthe-

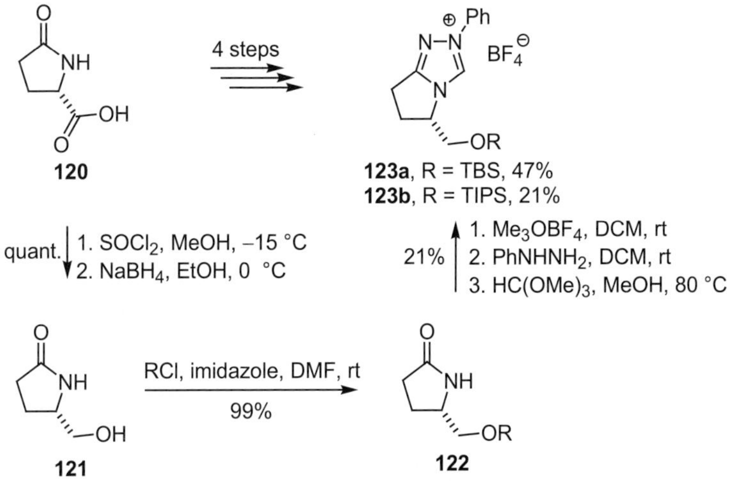

Scheme 31. Preparation of triazolium salts **123** derived from pyroglutamic acid

sis of **123** also gave access to the chiral tetracyclic triazolium salt **127**, which was obtained as a solid in 59% yield.

The asymmetric intramolecular benzoin condensation with model substrate **106** (R = Me) and the chiral triazolium salt **119** as precatalyst gave rise to the desired acyloin **107** in good yields by utilizing toluene as solvent and diazabicycloundecane (DBU) or KO*t*-Bu as the base (Scheme 33). Unfortunately only moderate enantiomeric excess (37%–48%) could be achieved, even at 5°C. The use of 10 mol% of the TBS-substituted catalyst **123** and stoichiometric amounts of DBU in toluene at room temperature enabled the methyl-substituted acyloin **106** to be obtained in high yield (92%) but with only moderate enantioselectivity (61%). Application of the TIPS substituted catalyst **123** under the same conditions resulted in an increased enantiomeric excess of 77%, which could be further improved to 84% with almost unchanged yields by performing the reaction at 5°C. The reactions with the tetracyclic catalyst **127** were conducted in tetrahydrofuran for better solubility. Furthermore, DBU was found to cause side-reactions that could be suppressed when KO*t*-Bu was used in substoichiometric amounts (9

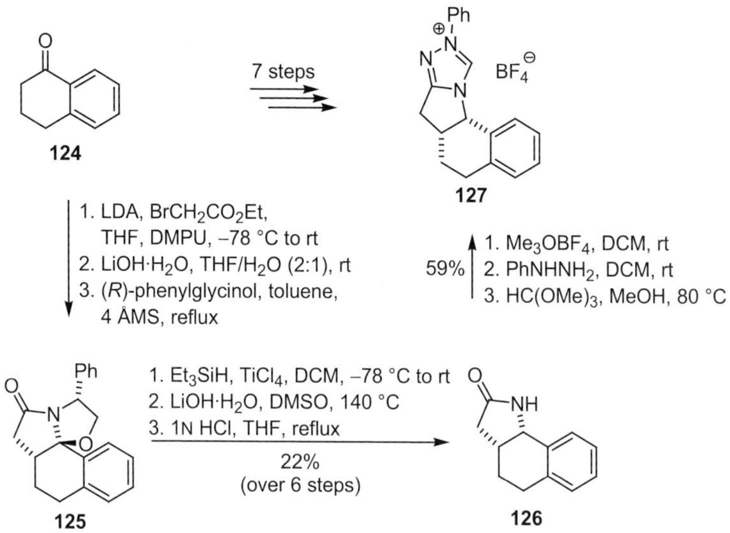

Biomimetic Organocatalytic C–C-Bond Formations

Scheme 32. Synthesis of the tetracyclic triazolium salt **127**

mol%). Product **107** (R = Me) could be obtained in high yield (93%) and with an excellent enantiomeric excess of 94%. The increased steric demand at the ketone function of the substrates **106** resulted in the reaction rate being much lower. We were pleased to see that the steric bulk of the ketone function had a significant influence on the enantiomeric excess, and almost complete inductions could be achieved (up to 98% *ee*). In general, the TIPS-substituted triazolium salt **123** derived from pyroglutamic acid delivered lower enantiomeric excesses than the tetracyclic carbene precursor **127**. However, the ease of its preparation and the low price of l-pyroglutamic acid also make it an attractive catalyst.

The absolute configuration of the quaternary stereocenter of the acyloin **107** (R = Me) produced was determined to be *S* by comparison of the measured optical rotation value with the corresponding literature data (Davis and Weismiller 1990). This stereochemical outcome might be explained by the transition state shown in Scheme 33, which is an adaptation of the transition state proposed by Dudding and Houk on the basis of computational calculations (Dudding and Houk 2004). The *Si*

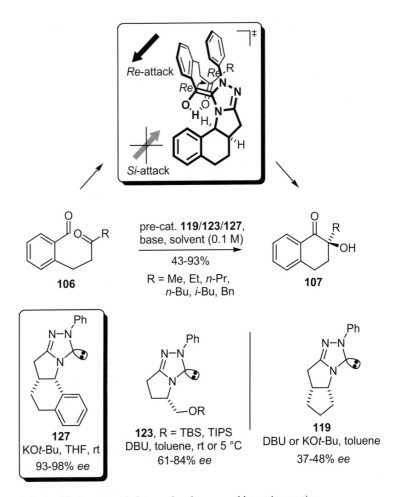

Scheme 33. Asymmetric intramolecular crossed benzoin reaction

face of the Breslow intermediate (Breslow 1958), which is formed as its *E* isomer in the hypothetical catalytic cycle, would be sterically shielded by the tetrahydro naphthalene residue of the tetracyclic catalyst. The *Re* face of the intermediate would attack the ketone function at its *Re* face

($R \neq Bn$). Furthermore, a favorable pre-arrangement for the formation of the C–C bond might be caused by the activation of the ketone function by an intramolecular H bridge. Thus, the S configuration of the new stereocenter would be preferred, which in fact is observed.

Investigations to widen the scope of the asymmetric intramolecular benzoin reaction by utilizing aldehyde-ketone **108a**, where the aldehyde and the ketone function are interchanged relative to **106**, show a promising 67% ee for the resulting acyloin **109a** as well as high yields (68%). Utilizing substrate **106b** to synthesize the five-membered cyclic acyloin **107b** resulted in an excellent yield and a good enantiomeric excess (74%). When the reaction was carried out at 5°C, the enantiomeric excess could be slightly increased to 75% (Scheme 34).

With the enantioselective intramolecular benzoin reaction established as a synthetic tool, and in combination with our efforts in the synthesis of bioactive natural products bearing a quaternary α-hydroxy ketone unit (Davis and Weismiller 1990; Heller and Tamm 1981), such as the 4-chromanone derivative (*S*)-eucomol (Böhler and Tamm 1967; Crouch et al. 1999), a catalytic asymmetric synthesis of various 3-hydroxy-4-chromanones brought about by the chiral triazolium salts **127, 123b** and **102** as pre-catalysts was investigated (Enders et al. 2006d). The sterically different pre-catalysts were chosen in order to adjust the catalyst system to the steric and electronic properties of the substrates **128**. A screening of the reaction conditions indicated 10 mol% of the

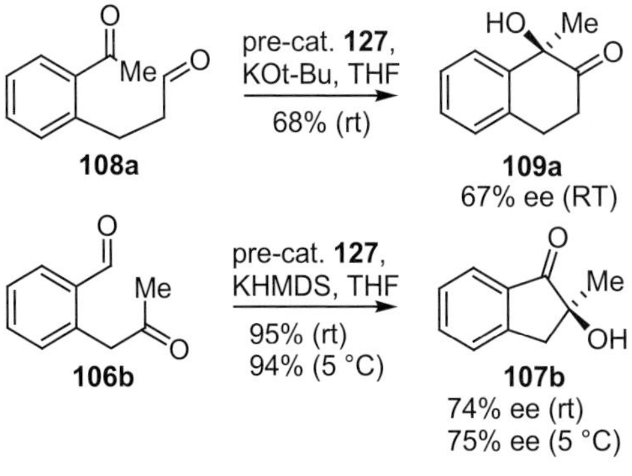

Scheme 34. Further scope of the intramolecular crossed benzoin reaction

tetracyclic pre-catalyst **127** and substoichiometric amounts of potassium hexamethyldisilazide (9 mol%) as base at room temperature and a higher dilution (0.05 M) to be suitable to suppress a competing aldol reaction. Again the tetracyclic catalyst **127** was shown to be suitable to a broad range of substrates, yielding the corresponding 4-chromanones with high enantiomeric excesses. The increased steric demand at the ketone function of substrates **128** resulted in an increase in the enantiomeric excesses up to 95%. Activated substrates (R^1 = 2,4-di-Br, 2-NO_2) required only short reaction times and could be converted with the bulkier catalyst **102**, resulting in most cases in better inductions (up to 99% *ee*). Sterically demanding functionalizations at the aromatic part

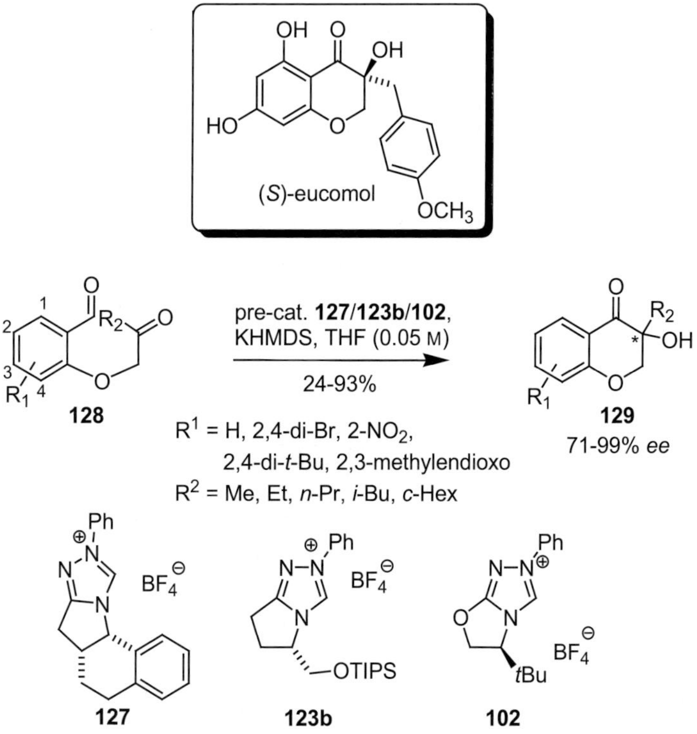

Scheme 35. Synthesis of 3-hydroxy-4-chromanones **129**

of substrates **128** (R^1 = 2,4-di-*t*-Bu) caused much lower conversions with pre-catalyst **127** even after long reaction times. By running the reaction at 45°C the conversion could be improved (Scheme 35).

The absolute configuration of chromanone **129** (R^1 = 2,4-di-Br, R^2 = *i*-Bu) was assigned by X-ray analysis of the corresponding (*S*)-

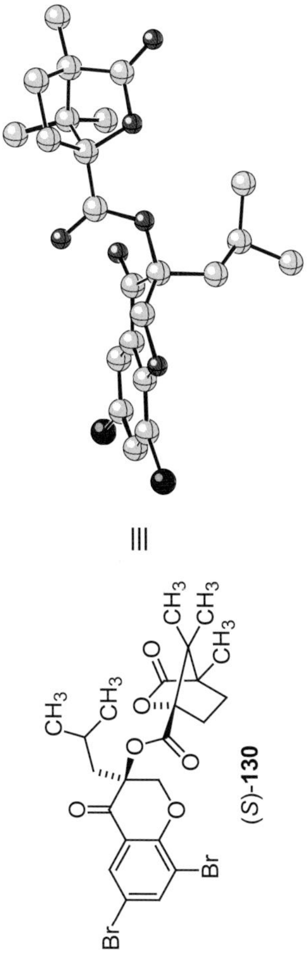

Fig. 14. X-ray analysis of (*S*)-camphanylester **130**

camphanylester [Fig. 14; CCDC 607797 (6a) and CCDC 607796 (6f) contain the supplementary crystallographic data, which can be obtained free of charge from the Cambridge Crystallographic Data Centre via www.ccdc.cam.ac.uk/data_request/cif]. Thus, the absolute configuration of the quaternary stereocenter was determined to be *S* utilizing precatalyst **127**, which is in accordance with our postulated transition state (Enders et al. 2006c; Takikawa et al. 2006).

3.3 The Stetter Reaction

3.3.1 Asymmetric Intermolecular Stetter Reactions

In the early 1970s Stetter and co-workers succeeded in transferring the concept of the thiazolium catalyzed nucleophilic acylation to the substrate class of Michael acceptors (Stetter 1976; Stetter and Schreckenberg 1973). Since then, the catalytic 1,4-addition of aldehydes **6** to an acceptor bearing an activated double bond **131** carries his name. The Stetter reaction enables a new catalytic pathway for the synthesis of 1,4-bifunctional molecules **132**, such as 1,4-diketones, 4-ketoesters and 4-ketonitriles (Stetter and Kuhlmann 1991; for a short review, see: Christmann 2005). The reaction can be catalyzed by a broad range of thiazolium salts. Stetter and co-workers found the benzyl-substituted thiazolium salt **86a** to give the best results for the addition of aliphatic aldehydes, whereas **86b** and **86c** were chosen for the addition of aromatic aldehydes. Any one of these three was found to be suitable for additions with heterocyclic aldehydes. Salt **86d** was utilized with α,β-unsaturated esters (Fig. 15).

This versatile method has found a broad application in the synthesis of organic key intermediates and diverse natural products. In the catalytic cycle, the aldehyde **6** is presumably activated by the carbene under generation of the Breslow intermediate **83** and subsequent nucle-

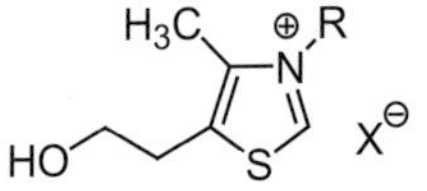

86a, R = Bn, X = Cl
86b, R = Me, X = I
86c, R = Et, X = Br
86d, R = $(CH_2)_2OEt$, X = Br

Fig. 15. Thiazolium salts **10** for the intermolecular Stetter reaction

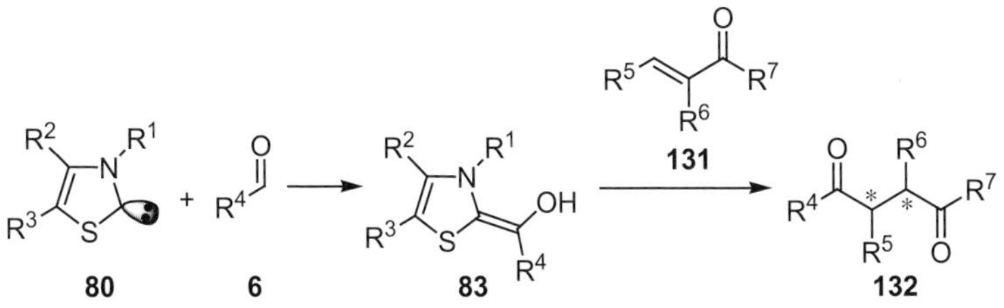

Scheme 36. The Stetter Reaction

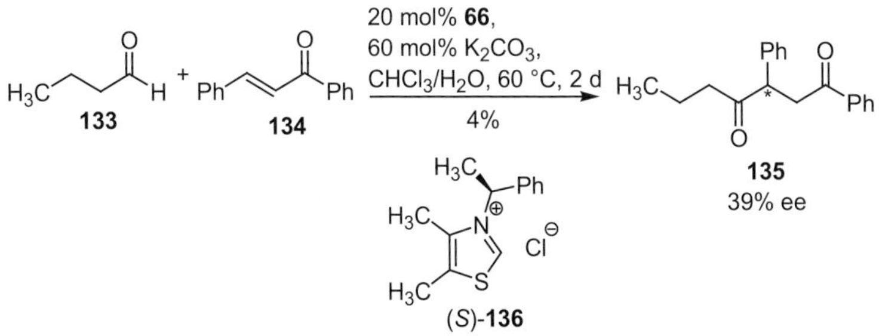

Scheme 37. First attempts of an asymmetric Stetter reaction

ophilic attack of the acyl anion equivalent to the Michael acceptor **131** (Scheme 36).

First attempts of an asymmetric Stetter reaction were made 1989 in our research group with the investigation of chiral thiazolium salts such as **136** as precatalysts. The reaction of *n*-butanal (**133**) with chalcone (**134**) in a two-phase system gave the 1,4-diketone **135** with an enantiomeric excess of 39%, but a low yield of only 4% (Scheme 37) (Tiebes 1990; Enders 1993; Enders et al. 1993b). The catalytic activity of thiazolium as well as triazolium salts in the Stetter reaction persisted at a rather low level. Triazolium salts have been shown to possess a catalytic activity in the non-enantioselective Stetter reaction (Stetter and Kuhlmann 1991), but in some cases stable adducts with Michael acceptors have been observed (Enders et al. 1996a), which might be a possible reason for their failure in catalysis.

3.3.2 Asymmetric Intramolecular Stetter Reactions

In 1995 Ciganek reported an intramolecular version of the Stetter reaction (Ciganek 1995). 2-Formyl phenoxycrotonates and -acrylates **137** have been shown to be highly active substrates for the Stetter reaction. The reactivity of the substrates was considerably enhanced may be due to entropic factors. The reaction also proceeds in the absence of triethylamine as base. The catalyst is presumably activated by DMF taking the place of the amine.

In 1996, our research group observed an activity of triazolium salts, for example (*S*, *S*)-**97** as precatalysts in this intramolecular reaction. The stereoselective synthesis of various 4-chromanones (*R*)-**138** via the first asymmetric intramolecular Stetter reaction was performed with enantiomeric excesses of 41%–74% and yields of 22%–73% (Scheme 38) (Enders et al. 1996c).

The observed absolute configuration of the products is in compliance with a simple transition state model where the phenyl group of the dioxane moiety shields the *Re* face of the intermediate formed by addition of the nucleophilic carbene to the aldehyde, therefore, directing the attack of the enoate Michael acceptor to occur with the less hindered face, that is the *Si* face of the enamine (Fig. 16). The electrophilic part of the intermediate bearing the activated C=C double bond is approached by

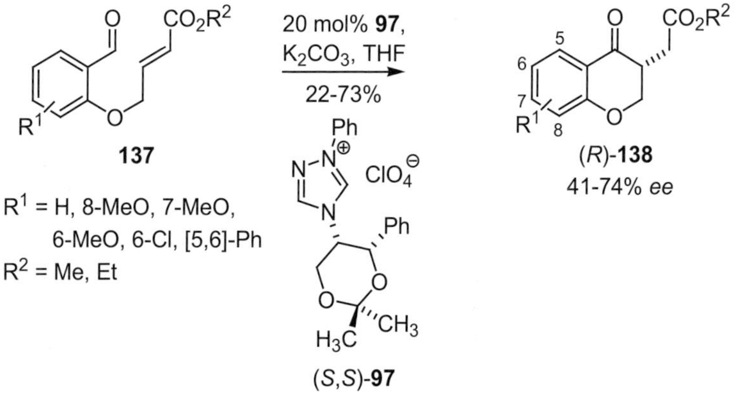

137

R^1 = H, 8-MeO, 7-MeO, 6-MeO, 6-Cl, [5,6]-Ph
R^2 = Me, Et

20 mol% **97**, K_2CO_3, THF
22-73%

(*S*,*S*)-**97**

(*R*)-**138**
41-74% ee

Scheme 38. Asymmetric intramolecular Stetter Reaction

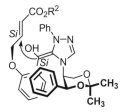

Fig. 16. Proposed transition state of the intramolecular Stetter reaction

the enamine β-C-atom (the part which is reminiscent of a Breslow intermediate) at the *Si* face, leading to an (*R*)-configuration at the newly formed stereogenic centre.

After these early results of our group not much attention has been paid to this important transformation. In the last years, however, Rovis et al. (Kerr et al. 2002; Kerr and Rovis 2003; Kerr and Rovis 2004; Moore et al. 2006; Read de Alaniz and Rovis 2005; Reynolds and Rovis 2005; Liu and Rovis 2006) and others (Nakamura et al. 2005; Mennen et al. 2005b; Murry et al. 2001; Mattson et al. 2004; Mennen et al. 2005a; Myers et al. 2005; Mattson et al. 2006a, b) have achieved significant progress.

4 Conclusion

This overview demonstrates that the concept of biomimetic C–C bond formations is applicable to a broad range of asymmetric reactions. The C_3+C_n concept using the DHA equivalent 2,2-dimethyl-1,3-dioxan-1-one leads to several carbohydrate derivatives via a proline-catalyzed aldol reaction. By using related Mannich reactions it was also possible to open up a flexible entry to amino sugars, carbasugars, polyoxamic acid and phytosphingosine derivatives. Furthermore, novel ulosonic acid precursors were obtained via an organocatalytic entry. The concept was extended to multicomponent cascade reactions leading to tetrasubstitued cyclohexene carbaldehydes. Within this process four stereogenic centers are generated in three consecutive carbon–carbon bond formations with high diastereo- and complete enantiocontrol. Starting from diene containing aldehyde substrates the organocatalytic domino process could be combined with an intramolecular Diels-Alder reaction in one pot providing tricyclic carbon frameworks under control of five carbon-carbon bonds and up to eight stereocenters. In addition N-heterocyclic carbenes were successfully used in asymmetric benzoin reactions. The first asymmetric intramolecular crossed aldehyde-ketone benzoin reaction was developed yielding α-hydroxy tetralone and -indanone derivatives. This new principle could be extended to the asymmetric synthesis 3-hydroxy 4-chromanones demonstrating an efficient entry to bioactive natural product synthesis possessing this structural motif. The fast developing field of organocatalysis will yield new routes and versatile synthetic methods, which can be applied in biomimetic processes.

References

Acevedo CM, Kogut EF, Lipton MA (2001) Synthesis and analysis of the sterically constrained L-glutamine analogues (3S,4R)-3,4-dimethyl-L-glutamine and (3S,4R)-3,4-dimethyl-L-pyroglutamic acid. Tetrahedron 57:6353

Lehninger AL (1993) Principles of biochemistry. Worth, New York

Alexander C, Rietschel ET (1999) Bakterielle Lipopolysaccharide-Hochaktive Stimulatoren der angeborenen Immunität. BIOspektrum (Heidelb) 4:275–281

Angata T, Varki A (2002) Chemical diversity in the sialic acids and related alpha-keto acids: an. evolutionary perspective. Chem Rev 102:439

Arduengo AJ 3rd (1999) Looking for stable carbenes: the difficulty in starting anew. Acc Chem Res 32:913

Arduengo AJ 3rd, Harlow RL, Kline M (1991) A stable crystalline carbene. J Am Chem Soc 113:361

Arduengo AJ 3rd, Dias HVR, Harlow RL, Kline M (1992) Electronic stabilization of nucleophilic carbenes. J Am Chem Soc 114:5530

Arduengo AJ 3rd, Kraftczyk R (1998) Auf der Suche nach Stabilen Carbenen. R Chem Unserer Zeit 32:6

Arend M, Westermann B, Risch N (1998) Modern variants of the Mannich reaction. Angew Chem Int Ed Engl 37:1044–1070

Atsumi S, Umezawa K, Iinuma H, Naganawa H, Nakamura H, Iitaka Y, Takeuchi T (1990a) Production, isolation and structure determination of a novel beta-glucosidase inhibitor, cyclophellitol, from Phellinus sp. J Antibiot (Tokyo) 43:49–53

Atsumi S, Iinuma H, Nosaka C, Umezawa K (1990b) Biological activities of cyclophellitol. J Antibiot (Tokyo) 43:1579–1585

Bach G, Breiding-Mack S, Grabley S, Hammann P, Hütter K, Thiericke R, Uhr H, Wink J, Zeeck A (1993) Gabosines, new carba-sugars from Streptomyces. Justus Liebigs Ann Chem 33:241–250

Banaszek A, Mlynarski J (2005) Recent advances in the chemistry of bioactive 3-deoxy-ulosonic acids. Stud Nat Prod Chem 30:419

Banwell M, De Savi C, Watson K (1998) Diastereoselective synthesis of (2)-N-acetylneuraminic acid (Neu5Ac) from a non-carbohydrate source. J Chem Soc [Perkin 1] 1998:2251–2252

Barenholz Y, Gatt S (1967) The utilization and degradation of phytosphingosine by rat liver. Biochem Biophys Res Commun 27:319–324

Beeson TD, MacMillan DWC (2005) Enantioselective organocatalytic alpha-fluorination of aldehydes. J Am Chem Soc 127:8826–8828

Berecibar A, Grandjean C, Siriwardena A (1999) Synthesis and biological activity of natural aminocyclopentitol glycosidase inhibitors: mannostatins, trehazolin, allosamidins, and their analogues. Chem Rev 99:779–844

Bobbio C, Gouverneur V (2006) Catalytic asymmetric fluorinations. Org Biomol Chem 4:2065

Böhler P, Tamm C (1967) The homo-isoflavones, a new class of natural product. Isolation and structure of eucomin and eucomol. Tetrahedron Lett 36:3479

Borysenko CW, Spaltenstein A, Straub JA, Whitesides GM (1989) The synthesis of aldose sugars from half-protected dialdehydes using rabbit muscle aldolase. J Am Chem Soc 111:9275

Bourissou D, Bertrand G (1999) The chemistry of phosphinocarbenes. Adv Organomet Chem 44:175

Bourissou D, Guerret O, Gabbaï FP, Bertrand G (2000) Stable carbenes. Chem Rev 100:39–92

Breslow R (1958) On the mechanism of thiamine action. IV. Evidence from studies on model systems. J Am Chem Soc 80:3719

Breuer K (1997) PhD thesis. RWTH Aachen University

Brodesser S, Sawatzki P, Kolter T (2003) Bioorganic chemistry of ceramide. Eur J Org Chem 11:2021–2034

Calvin M (1962) The path of carbon in photosynthesis. Angew Chem Int Ed Engl 1:65

Carter HE, Clemer WD, Lands WM, Muller KL, Tomizawa HH (1954) Biochemistry of the sphingolipids. VIII. Occurrence of a long chain base in plant phosphatides. J Biol Chem 206:613–623

Casiraghi G, Rassu G, Spanu P (1995) Stereoselective approaches to bioactive carbohydrates and alkaloids-with a focus on recent syntheses drawing from the chiral pool. Chem Rev 95:1677

Castells J, Geijo F, López Calahorra F (1980) The "formoin reaction": a promising entry to carbohydrates from formaldehyde. Tetrahedron Lett 21:4517

César V, Bellemin-Laponnaz S, Gade LH (2004) Chiral *N*-heterocyclic carbenes as stereodirecting ligands in asymmetric catalysis. Chem Soc Rev 33:619

Chambers RD (2004) Fluorine in organic chemistry. Blackwell, Oxford

Chapman CJ, Frost CG (2007) Tandem and domino catalytic strategies for enantioselective synthesis. Synthesis 2007:1–21

Christmann M (2005) New developments in the asymmetric Stetter reaction. Angew Chem Int Ed Engl 44:2632–2634

Chupak L, Luebbers T, Trost BM (1998) Total synthesis of (±)- and (+)-valienamine via a strategy derived from new palladium-catalyzed reactions. J Am Chem Soc 120:1732

Ciganek E (1995) Esters of 2,3-dihydro-3-oxobenzofuran-2-acetic acid and 3,4-dihydro-4-oxo-2H-1-benzopyran-3-acetic acid by intramolecular Stetter reactions. Synthesis 1995:1311–1314

Cookson R, Lane RM (1976) Conversion of dialdehydes into cyclic α-ketols by thiazolium salts: synthesis of cyclic 2-hydroxy-2-enones. J Chem Soc Chem Commun 1976:804

Crouch NR, Bangnai V, Mulholland DA (1999) Homoisoflavanones from three South African *Scilla* species. Phytochemistry 51:943

Danishefsky SJ, DeNinno MP, H Chen S (1988) Stereoselective total syntheses of the naturally occurring enantiomers of N-acetylneuraminic acid and 3-deoxy-D-manno-2-octulosonic acid. A new and stereospecific approach to sialo and 3-deoxy-D-manno-2-octulosonic acid conjugates. J Am Chem Soc 110:3929

Davis FA, Weismiller MC (1990) Enantioselective synthesis of tertiary α-hydroxy carbonyl compounds using [(8,8-dichlorocamphoryl)sulfonyl]oxaziridine. J Org Chem 55:3715

DeNinno MP (1991) The synthesis and glycosidation of N-acetylneuraminic acid. Synthesis 1991:583–593

Despagnet E, Gornitzka H, Rozhenko AB, Schoeller WW, Bourissou D, Bertrand G (2002) Stable non-push-pull phosphanylcarbenes: NMR spectroscopic characterization of a methylcarbene. Angew Chem Int Ed Engl 41:2835–2837

Díez-González S, Nolan SP (2005) Carbene and transition metal-mediated transformations. Annu Rep Prog Chem Sect B Org Che 101:171–191

Dimmock JR, Sidhu KK, Chen M, Reid RS, Allen TM, Kao GY, Truitt GA (1993) Anticonvulsant activities of some arylsemicarbazones displaying potent oral activity in the maximal electroshock screen in rats accompanied by high protection indices. Eur J Med Chem 36:2243–2252

Dixon DA, Dobbs KD, Arduengo AJ 3rd, Bertrand G (1991) Electronic structure of λ5-phosphaacetylene and corresponding triplet methylenes. J Am Chem Soc 113:8782

Dondoni A, Marra A, Merino P (1994) Installation of the pyruvate unit in glycidic aldehydes via a Wittig olefination-Michael addition sequence utilizing a thiazole-armed carbonyl ylide. A new stereoselective route to 3-deoxy-2-ulosonic acids and the total synthesis of DAH, KDN, and 4-epi-KDN. J Am Chem Soc 116:3324

Dudding T, Houk KN (2004) Computational predictions of stereochemistry in asymmetric thiazolium- and triazolium-catalyzed benzoin condensations. Proc Natl Acad Sci U S A 101:5770–5775

Dwek RA (1996) Glycobiology: toward understanding the function of sugars. Glycobiology. Chem Rev 96:683–720

Enders D (1993) Enzymemimetic C–C and C–N bond formations. In: Stereoselective synthesis. Springer-Verlag, Berlin Heidelberg New York, p 63

Enders D, Balensiefer T (2004) Nucleophilic carbenes in asymmetric organocatalysis. Acc Chem Res 37:534–541

Enders D, Bockstiegel B (1989) Enantioselective alkylation of 2,2-dimethyl-1,3-dioxan-5-one using the SAMP-/RAMP-hydrazone method. Synthesis 1989:493–496

Enders D, Breuer K (1999) In: Comprehensive asymmetric catalysis, vol 3. Springer-Verlag, Berlin Heidelberg New York, pp 1093–1102

Enders D, Gasperi T (2007) Proline organocatalysis as a new tool for the asymmetric synthesis of ulosonic acid precursors. Chem Commun 2007:88–90

Enders D, Grondal C (2005) Direct organocatalytic de novo synthesis of carbohydrates. Angew Chem Int Ed Engl 44:1210–1212

Enders D, Grondal C (2006) Direct asymmetric organocatalytic de novo synthesis of carbohydrates. Tetrahedron 62:329–337

Enders D, Hüttl MRM (2005) Direct organocatalytic α-fluorination of aldehydes and ketones. Synlett 2005:991

Enders D, Hüttl MRM (2006) Control of four stereocentres in a triple cascade organocatalytic reaction. Nature 441:861–863

Enders D, Kallfass U (2002) An efficient nucleophilic carbene catalyst for the asymmetric benzoin condensation. Angew Chem Int Ed Engl 41:1743–1745

Enders D, Müller-Hüwen A (2004) Asymmetric synthesis of 2-amino-1,3-diols and D-erythro-sphinganine. Eur J Org Chem 2004:1732

Enders D, Niemeier O (2004) Thiazol-2-ylidene catalysis in intramolecular crossed aldehyde-ketone benzoin reactions. Synlett 2004:2111–2114

Enders D, Oberbörsch S (2002) Asymmetric Mannich reactions with α-silylated trimethylsilyl enol ethers and *N*-alkoxycarbonyl Imines. Synlett 2002:471–473

Enders D, Potthoff M (1997) Regio- and enantioselective synthesis of alpha-fluoroketones by electrophilic fluorination of alpha-silylketone enolates with N-fluorobenzosulfonimide. Angew Chem Int Ed Engl 36:2362–2364

Enders D, Seki A (2002) Proline-catalyzed enantioselective Michael additions of ketones to nitrostyrene. Synlett 2002:26–28

Enders D, Vrettou M (2006) Asymmetric synthesis of (+)-polyoxamic acid via an efficient organocatalytic Mannich reaction as the key step. Synthesis 2155–2158 [and literature cited therein]

Enders D, Dyker H, Raabe G, Runsink J (1992) Enantio- and diastereoselective synthesis of 3-substituted cyclic hemiketals of ω-hydroxy-2-oxoesters. Synlett 1992:901–903

Enders D, Dyker H, Raabe G (1993a) Enantioselective aldol reactions with a phosphoenolpuryvate equivalent: asymmetric synthesis of 4-hydroxy-2-oxocarboxylic acid esters. Angew Chem Int Ed Engl 32:421–423

Enders D, Bockstiegel B, Dyker H, Jegelka U, Kipphardt H, Kownatka D, Kuhlmann H, Mannes D, Tiebes J, Papadopoulos K (1993b) Enzymmimetische C–C-Verknüpfungen. In: Dechema-Monographies, vol 129. VCH, Weinheim, p 209

Enders D, Whitehouse DL, Runsink J (1995a) Diastereo- and enantioselective synthesis of L-*threo*- and D-*erythro*-sphingosine. Chem Eur J 1:382

Enders D, Breuer K, Raabe G, Runsink J, Teles JH, P Melder J, Ebel K, Brode S (1995b) Preparation, structure, and reactivity of 1,3,4-triphenyl-4,5-dihydro-1H-1,2,4-triazol-5-ylidene, a new stable carbene. Angew Chem Int Ed Engl 34:1021–1023

Enders D, Breuer K, Runsink J, Teles JH (1996a) Chemical reactions of the stable carbene 1,3,4-triphenyl-4,5-dihydro-1H-1,2,4-triazol-5-ylidene. Liebigs Ann Chem 1996:2019

Enders D, Breuer K, Teles JH (1996b) A novel asymmetric benzoin reaction catalyzed by a chiral triazolium salt. Preliminary communication. Helv Chim Acta 79:1217

Enders D, Breuer K, Runsink J, Teles JH (1996c) The first asymmetric intramolecular Stetter reaction. Preliminary communication. Helv Chim Acta 79:1899

Enders D, Ward D, Adam J, Raabe G (1996d) Efficient regio- and enantioselective Mannich reactions. Angew Chem Int Ed Engl 35:981–984

Enders D, Breuer K, Teles JH, Ebel K (1997a) 1,3,4-triphenyl-4,5-dihydro-1H-1,2,4-triazol-5-ylidene: applications of a stable carbene in synthesis and catalysis. J Prakt Chem Chem-Ztg 339:397–399

Enders D, Breuer K, Raabe G, Simonet J, Ghanimi A, Stegmann HB, Teles JH (1997b) A stable carbene as π-acceptor. Electrochemical reduction to the radical anion. Tetrahedron Lett 38:2833

Enders D, Oberbörsch S, Adam J (2000) a-Silyl controlled asymmetric Mannich reactions of acyclic ketones with imines. Synlett 2000:644–646

Enders D, Faure S, Potthoff M, Runsink J (2001) Diastereoselective electrophilic fluorination of enantiopure α-silylketones using N-fluoro-benzosulfonimide: regio- and enantioselective synthesis of α-fluoroketones. Synthesis 2001:2307–2319

Enders D, Voith M, Ince SD (2002a) Preparation and reactions of 2,2-dimethyl-1,3-dioxan-5-one-SAMP-hydrazone: a versatile chiral dihydroxyacetone equivalent. Synthesis 2002:1775–1779

Enders D, Adam J, Oberbörsch S, Ward D (2002b) Asymmetric Mannich reactions by α-silyl controlled aminomethylation of ketones. Synthesis 2002:2737–2748

Enders D, Breuer K, Kallfass U, Balensiefer T (2003) Preparation and application of 1,3,4-triphenyl-4,5-dihydro-1H-1,2,4-triazol-5-ylidene, a stable carbene. Synthesis 2003:1292–1295

Enders D, Voith M, Lenzen A (2005a) The dihydroxyacetone unit—a versatile C3 building block in organic synthesis. Angew Chem Int Ed Engl 44:1304

Enders D, Grondal C, Vrettou M, Raabe G (2005b) Direct organocatalytic de novo synthesis of carbohydrates. Angew Chem Int Ed Engl 44:4079

Enders D, Paleek J, Grondal C (2006a) A direct organocatalytic entry to sphingoids: asymmetric synthesis of D-arabino- and L-ribo-phytosphingosine. Chem Commun 2006:655–657

Enders D, Grondal C, Vrettou M (2006b) Efficient entry to amino sugars and derivatives via asymmetric organocatalytic Mannich reactions. Synthesis 2006:3597

Enders D, Niemeier O, Balensiefer T (2006c) Asymmetric intramolecular crossed-benzoin reactions by N-heterocyclic carbene catalysis. Angew Chem Int Ed Engl 45:1463

Enders D, Niemeier O, Raabe G (2006d) Asymmetric synthesis of chromanones via N-heterocyclic carbene catalyzed intramolecular crossed-benzoin reactions. Synlett 2006:2431

Enders D, Grondal C, Hüttl MRM (2007a) Asymmetric Organocatalytic Domino Reactions. Angew Chem Weinheim Bergstr Ger 119:1590

Enders D, Hüttl MRM, Runsink J, Raabe G, Wendt B (2007b) Organocatalytic one pot asymmetric synthesis of functionalized tricyclic carbon skeletons via a triple cascade/Diels-Alder sequence. Angew Chem 119:471

Enders D, Balensiefer T, Niemeier O, Christmann M (2007c) Nucleophilic N-heterocyclic carbenes om asymmetric organocatalysis. In: Dalko PI (ed) Enantioselective organocatalysis—reactions and experimental procedures. Wiley-VCH, Weinheim, p 331

Ennis MD, Hoffman RL, Ghazal NB, Old DW, Mooney PA (1996) Asymmetric synthesis of *cis*-fused bicyclic pyrrolidines and pyrrolidinones via chiral polycyclic lactams. J Org Chem 61:5813

Garrison JC, Youngs WJ (2005) Ag(I) N-heterocyclic carbene complexes: synthesis, structure, and application. Chem Rev 105:3978–4008

Gerhards AU, Leeper FJ (1997) Synthesis of and asymmetric induction by chiral polycyclic thiazolium salts. Tetrahedron Lett 38:3615–3618

Grondal C (2006) Asymmetrische organokatalytische de novo Synthese von Kohlenhydraten, Phytosphingosinen und 1-*epi*-(+)MK7607. Dissertation. RWTH Aachen

Grondal C, Enders D (2006) A direct entry to carbasugars: asymmetric synthesis of 1-epi-(+)-MK7607. Synlett 2006:3507–3509

Guo HC, Ma JA (2006) Catalytic asymmetric tandem transformations triggered by conjugate additions. Angew Chem Int Ed Engl 45:354–366

Hachisu Y, Bode JW, Suzuki K (2003) Catalytic intramolecular crossed aldehyde–ketone benzoin reactions: a novel synthesis of functionalized preanthraquinones. J Am Chem Soc 125:8432–8433

Hachisu Y, Bode JW, Suzuki K (2004) Thiazolium ylide-catalyzed intramolecular aldehydeketone benzoin-forming reactions: substrate scope. Adv Synth Catal 346:1097–1100

Hahn FE (2006) Heterocyclic carbenes. Angew Chem Int Ed Engl 45:1348–1352

Hamashima Y, Sodeoka M (2006) Enantioselective fluorination reactions catalyzed by chiral palladium complexes. Synlett 2006:1467–1478

Hayashi Y, Gotoh H, Hayashi T, Shoji M (2005) Diphenylprolinol silyl ethers as efficient organocatalysts for the asymmetric Michael reaction of aldehydes and nitroalkenes. Angew Chem Int Ed Engl 44:4212–4215

Heller W, Tamm C (1981) Homoisoflavones and biogenetically related compounds. Fortschr Chem Org Naturst 40:105

Herrmann WA (2002) N-Heterocyclische Carbene: ein neues Konzept in der metallorganischen Katalyse. Angew Chem 114:1342

Herrmann WA, Koecher C (1997) N-Heterocyclic carbenes. Angew Chem Int Ed Engl 365:2162–2187

Holstein Wagner S, Lundt I (2001) Synthesis of carba sugars from aldonolactones. Part IV. Stereospecific synthesis of carbaheptopyranoses by radical-induced carbocyclisation of 2,3-unsaturated octonolactones. J Chem Soc [Perkin 1] 2001:780

Horii S, Iwasa T, Mizuta E, Kameda YJ (1971) Studies on validamycins, new antibiotics. VI. Validamine, hydroxyvalidamine and validatol, new cyclitols. J Antibiot (Tokyo) 24:59–63

Huang Y, Walji AM, Larsen CH, MacMillan DWC (2005) Enantioselective organo-cascade catalysis. J Am Chem Soc 127:15051–15053

Igau A, Grutzmacher H, Baceiredo A, Bertrand G (1988) Analogous α,α'-bis-carbenoid, triply bonded species: synthesis of a stable $\lambda 3$-phosphino carbene-$\lambda 5$-phosphaacetylene. J Am Chem Soc 110:6463

Igau A, Baceiredo A, Trinquier G, Betrand G (1989) Bis (diisopropylamino) phosphino] trimethylsilylcarbene: a stable nucleophilic carbene. Angew Chem Int Ed Engl 28:621

Ishikawa T, Shimizu Y, Kudoh T, Saito S (2003) Conversion of D-glucose to cyclitol with hydroxymethyl substituent via intramolecular silyl nitronate cycloaddition reaction: application to total synthesis of (+)-cyclophellitol. Org Lett 5:3879–3882

Isogai A, Sakuda S, Nakayama J, Watanabe S, Suzuki S (1987) Isolation and structural elucidation of a new cyclitol derivative, streptol, as a plant growth regulator. Agric Biol Chem 51:2277

Isono K, Asahi K, Suzuki S (1969) Studies on polyoxins, antifungal antibiotics. 13. The structure of polyoxins. J Am Chem Soc 91:7490–7505

Jordan F (2003) Current mechanistic understanding of thiamine diphosphate-dependent enzymatic reactions. Nat Prod Rep 20:184–201

Kameda Y, Horii SJ (1972) The unsaturated cyclitol part of the new antibiotics, the validam. J Chem Soc Chem Commun 1972:746

Kameda Y, Asano N, Yoshikawa M, Takeuchi M, Yamaguchi T, Matsui K, Horii S, Fukase H (1984) Valiolamine, a new alpha-glucosidase inhibiting aminocyclitol produced by Streptomyces hygroscopicus. J Antibiot (Tokyo) 37:1301–1307

Kamitakahara H, Suzuki T, Nishigori N, Suzuki Y, Kamie O, Wong C (1998) Ein Lysogangliosid/Poly-L-glutaminsäure-Konjugat als picomolarer Inhibitor von Influenza-Hämagglutinin. Angew Chem Weinheim Bergstr Ger 110:1607

Karlsson KA, Samuelsson BE, Steen GO (1968) Structure and function of sphinolipids. 1. Differences in sphingolipid long-chain base pattern between kidney cortex, medulla, and papillae. Acta Chem Scand 22:1361–1364

Katz L (1997) Manipulation of modular polyketide synthases. Chem Rev 97:2557–2576

Kauer Zinn F, Viciu MS, Nolan SP (2004) Carbenes: reactivity and catalysis. Annu Rep Prog Chem Sect B Org Che 100:231–249

Kawano Y, Higuchi R, Isobe R (1988) Biologically active glycosides from asteroidea, XIII. Glycosphingolipids from the starfish *Acanthaster planci*, 2. Isolation and structure of six new cerebrosides. T Komori Liebigs Ann Chem 1988:19

Kerr MS, Rovis T (2003) Effect of the Michael Acceptor in the asymmetric intramolecular Stetter reaction. Synlett 2003:1934–1936

Kerr MS, Rovis T (2004) Enantioselective synthesis of quaternary stereocenters via a catalytic asymmetric Stetter reaction. J Am Chem Soc 126:8876–8877

Kerr MS, Read de Alaniz J, Rovis T (2002) A highly enantioselective catalytic intramolecular Stetter reaction. J Am Chem Soc 124:10298

Kerr MS, Read de Alaniz J, Rovis T (2005) An efficient synthesis of achiral and chiral 1,2,4-triazolium salts: bench stable precursors for N-heterocyclic carbenes. J Org Chem 70:5725–5728

Kiefel MJ, von Itzstein M (2002) Recent advances in the synthesis of sialic acid. Chem Rev 102:471

Kleemann A, Engel J (1982) Pharmazeutische Wirkstoffe: Synthese, Patente, Anwendungen. Thieme, Stuttgart

Knight RL, Leeper FJ (1997) Synthesis of and asymmetric induction by chiral bicyclic thiazolium salts. Tetrahedron Lett 38:3611

Knight RL, Leeper FJ (1998) Comparison of chiral thiazolium and triazolium salts as asymmetric catalysts for the benzoin condensation. J Chem Soc [Perkin 1] 1998:1891

Kobayashi S, Ishitani H (1999) Catalytic enantioselective addition to imines. Chem Rev 99:1069–1094

Kobayashi S, Furuta T, Hayashi T, Nishijima M, Hanada K (1998) Catalytic asymmetric syntheses of antifungal sphingofungins and their biological activity as potent inhibitors of serine palmitoyltransferase (SPT). J Am Chem Soc 120:908

Kober R, Papadopoulos K, Miltz W, Enders D, Steglich W, Reuter H, Puff H (1985) Synthesis of diastereo- and enantiomerically pure α-amino-γ-oxo acid esters by reaction of acyliminoacetates with enamines derived from 6-membered ketones. Tetrahedron 41:1693

Kolter T (2004) Conformational restriction of sphingolipids. In: Schmuck C, Wennemers H (eds) Highlights in bioorganic chemistry: methods and applications. Wiley-VCH, Weinheim, p 48

Kolter T, Sandhoff K (1999) Sphingolipids—their metabolic pathways and the pathobiochemistry of neurodegenerative diseases. Angew Chem Int Ed Engl 38:1532

Korotkikh NI, Shvaika OP, Rayenko GF, Kiselyov AV, Knishevitsky AV, Cowley AH, Jones JN, Macdonald CLB (2005) Stable heteroaromatic carbenes of the benzimidazole and 1,2,4-triazole series. Arkivoc 8:10–43

Khosla C (1997) Harnessing the biosynthetic potential of modular polyketide synthases. Chem Rev 97:2577

Khosla C, Gokhale RS, Jacobsen JR, Cane DE (1999) Tolerance and specificity of polyketide synthases. Annu Rev Biochem 68:219

Lapworth A (1903) Qualitative study on the formation of cyanohydrin in water. J Chem Soc 83:995

Li H, Matsunaga S, Fusetani N (1995) Halicylindrosides, antifungal and cytotoxic cerebrosides from the marine sponge Halichondria cylindrata. Tetrahedron 51:2273

Li YT, Hirabayashi Y, DeGasperi R, Yu RK, Ariga T, Koerner TAW, C Li S (1984) Isolation and characterization of a novel phytosphingosine-containing GM2 ganglioside from mullet roe (Mugil cephalus). J Biol Chem 259:8980–8985

Li YW, Zhu LY, Huang L (2004) Studies on the total synthesis of hainanolide (VIII)-introducing C_4-methoxy group, and forming the ring E (lactone). Chin Chem Lett 15:397

Liao J, Tao J, Lin G, Liu D (2005) Chemistry and biology of sphingolipids. Tetrahedron 61:4715

List B (2000) The direct catalytic asymmetric three-component Mannich reaction. J Am Chem Soc 122:9336

List B, Lerner RA, Barbas CF 3rd (2000) Proline-catalyzed direct asymmetric aldol reactions. J Am Chem Soc 122:2395

List B, Pojarliev P, Martin HJ (2001) Efficient proline-catalyzed Michael additions of unmodified ketones to nitro olefins. Org Lett 3:2423–2425

List B, Pojarliev P, Biller TW, Martin HJ (2002) The proline-catalyzed direct asymmetric three-component Mannich reaction: scope, optimization, and application to the highly enantioselective synthesis of 1,2-amino alcohols. J Am Chem Soc 124:827–833

Liu Q, Rovis T (2006) Asymmetric synthesis of hydrobenzofuranones via desymmetrization of cyclohexadienones using the intramolecular Stetter reaction. J Am Chem Soc 128:2552–2553

Lubineau A, Billault I (1998) New access to unsaturated keto carba sugars (gabosines) using an intramolecular Nozaki-Kishi reaction as the key step. J Org Chem 63:5668

Ma JA, Cahard D (2004) Asymmetric fluorination, trifluoromethylation, and perfluoroalkylation reactions. Chem Rev 104:6119

Mann J (1999) Chemical aspects of biosynthesis. Oxford Chemistry Primers. Oxford University Press, Oxford

Mannich C, Krösche W (1912) Ueber ein Kondensationsprodukt aus Formaldehyd, Ammoniak und Antipyrin. Arch Pharm 250:647–667

Marigo M, Fielenbach D, Braunton A, Kjoersgaard A, Jørgensen KA (2005a) Enantioselective formation of stereogenic carbon-fluorine centers by a simple catalytic method. Angew Chem 117:3769

Marigo M, Schulte T, Franzén J, Jørgensen KA (2005b) Asymmetric multicomponent domino reactions and highly enantioselective conjugated addition of thiols to alpha,beta-unsaturated aldehydes. J Am Chem Soc 127:15710–15711

Matsumoto T, Inoue S (1983) Selective formation of triose from formaldehyde catalysed by ethylbenzothiazolium bromide. J Chem Soc Chem Commun 1983:171–172

Matsumoto T, Yamamoto H, Inoue S (1984) Selective formation of triose from formaldehyde catalyzed by thiazolium salt. J Am Chem Soc 106:4829

Mattson AE, Bharadwaj AR, Scheidt KA (2004) The thiazolium-catalyzed Sila-Stetter reaction: conjugate addition of acylsilanes to unsaturated esters and ketones. J Am Chem Soc 126:2314–2315

Mattson AE, Bharadwaj AR, Zuhl AM, Scheidt KA (2006a) Thiazolium-catalyzed additions of acylsilanes: a general strategy for acyl anion addition reactions. J Org Chem 71:5715–5724

Mattson AE, Zuhl AM, Reynolds TE, Scheidt KA (2006b) Direct nucleophilic acylation of nitroalkenes promoted by a fluoride anion/thiourea combination. J Am Chem Soc 128:4932–4933

McCasland GE, Furuta S, Durham LJ (1966) Alicyclic carbohydrates. XXIX. The synthesis of a pseudo-hexose (2,3,4,5-tetrahydroxycyclohexanemethanol). J Org Chem 31:1516

Mehta G, Lakshminath S (2000) A norbornyl route to cyclohexitols: stereoselective synthesis of conduritol-E, *allo*-inositol, MK 7607 and gabosines. Tetrahedron Lett 41:3509

Mennen MS, Gipson JD, Kim YR, Miller SJ (2005a) Thiazolylalanine-derived catalysts for enantioselective intermolecular aldehyde-imine cross-couplings. J Am Chem Soc 127:1654

Mennen SM, Blank JT, Tran-Dubé MB, Imbriglio JE, Miller SJ (2005b) A peptide-catalyzed asymmetric Stetter reaction. Chem Commun 2005:195–197

Mizuhara S, Tamura R, Arata H (1951) The mechanism of thiamine action II. Proc Jpn Acad 27:302

Moore JL, Kerr MS, Rovis T (2006) Enantioselective formation of quaternary stereocenters using the catalytic intramolecular Stetter reaction. Tetrahedron 62:11477–11482

Mukaiyama T, Suzuki K, Yamada T, Tabusa F (1990) 4-O-Benzyl-23-O-isopropylidene-L-threose: a useful building block for stereoselective synthesis of monosaccharides. Tetrahedron 46:265

Muñiz K (2003) Improving enantioselective fluorination reactions: chiral N-fluoro ammonium salts and transition metal catalysts. In: Schmalz HG, Wirth T (eds) Organic synthesis highlights. Wiley-VCH, Weinheim

Murry JA, Frantz DE, Soheili A, Tillyer R, Grabowski EJJ, Reider PJ (2001) Synthesis of alpha-amido ketones via organic catalysis: thiazolium-catalyzed cross-coupling of aldehydes with acylimines. J Am Chem Soc 123:9696–9697

Musser JH (1992) Carbohydrates as drug discovery leads. Annu Rep Med Chem 27:301

Myers MC, Bharadwaj AR, Milgram BC, Scheidt KA (2005) Catalytic conjugate additions of carbonyl anions under neutral aqueous conditions. J Am Chem Soc 127:14675–14680

Nakamura T, Hara O, Tamura T, Makino K, Hamada Y (2005) A facile synthesis of chroman-4-ones and 2,3-dihydroquinolin-4-ones with quaternary carbon using intramolecular Stetter reaction catalyzed by thiazolium salt. Synlett 2005:155–157

Naroti T, Morita M, Akimoto K, Koezuka Y (1994) Agelasphins, novel antitumor and immunostimulatory cerebrosides from the marine sponge *Agelas mauritianus*. Tetrahedron 50:2771

Nicolaou KC, Montagnon T, Snyder SA (2003) Tandem reactions, cascade sequences, and biomimetic strategies in total synthesis. Chem Commun 5:551

Nicolaou KC, Edmonds DJ, Bulger PC (2006) Cascade reactions in total synthesis. Angew Chem Int Ed Engl 45:7134–7186

Nieman JA, Ennis MD (2000) Enantioselective synthesis of the pyrroloquinoline core of the martinellines. Org Lett 2:1395–1397

Niemeier O (2006) PhD thesis. RWTH Aachen University

Nilsson U, Meshalkina L, Lindqvist Y, Schneider G (1997) Examination of substrate binding in thiamin diphosphate-dependent transketolase by protein crystallography and site-directed mutagenesis. J Biol Chem 272:1864–1869

Notz W, List B (2000) Catalytic asymmetric synthesis of anti-1,2-diols. J Am Chem Soc 122:7386

Oda T (1952) Compounds of penicillin-producing molds. IV. Fungus cerebrin. J Pharm Soc Jpn 72:142

Ogawa S (1988) Synthetic studies on glycosidase inhibitors composed of 5a-carba-sugars. In: Chapleur Y (ed) Carbohydrate mimics, concepts and methods. Wiley-VCH, Weinheim, pp 87

Ogawa S, Tsunoda H (1992) Pseudo-sugars. XXXI. New synthesis of 2-amino-5a-carba-2-deoxy-α-DL-glucopyranose and its transformation into valienamine and valiolamine analogues. Liebigs Ann Chem 6:637–641

Okabe K, Keenan RW, Schmidt G (1968) Phytosphingosine groups as quantitatively significant components of the sphingolipids of the mucosa of the small intestines of some mammalian species. Biochem Biophys Res Commun 31:137–143

Omar F, Frahm AW (1989) Asymmetrische reduktive Aminierung von Cycloalkanonen, 9. Mitt.: Die asymmetrische Synthese GABA-verwandter cycloaliphatischer Aminosäuren Arch Pharm 322:461

Omar F, Frahm AW (1990) Asymmetrische reduktive Aminierung von Cycloalkanonen, 10. Mitt.: EPC-Synthese *cis*-bicyclischer Lactame und Amine. Arch Pharm 323:923

Pellisier H (2006) Asymmetric domino reactions. Part B: Reactions based on the use of chiral catalysts and biocatalysts. Tetrahedron 62:2143

Pellissier H (2006) Asymmetric domino reactions. Part A: Reactions based on the use of chiral auxiliaries. Tetrahedron 62:1619

Perry MC, Burgess K (2003) Chiral N-heterocyclic carbene-. transition metal complexes in asymmetric catalysis. Tetrahedron Asymmetry 14:951–961

Piers E, Romero MA (1993) Total synthesis of amphilectane-type diterpenoids: (±)-8-isocyano-10,14-amphilectadiene. Tetrahedron 49:5791

Pihko PM (2006) Enantioselective alpha-fluorination of carbonyl compounds: organocatalysis or metal catalysis? Angew Chem Int Ed Engl 45:544–547

Prakash GK, Beier P (2006) Construction of asymmetric fluorinated carbon centers. Angew Chem Int Ed Engl 45:2172–2174

Prieto A, Halland N, Jørgensen KA (2005) Novel imidazolidine-tetrazole organocatalyst for asymmetric conjugate addition of nitroalkanes. Org Lett 7:3897–3900

Raabe G, Breuer K, Enders D (1996) The role of conjugate interaction in stable carbenes of the 1,2,4-triazol-5-ylidene type and their energy of dimerisation. An ab initio study. Z Naturforsch [B] 51a:95–101

Ramón DJ, Yus M (2005) Asymmetric multicomponent reactions (AMCRs): the new frontier. Angew Chem Int Ed Engl 44:1602–1634

Rassu G, Auzzas D, Pinna L, Battistini L, Zanardi F, Marzocchi L, Acquotti D, Casiraghi G (2000) Variable strategy toward carbasugars and relatives. 1. Stereocontrolled synthesis of pseudo-beta-D-gulopyranose, pseudo-beta-D-xylofuranose, (pseudo-beta-D-gulopyranosyl)amine, and (pseudo-beta-D-xylofuranosyl)amine. J Org Chem 65:6307–6318

Read de Alaniz J, Rovis T (2005) A highly enantio- and diastereoselective catalytic intramolecular Stetter reaction. J Am Chem Soc 127:6284–6289

Reynolds NT, Rovis T (2005) The effect of pre-existing stereocenters in the intramolecular asymmetric Stetter reaction. Tetrahedron 61:6368–6378

Li LS, Wu YL (2002) Synthesis of 3-deoxy-2-ulosonic acid KDO and 4-epi-KDN, a highly efficient approach of 3-C homologation by propargylation and oxidation. Tetrahedron 58:9049

Schauer R (1982) Chemistry, metabolism, and biological functions of sialic acids. Adv Carbohydr Chem Biochem 40:131

Schauer R (ed) (1982) Sialic acids—chemistry, metabolism, function. Cell biology monographs, vol 10. Springer-Verlag, Berlin Heidelberg New York

Schoerken U, Sprenger GA (1998) Thiamin-dependent enzymes as catalysts in chemoenzymatic syntheses. Biochim Biophys Acta 1385:229–243

Sears P, Wong CH (1998) The role of carbohydrates in biologically active natural products. Cell Mol Life Sci 54:223 252

Seebach D (1979) Methoden der Reaktivitätsumpolung. Angew Chem Weinheim Bergstr Ger 91:259–278

Shenbagamurthi P, Smith HA, Becker JM, Steinfeld A, Naider F (1983) Design of anticandidal agents: synthesis and biological properties of analogues of polyoxin L. J Med Chem 26:1518–1522

Shimizu M, Hiyama T (2005) Modern synthetic methods for fluorine-substituted target molecules. Angew Chem Int Ed Engl 44:214–231

Silvestri MG, Desantis G, Mitchell M, Wong CH (2003) Asymmetric aldol reactions using aldolases. Top Stereochem 23:267

Soleilhavoup M, Baceiredo A, Treutler O, Ahlrichs R, Nieger M, Bertrand G (1992) Synthesis and X-ray crystal structure of [(iso-Pr2N)2P(H)CP(N-iso-Pr2)2]+CF3SO3–: a carbene, a cumulene, or a phosphaacetylene?. J Am Chem Soc 114:10959

Sollogoub M, Sinay P (2006) From sugars to carba-sugars. In: Levy DE, Fügedi P (eds) The organic chemistry of sugars, chap 8. CRC Press, Boca Raton

Soloshonok VA (ed) (1999) Enantiocontrolled synthesis of fluoro-organic compounds. Stereochemical challenges and biomedicinal targets. Wiley, New York

Song C, Jiang S, Singh G (2001) Syntheses of (–)-MK 7607 and other carba-sugars from(–)-shikimic acid. Synlett 12:1983

Sprenger GA, Pohl M (1999) Synthetic potential of thiamin. diphosphate-dependent enzymes. J Mol Catal B: Enzymatic 6:145–159

Staunton J, Weissmann KJ (2001) Polyketide biosynthesis: a millennium review. Nat Prod Rep 18:380–416

Steiner DD, Mase N, Barbas CF 3rd (2005) Direct asymmetric—fluorination of aldehydes. Angew Chem Weinheim Bergstr Ger 117:3772–3776

Stetter H (1976) Die katalysierte Addition von Aldehyden an aktivierte Doppelbindungen–Ein neues Syntheseprinzip. Angew Chem Weinheim Bergstr Ger 88:695

Stetter H, Kuhlmann H (1991) The catalyzed nucleophilic addition of aldehydes to electrophilic double bonds. Org React 40:407–496

Stetter H, Schreckenberg M (1973) A new method for addition of aldehydes to activated double bonds. Angew Chem Int Ed Engl 12:81

Stetter H, Rämsch RY, Kuhlmann H (1976) Über die präparative Nutzung der Thiazoliumsalz-katalysierten Acyloin- und Benzoin-Bildung, I. Herstellung von einfachen Acyloinen und Benzoinen. Synthesis 1976:733–735

Stryer L (1995) Biochemistry, 4th edn. WH Freedman and Company, New York

Suami T (1987) Synthetic ventures in pseudo-sugar chemistry. Pure Appl Chem 59:1509

Suami T (1990) Chemistry of pseudo-sugars. Top Curr Chem 154:257

Suami T, Ogawa S (1990) Chemistry of carba-sugars (pseudo-sugars) and their derivatives. Adv Carbohydr Chem Biochem 48:21

Sugai T, J Shen G, Ichikawa Y, Wong CH (1993) Synthesis of 3-deoxy-D-manno-2-octulosonic acid (KDO) and its analogs based on KDO aldolase-catalyzed reactions. J Am Chem Soc 115:413

Sundström M, Lindqvist Y, Schneider G, Hellman U, Ronne H (1993) Yeast TKL1 gene encodes a transketolase that is required for efficient glycolysis and biosynthesis of aromatic amino acids. J Biol Chem 268:24346–24352

Takamatsu K, Mikami M, Kiguschi K, Nozawa S, Iwamori M (1992) Structural characteristics of the ceramides of neutral glycosphingolipids in the human female genital tract–their menstrual cycle-associated change in the cervical epithelium and uterine endometrium, and their dissociation in the mucosa of the fallopian tube with the menstrual cycle. Biochim Biophys Acta 1165:177–182

Takikawa H, Hachisu Y, Bode JW, Suzuki K (2006) Catalytic enantioselective crossed aldehyde-ketone benzoin cyclization. Angew Chem Int Ed Engl 45:3492–3494

Taylor SD, Kotoris CC, Hum G (1999) Recent advances in electrophilic fluorination. Tetrahedron 55:12431

Tekavec TN, Louie J (2007) Transition metal-catalyzed reactions using N-heterocyclic carbene ligands (besides Pd- and Ru-catalyzed reactions). Top Organomet Chem 21:195

Teles JH, Melder JP, Ebel K, Schneider R, Gehrer E, Harder W, Brode S, Enders D, Breuer K, Raabe G (1996) The chemistry of stable carbenes. Part 2. Benzoin-type condensations of fromaldehyde catalyzed by stable carbenes. Helv Chim Acta 79:61–83

Teles JH, Breuer K, Enders D, Gielen H (1999) One pot synthesis of 3,4-disubstituted 1-alkyl-4H-1,2,4-triazol-1-ium salts. Synth Commun 29:1–9

Thorpe SR, Sweeley C (1967) Chemistry and metabolism of sphingolipids. On the biosynthesis of phytosphingosine by yeast. Biochemistry 6:887

Tiebes J (1990) Diploma thesis. RWTH Aachen University, Aachen
Tietze LF (1996) Domino reactions in organic synthesis. Chem Rev 96:115–136
Tietze LF, Beifuss U (1993) Sequential transformations in organic synthesis. A synthetic strategy with a future. Angew Chem Int Ed Engl 32:131
Tietze LF, Haunert F (2000) In: Vögtle F, Stoddart JF, Shibasaki M (eds) Stimulating concepts in chemistry. Wiley-VCH, Weinheim, S39
Tietze LF, Brasche G, Gerike K (2006) Domino reactions in organic chemistry. Wiley-VCH, Weinheim
Traxler P, Trinks U, Buchdunger E, Mett H, Meyer T, Müller M, Regenass U, Rösel J, Lydon N (1995) [(Alkylamino)methyl]acrylophenones: potent and selective inhibitors of the epidermal growth factor receptor protein tyrosine kinase. J Med Chem 38:2441–2448
Troy FA 2nd (1992) Polysialylation: from bacteria to brains. Glycobiology 2:5–23
Ukai T, Tanaka R, Dokawa T (1943) A new catalyst for acyloin condensation. J Pharm Soc Jpn 63:296 (Chem Abstr 1951, 45:5148)
Unger FM (1981) The chemistry and biological significance of 3-deoxy-D-nanno-2-octulosonic acid (KDO). Adv Carbohydr Chem Biochem 38:323
Vance DE, Sweeley CC (1967) Quantitative determination of the neutral glycosyl ceramides in human blood. J Lipid Res 8:621
Varki A (1992) Diversity in the sialic acids. Glycobiology 2:25
Varki A (1993) Biological roles of oligosaccharides: all of the theories are correct. Glycobiology 3:97–130
Voight EA, Rein C, Burke SD (2002) Synthesis of sialic acids via desymmetrization by ring-closing metathesis. J Org Chem 67:8489–8499
von Itzstein M, Kiefel MJ (1997) Static acid analogues as potential antimicrobial agents. In: Witczak ZJ, Nieforth KA (eds) Carbohydrates in drug design. Marcel Decker, New York, p 39
Wanzlick HW (1962) Nucleophile Carben-Chemie. Angew Chem 74:129
Wanzlick HW, Kleiner HJ (1964) Low-energy "carbenes". Angew Chem Int Ed Engl 2:65
Wasilke JC, Obrey SJ, Baker RT, Bazan GC (2005) Concurrent tandem catalysis. Chem Rev 105:1001–1020
Welch JT, Eswarakrishnan S (1991) Fluorine in bioorganic chemistry. Wiley, New York
Wenzel AG, Jacobsen EN (2002) Asymmetric catalytic Mannich reactions catalyzed by urea derivatives: enantioselective synthesis of beta-aryl-beta-amino acids. J Am Chem Soc 124:12964–12965
Wertz PW, Miethke MC, Long SA, Stauss JS, Downing DT (1985) The composition of the ceramides from human stratum corneum and from comedones. J Invest Dermatol 84:410–412

Weymouth-Wilson AC (1997) The role of carbohydrates in biologically active natural products. Nat Prod Rep 14:99–110
Witczak ZJ (1997) Carbohydrates: new and old targets of rational drug design. In: Carbohydrates in drug design. Marcel Dekker, New York, pp 1–37
Wöhler F, Liebig J (1832) Untersuchungen über das Radikal der Benzoesäure. Ann Pharm 3:249–287
Wong CH (2003) Carbohydrate based drug discovery, Chap 24. Wiley-VCH, Weinheim
Wong CH (ed) (2003) Carbohydrate based drug discovery, Chap 26.3. Wiley-VCH, Weinheim
Yang JW, Hechavarria Fonseca MT, List B (2005) Catalytic asymmetric reductive Michael cyclization. J Am Chem Soc 127:15036–15037
Yoshikawa N, Chiba N, Mikawa T, Ueno S, Harimaya K, Iwata M (1994) Mitsubishi Chemical Industries patent. Jpn Kokai Tokkyo Koho JP 0630600
Zeitler K (2005) Extending mechanistic routes in heterazolium catalysis—promising concepts for versatile synthetic methods. Angew Chem Int Ed Engl 44:7506–7510

Organocatalytic Syntheses of Bioactive Natural Products

M. Christmann(✉)

Institut für Organische Chemie, RWTH Aachen, Landoltweg 1, 52074 Aachen, Germany
email: *christmann@oc.rwth-aachen.de*

1	Introduction	125
2	UCS1025A	126
3	Lepidopteran Sex Pheromones	133
4	N-Heterocycles	135
5	Conclusion	136
	References	136

Abstract. The current scope and limitations of organocatalytic reactions and the consequences for the strategic planning of natural product syntheses are discussed. Examples from our group include the total synthesis of UCS1025A and lepidopteran sex pheromones.

1 Introduction

Natural product synthesis and medicinal chemistry exist in a symbiotic relationship with the development of synthesis methodology. Noyori's asymmetric hydrogenations, Sharpless' olefin oxidations, Grubbs' olefin metathesis, Buchwald-Hartwig couplings and Jacobsen's hydrolytic kinetic resolution are illustrious examples with many practical applications. The key to the success of the above-mentioned reactions is that they have provided *reliable* shortcuts to more traditional synthetic strategies. It is no exaggeration to dub catalytic reactions the cornerstones in the design of future sustainable chemical processes. Besides the field of transition metal-catalyzed processes and enzymatic reactions, organocatalysis has undoubtedly enriched the field of asymmetric catalysis. The stage is now set to examine its practicability and reliability. In a recent review (de Figueiredo and Christmann 2007), we have compiled applications of organocatalytic reactions in the synthesis of drugs and natural products. Here we discuss the role of organocatalysis in our synthesis (de Figueiredo et al. 2007) of the telomerase inhibitor UCS1025A from the planning perspective. In addition, applications to the synthesis of other bioactive molecules are presented.

2 UCS1025A

In a screening program for new fungal metabolites (Nakai et al. 2006), Yamashita et al. isolated UCS1025A (**1**) from *Acremonium* sp. KY4917 (Nakai et al. 2000). The absolute and relative configuration was elucidated by Agatsuma and Kanda using two-dimensional NMR and X-ray crystallographic analysis (Agatsuma et al. 2002). UCS1025A possesses a pentacyclic framework with a dense array of eight stereocenters and an unprecedented furopyrrolizidine (Scheme 1). More interestingly, **1** comprises a new natural product telomerase inhibitor (Shin-ya et al. 2001; Warabi et al. 2003, 2005; Rezler et al. 2005; Doi et al. 2006; Fürstner et al. 2006) and many groups (Snider and Neubert 2004) have embarked on its synthesis. In early 2006, Danishefsky (Lambert and Danishefsky 2006) and Hoye (Hoye and Dvornikovs 2006) succeeded. Our retrosynthetic analysis disconnects UCS1025A into the tentative pyrrolizidine-derived enolate **2** and the decalin aldehyde **3** (Scheme 1).

For the synthesis of the pyrrolizidine fragment (**2**), we identified maleic anhydride (**5**, 7 €/kg), aminobutyric acid (**6**, 120 €/kg) and trienal **7** as readily available starting materials. In the most ambitious disconnection, the racemic pyrrolizidine carboxylic acid **4a**, could be assembled, in principle, in a single step from 4-aminobutyric acid (**6**) and maleic anhydride (**5**). A subsequent kinetic resolution via oxa-Michael addition would then generate an enantiomerically enriched enolate equivalent, which could in turn add to the aldehyde **3**. However,

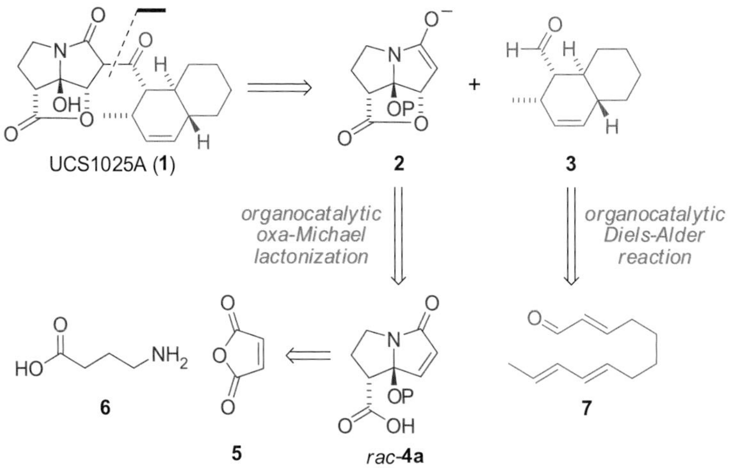

Scheme 1. Retrosynthetic disconnection of UCS1025A

such an approach defies some general assumptions concerning the activation of carboxylic acids. First, it is assumed that free carboxylic acids are poor precursors for the generation of active enolate equivalents. Second, despite the enantioselective addition of a plethora of nucleophiles to Michael acceptors, an enantioselective addition of carboxylic acids has not been reported. The key challenge in this route is the unprecedented enantioselective addition of a carboxylic acid to a Michael acceptor (**4a**→**2**). For the decalin part (**3**), we envisioned an organocatalytic intramolecular Diels–Alder reaction of triene **7**. Inspired by the work of Rudler (Rudler et al. 2005) and Langer (Ullah et al. 2005) and in combination with the soft-enolization methodology developed by Hoye and co-workers (Hoye et al. 2006), we speculated that bis-silylketene acetals formed in situ would provide an acceptable compromise between the reactivity and the desired lability of the ester products. This strategy sidesteps 4-maleimidoalkyl ester intermediates and the harsh basic conditions required for their hydrolysis. Toward this end, a solution of 4-aminobutyric acid (**6**) and maleic anhydride (**5**) was treated with triethylamine and TBSOTf (Scheme 2). Upon quenching with 1 N HCl, the pyrrolizidine carboxylic acid silyl ester **4b** was

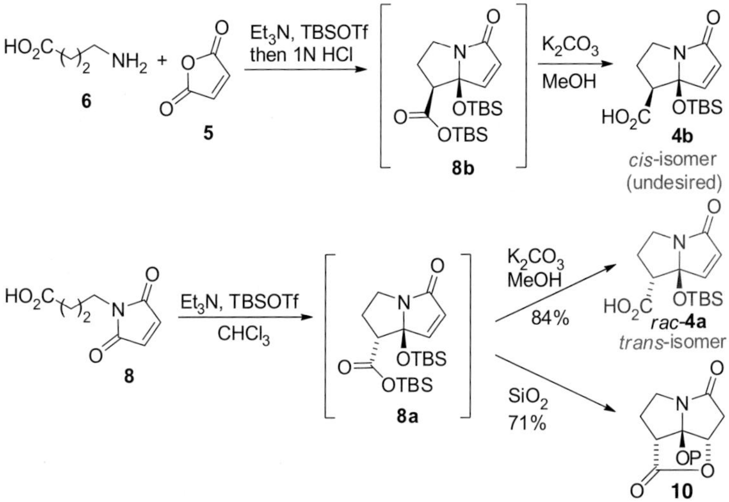

Scheme 2. One-step synthesis of the desired pyrrolizidine carboxylic acid *rac*-**4a**

obtained. Work-up with potassium carbonate in methanol afforded the carboxylic acid **4b**. We were surprised to see that the undesired *cis*-diastereomer **4b** was formed with high selectivity. This result also suggests that the reaction does not proceed via a maleimide. When commercially available maleimide **8** was submitted to the same conditions, the desired silylated *trans*-pyrrolizidine carboxylic acid *rac*-**4a** was formed. Depending on the work-up, either tricyclic lactone **10** or carboxylic acid **4a** was obtained.

We then turned our attention to the kinetic resolution of *rac*-**4a** via an oxa-Michael lactonization. Following the seminal work of Wynberg (Wynberg and Hiemstra 1981), many groups have contributed to the cinchona alkaloid-catalyzed conjugate addition to Michael acceptors. Among the nucleophiles used, soft anions usually prevail, although progress has been reported in the intramolecular oxa-Michael addition of alcohols to α,β-unsaturated esters (Sekino et al. 2004). Interestingly, no enantioselective addition of a carboxylic acid to a Michael acceptor has yet been achieved. When *rac*-**4a** and quinine (2:1) are dissolved, we

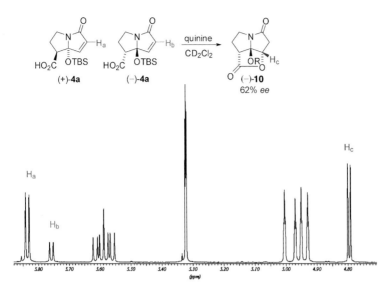

Scheme 3. Kinetic resolution of *rac*-**4a**

observed a kinetic resolution process in the NMR tube. In solution, quinine (like a shift reagent) and the enantiomers of **4a** form diastereomeric ion pairs, which lead to isolated resonances (H_a, H_b and H_c) indicative of the reaction progress. As shown in Scheme 3, (–)-**4a** cyclizes faster than its enantiomer to form the enantiomerically enriched lactone **10**. The enantiomeric ratio of the carboxylic acids and the conversion can be easily determined by integration of the respective resonances. It is important to note that this system is suited to contribute to the understanding of cinchona alkaloid-catalyzed processes, since solvent effects, catalyst performance and kinetics are directly observable.

Recently, Blackmond (Klussmann et al. 2006) and Hayashi (Hayashi et al. 2006) investigated the solid–liquid equilibrium of scalemic amino acids in different solvents and the consequences for catalysis. The most striking example, serine, has its eutectic at >99% *ee*, with the practical consequence that an almost racemic sample can provide an almost enantiopure solution in the solid–liquid equilibrium. It is also known that trituration can influence enantiomeric and diastereomeric compo-

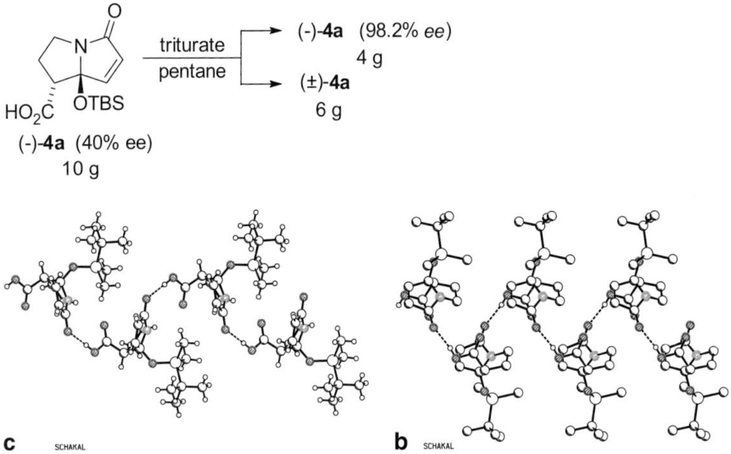

Scheme 4. Trituration enrichment at a high eutectic *ee* and X-ray structures of (±)-**4a** (*left*) and (–)-**4a** (*right*)

sition of scalemic and diastereomeric mixtures. In most cases, trituration is used to remove minor impurities (Myers et al. 1997; Srinivasan et al. 2005), as in our case where we were just aiming separate lactone **10** from the carboxylic acid **4a**. Interestingly, we found that when a weakly enriched scalemic mixture of **4a** is triturated in hot *n*-pentane, (–)-**4a** is readily dissolved, while racemic **4a** remains as a solid residue (Scheme 4).

After having developed a reliable route to the enantiopure carboxylic acid **4a**, we next turned our attention to gain access to the enantiopure decalin fragment. One of the most efficient approaches to this structural motif is the intramolecular Diels–Alder reaction. The first systematic studies in the racemic series were performed by Roush and co-workers in 1981 (Roush and Hall 1981). Evans (Evans et al. 1984, 1988) and Oppolzer (Oppolzer and Dupuis 1985) showed that chiral auxiliaries in combination with strong Lewis acids allowed the generation of the *endo* adducts with excellent diastereoselectivities (Scheme 5). Recently, chiral Lewis acid catalysts were successfully used by Evans (Evans et al. 1999) and Corey (Zhou et al. 2003).

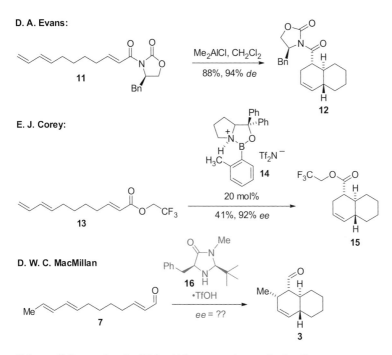

Scheme 5. Intramolecular Diels–Alder approaches to the decalin core

In our synthetic strategy, an approach using MacMillan's catalyst **16** would be most useful because it provides a direct coupling partner (**3**), while the other approaches require further functionalizations of the decalin. However, Evans' auxiliary-mediated method provides an attractive handle to upgrade the enantiomeric excess by separating *diastereomers* prior to the cleavage of the auxiliary. We therefore kept the Evans approach as a safety backup. From the work of Hagiwara we knew that the corresponding decalin alcohol **19** is a solid, which could be used for an eventual improvement of the enantiomeric excess by crystallization. In our synthesis, aldehyde **17** (Larsen and Grieco 1994) was reacted with commercially available Wittig reagent **18** to give the desired trienal **7** in 91% yield (Scheme 6). This compound was subjected to an organocatalytic Diels–Alder reaction using MacMillan's catalyst **16**.

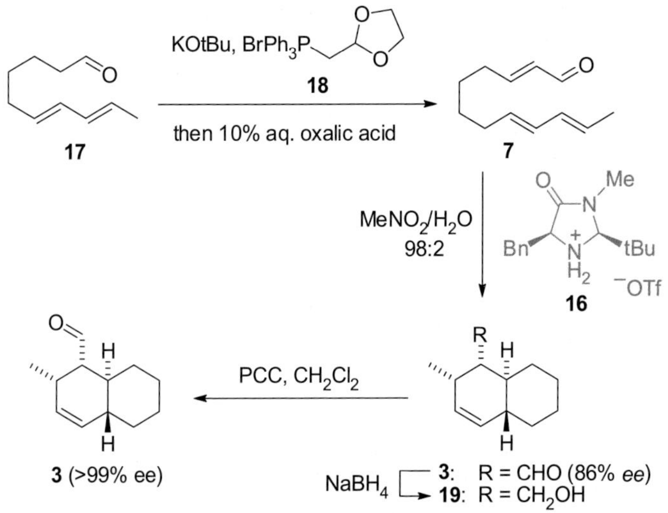

Scheme 6. Enantioselective synthesis of the decalin **3**

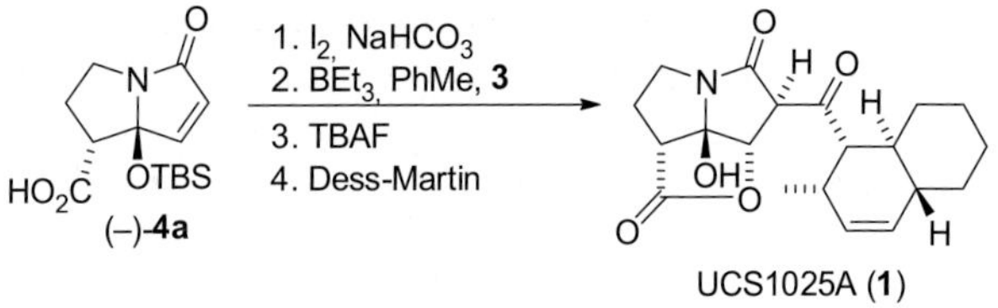

Scheme 7. Completion of the synthesis

We were able to halve the catalyst loading to 10 mol% using nitromethane as the solvent and obtained the Diels–Alder adduct **3** in 74% yield and 86% *ee*. The *ee* of **3** was upgraded to >99% by recrystallization of the corresponding alcohol (**19**) followed by reoxidation (de Figueiredo et al. 2007).

Using Danishefsky's elegant fragment assembly (Lambert and Danishefsky 2006), **4a** and **3** were joined to UCS1025A in just four further steps (Scheme 7), which now allows for a rapid generation of analogs, which supports our SAR-studies for this important telomerase inhibitor.

Four recent developments in reaction methodology—two of them being organocatalytic reactions—now allow for a very efficient access to UCS1025A. Hoye's soft enolization cyclization (Hoye et al. 2006), our oxa-Michael lactonization/trituration (de Figueiredo et al. 2007), MacMillan's organocatalytic Diels–Alder reaction (Wilson et al. 2005) and Danishefsky's BEt$_3$-mediated coupling (Lambert and Danishefsky 2006) have made UCS1025A available in gram scale for biological studies.

3 Lepidopteran Sex Pheromones

During the past decade, the horse chestnut leafminer *Cameraria ohridella* (Deschka and Dimic 1986) has spread from Macedonia over Europe and infested most of the horse chestnut trees (*Aesculus hippocastanum*). The insect's life cycle begins with oviposition onto the leaf surface. The larva then eats itself through the leaf which leads to brown spots, giving the tree an autumnal appearance. The cycle repeats several times per year. Lepidoteran sex pheromones are rather simple straight-chain hydrocarbons terminated by hydroxyl, aldehyde or acetate groups. The hydrocarbon chain is often decorated with different degrees of unsaturation with the *E,Z*-diene moiety being a frequent motif. We have developed a divergent approach to 8*E*,10*Z*-tetradeca-8,10-dienal (**20**) (*Cameraria ohridella*) and other members of this class of lepidopteran pheromones, which takes advantage of the recently discovered organocatalytic transfer hydrogenation of enals (Scheme 8) (Yang et al. 2004, 2005; Ouellet et al. 2005).

A double Wittig homologation of the hexanedial **21** (López et al. 2005) with the phosphonium salt **18** affored decadienedial **22** (Iriye et al. 1988) in good yield. It is noteworthy that the isomerization to thermodynamically more stable *E*-enal occurs upon acidic hydrolysis. A subsequent Wittig reaction with one equivalent of butyltriphenylphosphonium bromide afforded trienal **23** in good selectivity (*Z*:*E*>9:1) albeit in low yield. This material was subjected to the conjugate reduction protocol developed by List to afford 8*E*,10*Z*-tetradeca-8,10-dienal **20**.

Our strategy can be easily extended to the synthesis of other lepidopteran pheromones just by altering the chain length of starting alde-

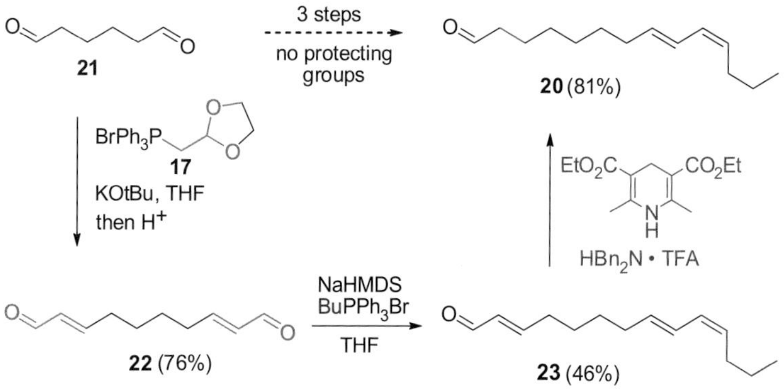

Scheme 8. Synthesis of 8*E*,10*Z*-tetradeca-8,10-dienal (**19**) (*Cameraria ohridella*)

hyde as well as the alkylphosphonium salt. Scheme 9 illustrates some lepidopteran pheromones (**24–30**) synthesized during this study. It should be noted that our approach is maybe not the best for large-scale pheromone synthesis. However, it is significantly shorter and therefore more rapid than the syntheses reported previously. As evaluation and biological testing requires only minute amounts of material, this route also allows for a diversity oriented pheromone synthesis. Elaboration of the aldehyde functionality into alcohol and acetate groups may give rise to other lepidopteran pheromones as exemplified by the synthesis of bombykol and two other pheromones. The aldehyde **28** was reduced to **29** (*Lobesia botrana*) (El-Sayed et al. 1999) using $NaBH_4$ in methanol. Subsequently, **29** was acetylated with acetic anhydride in pyridine to yield **30** (*Matsumuraeses falcana*) in 95% yield (Negishi et al. 1999).

We have developed a straightforward strategy for the synthesis of an important class of lepidopteran sex pheromones starting from simple dialdehydes. The combination of a Wittig reaction and an organocatalytic reduction represents a useful sequence for the nontrivial two-carbon homologation of aldehydes.

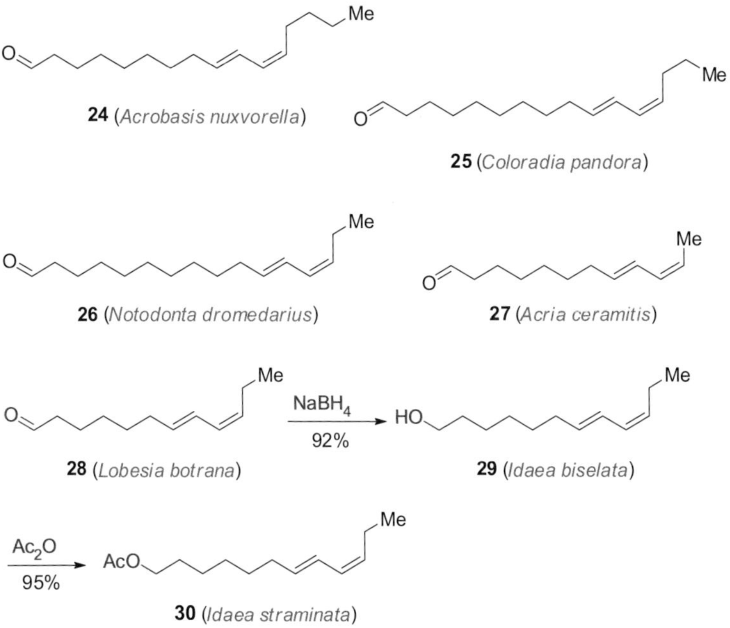

Scheme 9. Synthesis of a collection of lepidopteran pheromones

4 *N*-Heterocycles

In the early stages of the UCS1025A campaign, we developed a novel two-step approach to small *N*-heterocycles using an organocatalytic Mannich reaction and a novel CDI-mediated dehydrative cyclization (Münch et al. 2004). In the first step, we either used two identical aldehydes as the donor and the acceptor (Mannich dimerization) or an enolizable and a non-enolizable aldehyde (three-component Mannich reaction) followed by *in situ* reduction of the aldehyde function. In the second step, the hydroxyl group was displaced by the amino group to give the desired azetidines. As an example, we synthesized an analogue of the cholesterol reducing drug ezetimibe (de Figueiredo et al. 2006).

5 Conclusion

We have shown organocatalysis to be a powerful tool for the synthesis of drugs and natural products. Our work and that of others have clearly demonstrated that organocatalytic strategies can cut down the total number of synthetic operations. As organocatalysts are usually non-toxic, air- and moisture-stable and often available from renewable sources, they are destined to have an impact on the development of future sustainable chemical processes.

Acknowledgements. I would like to thank the Fonds der Chemischen Industrie for a Liebig fellowship, the Deutsche Forschungsgemeinschaft for funding (SPP1179 Organokatalyse) and Prof. Dieter Enders for his encouragement and support.

References

Agatsuma T, Akama T, Nara S, Matsumiya S, Nakai R, Ogawa H, Otaki S, Ikeda S, Saitoh Y, Kanda Y (2002) UCS1025A and B, new antitumor antibiotics from the fungus *Acremonium* species. Org Lett 4:4387–4390

de Figueiredo RM, Christmann M (2007) Organocatalytic synthesis of drugs and bioactive natural products. Eur J Org Chem 2575–2600

de Figueiredo RM, Fröhlich R, Christmann M (2006) N,N'-Carbonyldiimidazole-mediated cyclization of amino alcohols to substituted azetidines and other N-heterocycles. J Org Chem 71:4147–4154

de Figueiredo RM, Berner R, Julis J, Liu T, Türp D, Christmann M (2007) Bidirectional, organocatalytic synthesis of lepidopteran sex pheromones. J Org Chem 72:640–642

de Figueiredo RM, Fröhlich R, Christmann M (2007a) Efficient Synthesis and Resolution of Pyrrolizidines. Angew Chem Int Ed 46:2883–2886

de Figueiredo RM, Voith M, Fröhlich R, Christmann M (2007b) Synthesis of a Malimide Analogue of the Telomerase Inhibitor UCS1025A Using a Dianionic Aldol Strategy. Synlett 391–394

Deschka G, Dimic N (1986) Cameraria ohridella sp. n. (Lep., Lithocolletidae) from Macedonia, Yugoslavia. Acta Ent Jugosl 22:11–23

Doi T, Yoshida T, Shin-ya K, Takahashi T (2006) Total Synthesis of (R)-Telomestatin. Org Lett 8:4165–4167

El-Sayed AM, Gödde J, Witzgall P, Arn H (1999) Characterization of pheromone blend for grapevine moth, *Lobesia botrana* by using flight track recording. J Chem Ecol 25:389–400

Evans DA, Barnes DM, Johnson JS, Lectka T, von Matt P, Miller SJ, Murry JA, Norcross RD, Shaughnessy EA, Campos KR (1999) Bis(oxazoline) and bis(oxazolinyl)pyridine copper complexes as enantioselective Diels–Alder catalysts: reaction scope and synthetic applications. J Am Chem Soc 121:7582–7594

Evans DA, Chapman KT, Bisaha J (1984) New asymmetric Diels–Alder cycloaddition reactions. Chiral α,β-unsaturated carboximides as practical chiral acrylate and crotonate dienophile synthons. J Am Chem Soc 106:4261–4263

Evans DA, Chapman KT, Bisaha J (1988) Asymmetric Diels–Alder cycloaddition reactions with chiral α,β-unsaturated N-acyloxazolidinones. J Am Chem Soc 110:1238–1256

Fürstner A, Domostoj MM, Scheiper B (2006) Total syntheses of the telomerase inhibitors dictyodendrin B, C, and E. J Am Chem Soc 128:8087–8094

Hayashi Y, Matsuzawa M, Yamaguchi J, Yonehara S, Matsumoto Y, Shoji M, Hashizume D, Koshino H (2006) Large nonlinear effect observed in the enantiomeric excess of proline in solution and that in the solid state. Angew Chem Int Ed 45:4593–4597

Hiemstra H, Wynberg H (1981) Addition of aromatic thiols to conjugated cycloalkenones, catalyzed by chiral β-hydroxy amines. A mechanistic study of homogeneous catalytic asymmetric synthesis. J Am Chem Soc 103:417–430

Hoye TR, Dvornikovs V (2006) Comparative Diels–Alder reactivities within a family of valence bond isomers: a biomimetic total synthesis of (±)-UCS1025A. J Am Chem Soc 128:2550–2551

Hoye TR, Dvornikovs V, Sizova E (2006) Silylative dieckmann-like cyclizations of ester-imides (and diesters). Org Lett 8:5191–5194

Iriye R, Toya T, Makino J, Aruga R, Doi Y, Handa S, Tanaka H (1988) Synthesis of aliphatic dienedials, ω-hydroxydienals, ω-hydroxy-2-alkenals, vinylketones, an enon-enal and aliphatic dienediones. Agric Biol Chem 52:989–996

Klussmann M, Iwamura H, Mathew S, Wells Jr. DH, Pandya U, Armstrong A, Blackmond DG (2006) Thermodynamic control of asymmetric amplification in amino acid catalysis. Nature 441:621–623

Lambert TH, Danishefsky SJ (2006) Total Synthesis of UCS1025A. J Am Chem Soc 128:426–427

Larsen SD, Grieco PA (1985) Aza Diels–Alder reactions in aqueous solution: cyclocondensation of dienes with simple iminium salts generated under Mannich conditions. J Am Chem Soc 107:1768–1769

López S, Fernández-Trillo F, Midón P, Castedo L, Saá C (2005) First stereoselective syntheses of (-)-siphonodiol and (-)-tetrahydrosiphonodiol, bioactive polyacetylenes from marine sponges. J Org Chem 70:6346–6352

Münch A, Wendt B, Christmann M (2004) A Mannich-cyclization approach for the asymmetric synthesis of saturated N-heterocycles. Synlett 2751–2755

Myers AG, Gleason JL, Yoon T, Kung DW (1997) Highly practical methodology for the synthesis of D- and L-α-amino acids, N-protected α-amino acids, and N-Methyl-α-amino acids. J Am Chem Soc 119:656–673

Nakai R, Ishida H, Asai A, Ogawa H, Yamamoto Y, Kawasaki H, Akinaga S, Mizukami T, Yamashita Y (2006) Telomerase inhibitors identified by a forward chemical genetics approach using a yeast strain with shortened telomere length. Chem Biol 13:183–190

Nakai R, Ogawa H, Asai A, Ando K, Agatsuma T, Matsumiya S, Akinaga S, Yamashita Y, Mizukami T (2000) UCS1025A, a novel antibiotic produced by Acremonium sp. J Antibiot 53:294–296

Negishi E, Yoshida T, Abramovitch A, Lew G, Williams RM (1991) Highly stereoselective syntheses of conjugated E,Z- and Z,Z-dienes, Z-enynes and Z-1,2,3-butatrienes via alkenylborane derivatives. Tetrahedron 47:343–356

Oppolzer W, Dupuis D (1985) Asymmetric intramolecular Diels–Alder reactions of N-acyl-camphor-sultam trienes. Tetrahedron Lett 26:5437–5440

Ouellet SG, Tuttle JB, MacMillan DWC (2005) Enantioselective organocatalytic hydride reduction. J Am Chem Soc 127:32–33

Rezler EM, Seenisamy J, Bashyam S, Kim MY, White E, Wilson WD, Hurley LH (2005) Telomestatin and diseleno sapphyrin bind selectively to two different forms of the human telomeric G-quadruplex structure. J Am Chem Soc 127:9439–9447

Roush WR, Hall SE (1981) Studies on the total synthesis of chlorothricolide: stereochemical aspects of the intramolecular Diels–Alder reactions of methyl undeca-2,8,10-trienoates. J Am Chem Soc 103:5200–5211

Rudler H, Denise B, Xu Y, Parlier A, Vaissermann J (2005) Bis(trimethylsilyl)-ketene acetals as C,O-dinucleophiles: one-pot formation of polycyclic γ- and δ-lactones from pyridines and pyrazines. Eur J Org Chem 3724–2744

Sekino E, Kumamoto T, Tanaka T, Ikeda T, Ishikawa T (2004) Concise synthesis of anti-HIV-1 Active (+)-inophyllum B and (+)-calanolide A by application of (-)-quinine-catalyzed intramolecular oxo-michael addition. J Org Chem 69:2760–2767

Shin-ya K, Wierzba K, Matsuo K, Ohtani T, Yamada Y, Furihata K, Hayakawa Y, Seto H (2001) Telomestatin, a novel telomerase inhibitor from *Streptomyces anulatus*. J Am Chem Soc 123:1262–1263

Snider BB, Neubert BJ (2004) A novel biomimetic route to the 3-acyl-5-hydroxy-3-pyrrolin-2-one and 3-Acyl-3,4-epoxy-5-hydroxypyrrolidin-2-one ring systems. J Org Chem 69:8952–8955

Srinivasan JM, Burks HE, Smith CR, Viswanathan R, Johnston JN (2005) Free radical-mediated aryl amination: a practical synthesis of (R)- and (S)-7-azaindoline α-amino acid. Synth 330–333

Ullah E, Rotzoll S, Schmidt A, Michalik D, Langer P (2005) Synthesis of 7,8-benzo-9-aza-4-oxabicyclo[3.3.1]nonan-3-ones by sequential 'condensation–iodolactonization' reactions of 1,1-bis(trimethylsilyloxy)ketene acetals with isoquinolines. Tetrahedron Lett 46:8997–8999

Warabi K, Hamada T, Nakao Y, Matsunaga S, Hirota H, van Soest RWM, Fusetani N (2005) Axinelloside A, an Unprecedented Highly Sulfated Lipopolysaccharide Inhibiting Telomerase, from the Marine Sponge, *Axinella infundibula*. J Am Chem Soc 127:13262–13270

Warabi K, Matsunaga S, van Soest RWM, Fusetani N (2003) Dictyodendrins A-E, the First Telomerase-Inhibitory Marine Natural Products from the Sponge *Dictyodendrilla verongiformis*. J Org Chem 68:2765–2770

Wilson RM, Jen WS, MacMillan DWC (2005) Enantioselective organocatalytic intramolecular Diels–Alder reactions. The asymmetric synthesis of solanapyrone D. J Am Chem Soc 127:11616–11617

Yang JW, Hechavarria Fonseca MT, List B (2004) A Metal-Free Transfer Hydrogenation: Organocatalytic conjugate reduction of α,β-unsaturated aldehydes. Angew Chem Int Ed 43:6660–6662

Yang JW, Hechavarria Fonseca MT, Vignola N, List B (2005) Metal-free, organocatalytic asymmetric transfer hydrogenation of α,β-unsaturated aldehydes. Angew Chem Int Ed 44:108–110

Zhou G, Hu QY, Corey EJ (2003) Useful Enantioselective Bicyclization Reactions Using an N-Protonated Chiral Oxazaborolidine as Catalyst. Org Lett 5:3979–3982

Asymmetric Organocatalysis on a Technical Scale: Current Status and Future Challenges

H. Gröger

Department of Chemistry and Pharmacy, University of Erlangen-Nuremberg, Henkestr. 42, 91054 Erlangen, Germany
email: *harald.groeger@chemie.uni-erlangen.de*

1	Introduction	141
2	Industrially Relevant Advantages of Organocatalysis	142
3	Organocatalytic Transformations of Industrial Relevance	143
3.1	Overview	143
3.2	Intramolecular Aldol Reaction: Hajos–Parrish–Eder–Wiechert–Sauer Reaction	144
3.3	Alkylation of Cyclic Ketones	145
3.4	Alkylation of Glycinates for the Synthesis of Optically Active α-Amino Acids	146
3.5	Strecker Reaction	149
3.6	Epoxidation/I: Julia–Colonna–Type Epoxidation	150
3.7	Epoxidation/II: Shi Epoxidation	153
3.8	Other Reactions	153
4	Conclusion and Outlook	154
References		155

1 Introduction

The development of methodologies for the production of chiral building blocks is of crucial importance, as such enantiomerically pure molecules are required as key intermediates in the synthesis of drugs. Due to the increasing tendency to use enantiomerically pure molecules rather than racemates as chiral drugs, there is an increasing interest in developing efficient synthetic technologies. Among conceivable approaches such as multi-step syntheses starting from chiral pool molecules, resolution processes, and asymmetric catalytic technologies, the latter represents the most attractive access for most cases. During recent decades an increasing tendency in industry was observed to apply asymmetric catalytic processes (review: Blaser and Schmidt 2004). Supplementing the established catalytic technologies 'Metal Catalysis' (reviews: Katsuki 1999; Jacobsen and Wu 1999) and 'Biocatalysis' (review: Drauz and Waldmann 2002), recently a third technology type emerged with 'Organocatalysis' (review: Berkessel and Gröger 2005). Seeking for new and innovative technology platforms, the chemical industry shows an increasing interest in organocatalytic reactions as a potential solution for large-scale applications.

2 Industrially Relevant Advantages of Organocatalysis

Organocatalysis offers several advantages not only with respect to its synthetic range. Among "typical" advantages of organocatalysis, in particular with respect to large-scale applications, are favorable economic data of many organocatalysts, the stability of organocatalysts as well as the potential for an efficient recovery (Berkessel and Gröger 2005).

Many organocatalysts are easily available from cheap raw materials from the 'chiral pool' or simple derivatives thereof (e.g., alkaloids and L-proline). In addition, for the majority of organocatalysts there are no concerns regarding moisture sensitivity (which can represent a serious issue in the case of chiral metal complexes used as Lewis acid catalysts). Thus, special equipment for handling organocatalysts is often not required. Recovery of organocatalysts after downstream processing for re-use has also already been reported for organocatalysts in several cases. Furthermore, immobilization represents a popular approach to simplify separation of the catalyst from the reaction mixture. In contrast to immobilized metal complexes (via a solid support-bound ligand), leaching problems are not a critical issue when using organocatalysts immobilized by forming covalent bonds with the solid support. Several immobilized organocatalysts have already been recycled efficiently.

Furthermore, many organocatalytic reactions are already known that proceed with both high conversion and enantioselectivity. There is a range of organocatalytic reactions known to give the desired products with excellent enantioselectivities of more than 99% *ee* (Berkessel and Gröger 2005).

3 Organocatalytic Transformations of Industrial Relevance

3.1 Overview

The suitability of organocatalytic reactions for larger-scale production processes of chiral building blocks has also already been demonstrated in some cases. Notably, different types of bond formation have been reported, comprising several carbon-carbon bond formations as well as oxidation processes. An overview about asymmetric organocatalytic processes with an industrial impact is given in Table 1. These syntheses comprise asymmetric organocatalytic reactions which have been scaled

Table 1 Organocatalytic processes of industrial relevance

Asymmetric organocatalytic reaction	Company	Developed at	Catalyst
Intramolecular aldol reaction	Schering AG Hoffm.-LaRoche	in house	L-proline
Alkylation of indanone derivative	Merck	in house	alkaloid-deriv. cat.
Alkylation of glycinates	Nagase	Maruoka group	phase-transfer-cat.
Strecker reaction	Rhodia ChiRex	Jacobsen group	(thio-)urea cat.
Protonation	Firmenich	in house	amino alcohol
Epoxidation of chalone and derivatives	Bayer AG Degussa AG	Julia/ Colonna group	poly-/oligo-Leu cat.
Epoxidation of alkenes	DSM	Shi group	chiral ketone

up already or which represent process technology solutions ready to be scaled up.

These examples underline the potential of organocatalytic reactions for commercial scale applications. The scale up of the corresponding reactions ranges from L-scale applications to applications on a (pilot) production scale. In the following, several types of these reactions are discussed.

3.2 Intramolecular Aldol Reaction: Hajos–Parrish–Eder–Wiechert–Sauer Reaction

The Hajos–Parrish–Eder–Wiechert–Sauer reaction certainly represents a historical landmark in the field of (asymmetric) organocatalysis. This asymmetric intramolecular aldol reaction was developed in the early 1970s independently by two industrial groups at Schering and Hoffmann-LaRoche, being one of the first major contributions to organocatalysis in general (Hajos and Parrish 1971, 1974a,b; Eder et al. 1971a,b). The target molecules **5** and **6** represent valuable intermediates for the asymmetric synthesis of steroids, and were envisaged as alternatives for the access to steroids instead of rare natural sources. As an organocatalyst, L-proline was used by both groups. At Hoffmann-LaRoche, Hajos and Parris showed that triketones **1** and **2** give, in an intramolecular aldol reaction, the aldol products **3** and **4**, which can subsequently be transformed in to the desired target products **5** and **6** (Hajos and Parrish 1971, 1974a,b). In the presence of 3 mol% of L-proline only, the intramolecular aldol reaction proceeds with enantioselectivities of 74%–93% *ee* (Scheme 1).

The Schering chemists Eder, Wiechert and Sauer demonstrated that the synthesis of the target molecules **5** and **6** can also be done as a one-

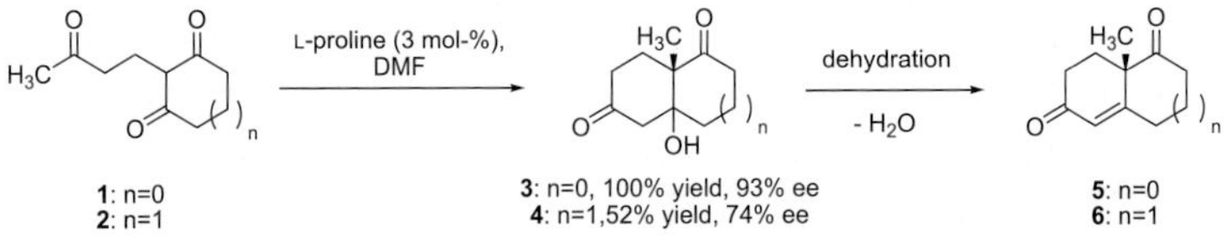

1: n=0
2: n=1

3: n=0, 100% yield, 93% ee
4: n=1,52% yield, 74% ee

5: n=0
6: n=1

Scheme 1. Proline-catalyzed intramolecular aldol reaction

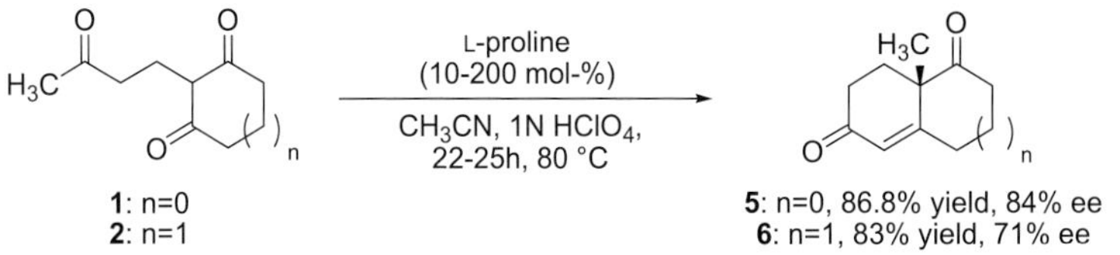

1: n=0
2: n=1

L-proline
(10–200 mol-%)
CH$_3$CN, 1N HClO$_4$,
22–25h, 80 °C

5: n=0, 86.8% yield, 84% ee
6: n=1, 83% yield, 71% ee

Scheme 2. Organocatalytic one-pot synthesis of steroid intermediates

pot reaction with enantioselectivities of up to 84% *ee* when using proline with a catalytic amount of 10–200 mol% (Scheme 2) (Eder et al. 1971a,b). Due to the easy access to the steroid precursors **1** and **2** starting from readily available raw materials, and the use of the economically attractive catalyst L-proline, this intramolecular aldol reaction has gained commercial attention. At Schering, the application of this L-proline catalysis has been carried out on a multikilogram scale (Berkessel and Gröger 2005).

3.3 Alkylation of Cyclic Ketones

A further strength of organocatalysis is its use for efficient carbon–carbon bond formation by means of alkylation processes. In the mid-1980s, Merck chemists developed an asymmetric alkylation of a cyclic ketone in the presence of a simple chinchona alkaloid (Dolling et al. 1984; Hughes et al. 1987; for an exciting review about process research at Merck, see Grabowski 2004). The resulting product **9**, bearing a quaternary stereogenic center, is an intermediate in the synthesis of indacrinone **10**. Notably, this impressive contribution from Merck chemists not only represents the first example of a highly asymmetric phase-transfer catalyst (PTC)-catalyzed alkylation, but also one of the first asymmetric organocatalytic syntheses applied on a larger scale.

Starting with enantioselectivities of below 10% *ee* at the beginning, a subsequent increase of the asymmetric induction was achieved when using individually made chinchona-derived quaternary ammonium salts. While *N*-benzylchinchonium reached approximately 30% *ee*, the use of analogue *p*-substituted derivatives led to enantioselecivities of up to 60% *ee*. Subsequent process development led to an efficient enantioselective alkylation process with enantioselectivities of up to 94% *ee*

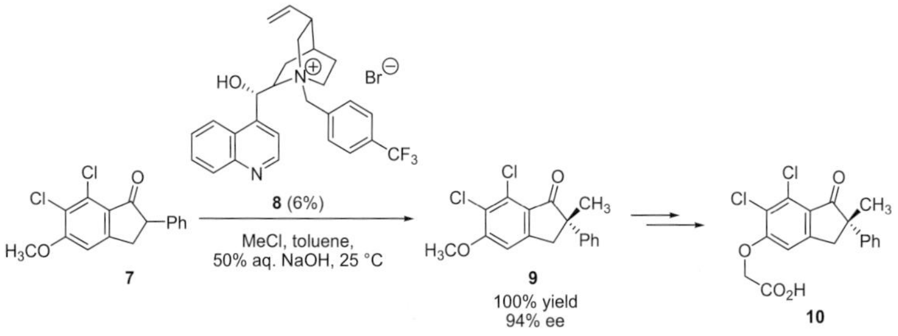

Scheme 3. Organocatalytic alkylation of a cyclic ketone

(Scheme 3) (Grabowski 2004; Dolling et al. 1984; Hughes et al. 1987). The yield of the desired product was 100%, and the required catalytic amount was just 6%. The large-scale feasibility of this process has been demonstrated successfully on a pilot plant scale (Grabowski 2004). Thus, this methodology belongs to the largest-scale organocatalytic reactions applied so far. By means of this methodology, the drug supply of this program has been realized until the demise of the candidate for toxicity reasons (Grabowski 2004). This phase-transfer method also shows advantageous economic data. It was reported that the cost of producing the desired (*S*)-enantiomer based on the asymmetric organocatalytic alkylation route using a catalytic amount below 10 mol% was significantly lower than the costs of producing the (*S*)-enantiomer by a resolution process (Grabowski 2004).

3.4 Alkylation of Glycinates for the Synthesis of Optically Active α-Amino Acids

Further great advancements in the field of asymmetric alkylation reactions have been made by several groups for the chiral phase transfer-catalyzed alkylation of glycinates. This type of reaction offers attractive access to enantiomerically pure, particularly nonproteinogenic α-amino acids. A pioneer in this field is the O'Donnell group (O'Donnell et al. 1989; for an excellent recent review, see O'Donnell 2001) who developed the first α-amino acid ester synthesis by means of this methodology. Notably, this group also reported a first scale up of the synthesis in

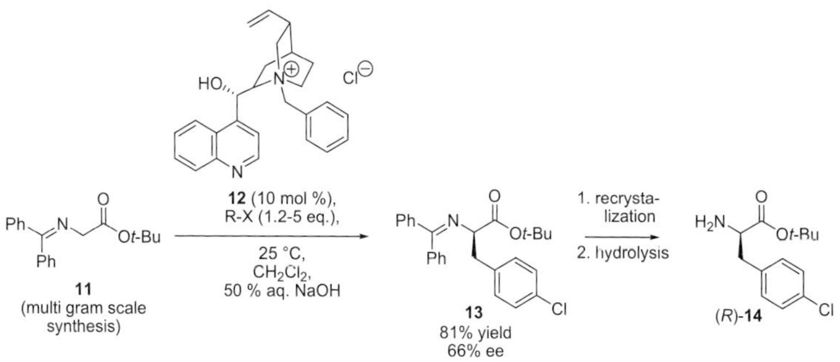

Scheme 4. Organocatalytic alkylation of a glycinate

a multigram-scale synthesis of the α-amino acid D-*p*-chlorophenylalanine, (*R*)-**14** (O'Donnell et al. 1989). The asymmetric alkaloid-catalyzed alkylation with a *p*-chlorobenzyl halide proceeds under formation of the glycinate **13** in 81% yield and with 66% *ee* when using a catalytic amount of 10 mol% of the chiral phase-transfer catalyst **12** (Scheme 4). Recrystallization, and subsequent hydrolysis afforded an enantiomerically pure sample of 6.5 g of the 'free' amino acid D-*p*-chlorophenylalanine, (*R*)-**14** (O'Donnell et al. 1989).

Besides the O'Donnell group, further important contributions in the field of asymmetric alkylation have been made by the groups of Lygo, Corey, Maruoka, Shiori, Kim, as well as Jew and Park (Berkessel and Gröger 2005). The latter group (Park et al. 2002; Jew et al. 2001) also applied their alkaloid-based PTC-catalyst on a 150-g-scale for the synthesis of a *p*-substituted phenylalanine derivative (H.-G. Park, personal communication). In addition, several patent applications describe the use of glycinate alkylation with alkaloid-type organocatalysts for the preparation of commercially interesting target molecules (Mulholland et al. 2002; Jew et al. 2002; Fujita et al. 2003; Jew et al. 2003).

Following the great achievements in alkaloid-type asymmetric alkylation of glycinates that have been made over the years, this methodology has recently been applied on larger scale for the preparation of particularly nonproteinogenic, optically active α-amino acids. A very successful application on the kilogram scale was reported by a GlaxoSmithKline research team for the preparation of 4-fluoro-β-(4-fluoro-

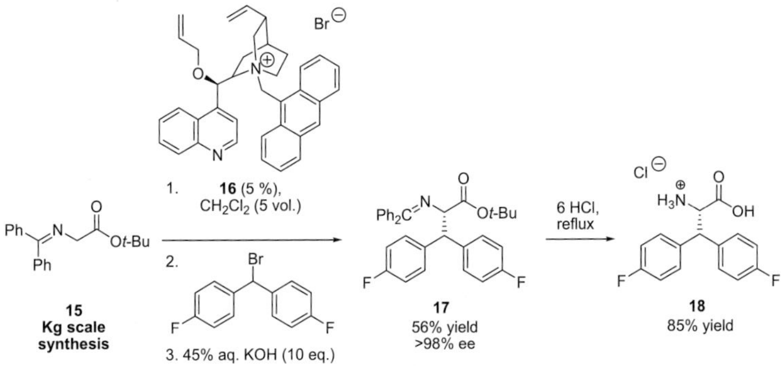

Scheme 5. Synthesis of (S)-4-fluoro-β-(4-fluorophenyl)-phenylalanine as its hydrochloride salt

phenyl)-phenylalanine using alkaloid-type phase-transfer organocatalyst **16** (Scheme 5; Patterson et al. 2006). In the presence of 5 mol% of **16** the reaction runs to completion within only 5 h, and gave the alkylated glycinate with an enantioselectivity of 60% ee. After work-up and recrystallization, the product **16** was obtained in 56% yield and with an enantiomeric excess of 98% ee. Subsequent hydrolysis in hydrochloric acid and work-up led to the amino acid 4-fluoro-β-(4-fluorophenyl)-phenylalanine as its hydrochloric acid salt (**17**) in 85% yield (Patterson et al. 2006).

Recently, the Maruoka group developed highly efficient phase transfer-organocatalysts, e.g., of type **20**, bearing a quaternary ammonium moiety for this type of reaction (Ooi et al. 1999; Ooi et al. 2003; review: Maruoka and Ooi 2003). The Maruoka organocatalysts show outstanding catalytic properties such as excellent enantioselectivities, high conversion and very low catalytic amounts in the range of 1 mol% or even below. Accordingly, the Maruoka organocatalysts also attracted industrial interest, and large-scale applications using the Maruoka organocatalyst have been carried out by Nagase Company synthesizing unnatural α-amino acids starting from glycine or alanine (Maruoka 2006; K. Maruoka, personal communication). Notably, the use of alanine (**19**) instead of glycine as a raw material leads to of α-amino acids bearing a quaternary stereogenic center. Representative examples based on the

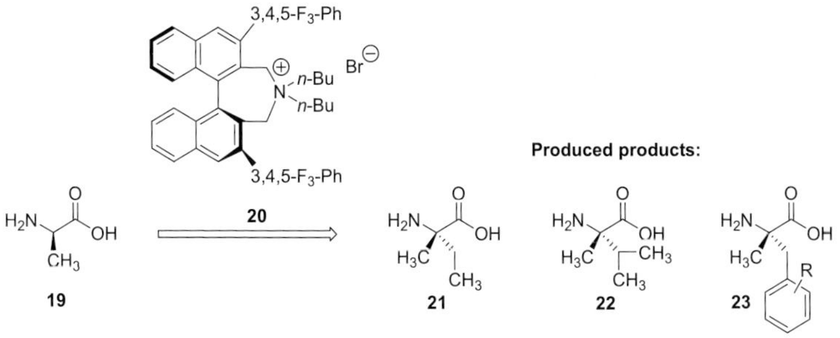

Scheme 6. Asymmetric synthesis of α-amino acids bearing a quaternary stereogenic center

use of alanine as a starting material are shown in Scheme 6 (Maruoka 2006; K. Maruoka, personal communication).

Due to the high efficiency the "state of the art" of this methodology reached, an increasing number of commercial applications thereof can be expected for the synthesis of in particular non-proteinogenic amino acids the future.

3.5 Strecker Reaction

The asymmetric catalytic Strecker reaction is another elegant approach for the synthesis of optically active α-amino acids. For this reaction the Jacobsen group developed excellent organocatalysts (Sigman and Jacobsen 1998a,b; Jacobsen and Sigman 1999; Sigman et al. 2000; Vachal and Jacobsen 2000), namely optically active urea or thiourea derivatives. Notably, these organocatalysts turned out to represent the most efficient catalyst type to date for the asymmetric hydrocyanation of imines. Due to its high efficiency the Jacobsen hydrocyanation technology has already been utilized for commercial purpose at Rodia ChiRex (http://www.rhodiachirex.com/techpages/amino_acid_technology.htm). The reaction concept is shown in Scheme 7. In the presence of the readily available organocatalyst **25** with a catalytic amount of 2 mol%, the asymmetric hydrocyanation proceeds with high conversion and enantioselectivity for a broad range of imines. The resulting α-amino nitrile products are conveniently isolated as the trifluoracetamides **26**, which

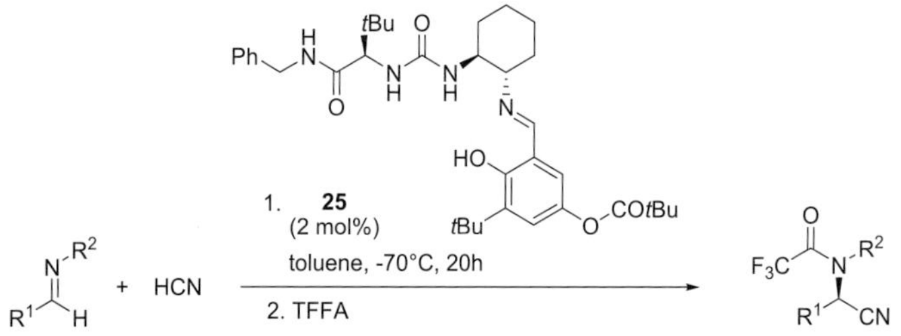

Scheme 7. Organocatalytic Strecker reaction

can be further converted into a range of enantiomerically pure building blocks, e.g., α-amino acids (http://www.rhodiachirex.com/techpages/amino_acid_technology.htm). In addition, minor variation of the catalyst in combination with immobilization on a resin support gave an analogue recyclable solid-supported organocatalyst (http://www.rhodiachirex.com/techpages/amino_acid_technology.htm).

3.6 Epoxidation/I: Julia–Colonna–Type Epoxidation

Epoxidation reactions belong to the most important (asymmetric) transformations. Besides asymmetric metal-catalyzed methodologies, analogous organocatalytic epoxidation has been known for a long time. In 1980, Julia et al. reported a simple asymmetric epoxidation catalyzed by polyamino acids (for the pioneering work, see: Julia et al. 1980, 1982). Subsequently, many groups contributed to the further development of this synthetic method, which turned out to be an efficient technology for the preparation of chalcone-derived epoxides (Berkessel and Gröger 2005). The epoxidation reaction utilizes hydrogen peroxide as an oxidant and proceeds under triphasic conditions. Among the main advantages of this epoxidation reaction are the use of an environmentally friendly organocatalyst, use of a cheap oxidant and base (NaOH), the potential recyclability of the organocatalyst, and the high enantioselectivities of up to 95% *ee*. However, with respect to technical applications, this method also shows some drawbacks (Bosch 2004). For example, a large excess of the catalyst with amounts of up to 200% (w/w) is needed. In addition, the catalyst has to be preactivated and the duration of this process is more than 6 h. Another drawback are the long reaction times of 1–5 days, which limits technical applicability significantly.

Researchers at Bayer AG addressed the critical issues with respect to technical application and developed successful solutions towards a technically applicable Julia–Colonna-type epoxidation (Bosch 2004; Geller et al. 2003, 2004a,b). Catalyst preparation has been improved by a straightforward synthesis of the poly-Leu-catalyst. Key features are cheap reagents and a shorter reaction time (Bosch 2004; Geller et al. 2003, 2004a). In particular the reaction time for the new polymerization process is only 3 h when the the process is carried out at 80 °C in toluene, compared with 5 days under classic reaction conditions (THF, room temperature). The catalyst prepared by the 'Bayer route' is also much more active, and does not require preactivation (Bosch 2004; Geller et al. 2003, 2004a). In parallel, the triphasic reaction system has been improved: A strongly enhanced reaction rate occurs in the presence of an achiral PTC as an additive (Bosch 2004; Geller et al. 2003, 2004a,b). Carrying out the epoxidation of chalcone **27** with 10 mol% of TBAB as an achiral PTC catalyst on a 100-g scale in the presence of a catalytic amount of 10%–20%(w/w) of the poly-L-Leu organocatalyst, equivalent to 0.25–0.7 mol%, led to a complete conversion within 12 h (Bosch 2004). The desired product **29** was obtained in 75% yield and with an enantiomeric excess of 97.6% *ee* (Scheme 8). Notably, on a smaller scale, a conversion of more than 99% was reached within only 7 min, whereas in the absence of TBAB the asymmetric poly-L-leucine catalyzed epoxidation gave only 2% conversion after 1.5 h. Furthermore, it was found that efficient stirring was essential for a complete conversion. Notably, the catalyst can be re-used without loss of reactivity and enantioselectivity (Bosch 2004).

Researchers at Degussa AG focused on an alternative solution with respect to a technical application of the Julia–Colonna epoxidation (Tsogoeva et al. 2002). Successful process development is based on the design of a continuously operated process in a chemzyme membrane reactor (CMR reactor). Therein, the epoxide and unconverted chalcone pass through the membrane whereas the polymer-enlarged organocatalyst is retained in the reactor by means of a nanofiltration membrane. The setup for this type of continuous epoxidation reaction is shown in Scheme 9.

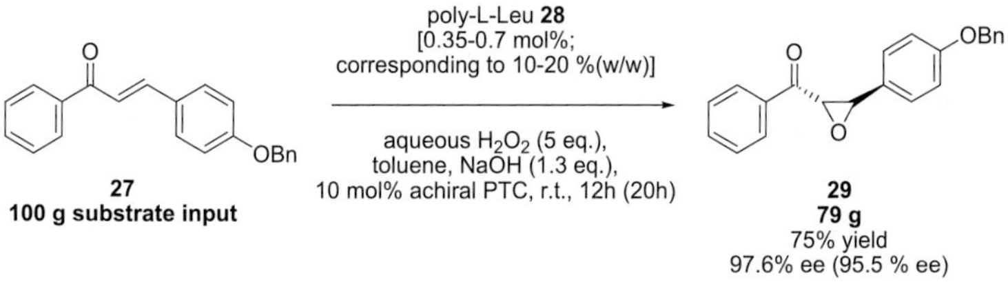

Scheme 8. Julia–Colonna-type epoxidation

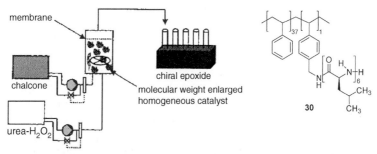

Scheme 9. Continuously operated epoxidation process in a chemzyme membran reactor (figure reprinted with permission form: Tsogoeva et al. 2002. Copyright 2002 Georg Thieme Verlag, Stuttgart, New York)

The chemzyme membrane reactor is based on the same continuous process concept as the efficient enzyme membrane reactor, which has already been applied for enzymatic α-amino acid resolution on industrial scale at a production level of hundreds of tons per year (Drauz and Waldmann 2002; Wandrey and Flaschel 1979; Wandrey et al. 1981; Gröger and Drauz 2004).

The prerequisite for this process, namely the availability of homogenous polymer-supported catalysts, have been fulfilled by Tsogoeva et al. developing, e.g., the oligo(L-Leu) catalyst **30** (Scheme 9; Tsogoeva et al. 2002). This catalyst has been used efficiently in the continuous CMR process with chalcone and urea-hydrogen peroxide as the oxidizing agent. The corresponding epoxidation reaction in the chemzyme membrane reactor with a volume of 10 mL furnished the epoxide product with enantioselectivities of up to 90%–95% throughout 50 residence times (Tsogoeva et al. 2002). Advantages of this CMR-concept with soluble polymer-supported oligo(L-Leu) are the high conversions in combination with an elegant catalyst recycling and separation, as well as high enantioselectivities (Tsogoeva et al. 2002).

3.7 Epoxidation/II: Shi Epoxidation

In addition to the Julia–Colonna epoxidation, also the Shi-epoxidation received commercial interest. This epoxidation has been developed by the Shi group in recent years and is based on the use of a D-fructose-derived enantiomerically pure ketone **32** as a catalyst (Tu et al. 1996; Wang et al. 1997; review: Shi 2004). Besides the simple approach to the catalyst starting from an easily available raw material, the high enantioselectivities and the very broad substrate range are further key advantages of this impressive epoxidation technology. Recently, DSM researchers jointly with Shi reported the application of this efficient Shi epoxidation technology on commercial scale for the synthesis of an epoxide derivative of **31** (Ager et al. 2007; Ager 2003). This compound is of interest as a key intermediate in the synthesis of lactone **33** (Ager et al. 2007). The multi-step synthesis is shown in Scheme 10. For the epoxidation key step, the overall yields were around 63%, and the resulting lactone **33** showed a chemical purity of 97% and an enantiomeric excess of 88% *ee*. This product quality was sufficient to be used for the subsequent step without further purification. The overall amount of produced lactone was greater than 100 kg (Ager et al. 2007) Notably, the overall synthesis of the lactone **33** was accomplished without isolation of any of the intermediates. It should be added that Ager et al. also reported a large-scale feasible and cost effective method for the synthesis of the organocatalyst, **32** (Ager et al. 2007).

3.8 Other Reactions

Several other examples in the field of (asymmetric) organocatalysis, which are not described in more detail in this contribution, have been reported as well. For example, Fehr and co-workers from Firmenich reported a very interesting asymmetric protonation reaction based on the use of an organocatalyst on technical scale (Fehr 2006; C. Fehr,

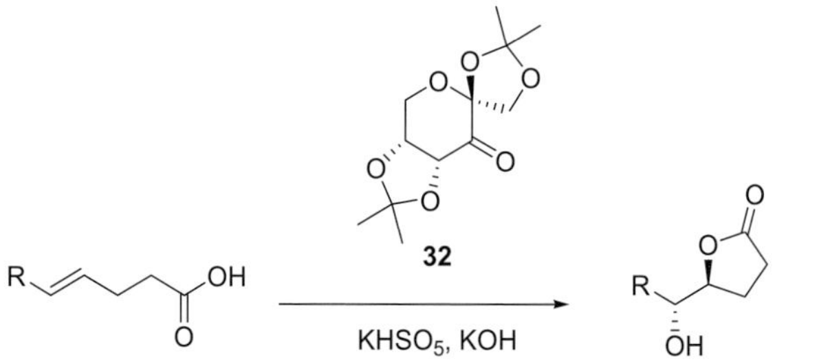

Scheme 10. Application of the Shi-type epoxidation

personal communication). In addition, organocatalytic transformation based on a thiourea organocatalyst, which were developed by the Takemoto group, was done successfully on kilogram scale (Y. Takemoto, personal communication).

4 Conclusion and Outlook

In conclusion, process development and scale-up work has been done already for several organocatalytic reactions. The results demonstrate that organocatalysis can represent a valuable tool for solutions on industrial scale also. Albeit the use of organocatalysis in industry is still limited, it can be expected that the broad variety of already developed efficient organocatalytic syntheses, in combination with further breakthroughs and new applications, will contribute to an increasing number of organocatalytic large-scale reactions in the future. Notably, many of the very recently developed syntheses in asymmetric organocatalysis fulfil already important criteria for an industrial process. In general, among key challenges for an increased use of organocatalysts will be the development of scalable procedures for catalyst preparation, further decrease of catalyst loading, and technically attractive downstream processing protocols. Another challenge is the development of highly atom economical attractive synthetic routes to the target molecules, e.g., by avoiding the use of protecting groups (in particular of those whose subsequent cleavage is difficult).

Acknowledgements. I am grateful for the support of many colleagues during the preparation of this contribution about large-scale organocatalysis. In this regard I would like to thank Dr. Dave Ager (DSM), Prof. Dr. Karlheinz Drauz, Dr. Ian Grayson (both Degussa AG), Dr. Charles Fehr (Firmenich), Prof. Dr. Eric Jacobsen (Harvard University), Prof. Dr. Jürgen Martens (University of Oldenburg), Prof. Dr. Kenji Maruoka (Kyoto University), Prof. Dr. Martin O'Donnell (University of Illinois), Prof. Dr. Yian Shi (Colorado State University), and Prof. Dr. Yoshiji Takemoto (Kyoto University) for personal communications. In addition, I would like to thank Prof. Dr. Albrecht Berkessel (University of Cologne), and Prof. Dr. Heribert Offermanns for many exciting discussions about asymmetric organocatalysis.

References

Ager D (2003) Homogeneous catalysis, from Coors to Heineken. In: 8th International Conference on Organic Process Research and Development, Conference presentation, Scientific Update Conference, Barcelona, 7–10 September 2003

Ager D, Anderson K, Oblinger E, Shi Y, VanderRoest J (2007) An epoxidation approach to a chiral lactone: application of the Shi-epoxidation. Org Process Res Dev 11:44

Berkessel A, Gröger H (2005) Asymmetric organocatalysis, from biomimetic concepts to synthetic applications. VCH-Wiley, Weinheim

Blaser HU, Schmidt E (eds) (2004) Asymmetric catalysis on industrial scale. Wiley-VCH, Weinheim

Bosch BE (2004) Scalable methods for functionalised chiral alcohols. In: Chiral Europe 2004 Conference Proceedings, Scientific Update Conferences, Mainz, 14–16 June 2004

Dolling UH, Davis P, Grabowski EJJ (1984) Efficient catalytic asymmetric alkylations, 1. Enantioselective synthesis of (+)-indacrinone via chiral phase-transfer catalysis. J Am Chem Soc 106:446

Drauz K, Waldmann H (eds) (2002) Enzyme catalysis in organic synthesis, vols 1–3, 2nd edn. Wiley-VCH, Weinheim

Eder U, Sauer G, Wiechert R (1971a) New type of asymmetric cyclization to optically active steroid CD partial structures. Angew Chem Int Ed Engl 10:496

Eder U, Wiechert R, Sauer G (1971b) Verfahren zur Herstellung optisch aktiver Bicycloalkan-Derivate. DE 2014757

Fehr C (2006) Synthetic applications of enantioselective protonation and case study for (S)-α-damascone. In: 45th Tutzing Symposiom Organocatalysis, Tutzing, Germany, 8–11 October 2006

Fujita K, Taguchi Y, Oishi A (2003) JP 3459986

Geller T, Krüger CM, Militzer HC (2003) Polyaminosäure-katalysiertes Verfahren zur enantioselektiven Epoxidierung von alpha,beta-ungesättigten Enonen und alpha,beta-ungesättigten Sulfonen. EP 1279671

Geller T, Gerlach A, Krüger CM, C Militzer H (2004a) Novel conditions for the Juliá–Colonna epoxidation reaction providing efficient access to chiral, nonracemic epoxides. Tetrahedron Lett 45:5065

Geller T, Krüger CM, C Militzer H (2004b) Scoping the triphasic/PTC conditions for the Juliá–Colonna epoxidation reaction. Tetrahedron Lett 45:5069

Grabowski EJ (2004) ACS Symposium Series, 870 (Chemical Process Research), 1

Gröger H, Drauz K (2004) Methods for the biocatalytic production of L-amino acids on industrial scale. In: Blaser HU, Schmidt E (eds) Asymmetric catalysis on industrial scale. VCH, Weinheim, p 131f

Hajos ZG, Parrish DR (1971) Asymmetric synthesis of optically active polycyclic organic compounds. DE 2102623

Hajos ZG, Parrish DR (1974a) Synthesis and conversion of 2-methyl-2-(3-oxobutyl)-1,3-cyclopentanedione to the isomeric racemic ketols of the [3.2.1]bicyclooctane and the perhydroindane series. J Org Chem 39:1612

Hajos ZG, Parrish DR (1974b) Asymmetric synthesis of bicyclic intermediates of natural product chemistry. J Org Chem 39:1615

http://www.rhodiachirex.com/techpages/amino_acid_technology.htm, July 2004

Hughes DL, H Dolling U, Ryan KM, Schoenewaldt EF, Grabowski EJJ (1987) Efficient catalytic asymmetric alkylations. 3. A kinetic and mechanistic study of the enantioselective phase-transfer methylation of 6,7-dichloro-5-methoxy-2-phenyl-1-indanone. J Org Chem 52:4745

Jacobsen EN, Sigman MS (1999) Parallel combinatorial approach to the discovery and optimization of catalysts and uses thereof. PCT Int Appl, WO9951546

Jacobsen EN, Wu MH (1999) Epoxidation of alkenes other than allylic alcohols. In: Jacobsen E, Pfaltz A, Yamamoto H (eds) Comprehensive asymmetric catalysis I–III. Springer, Berlin, Heidelberg, New York, p 649

Jew SS, Jeong BS, Yoo MS, Huh H, Park HG (2001) Chem Commun 1244

Jew SS, Park HG, Jeong BS, Yoo MS, Lee SH (2002) WO 2002083670

Jew SS, Park HG, Jeong BS, Yoo MS, Lee SH, Cho DH (2003) WO 2003045948

Julia S, Guixer J, Masana J, Rocas J, Colonna S, Annuziata R, Molinari HJ (1982) "Synthetic enzymes", highly stereoselective epoxidation of chalcone in a triphasic toluene-water-poly[(S)-alanine] system. J Chem Soc [Perkin 1]:1317

Julia S, Masana J, Vega J (1980) Angew Chem Int Ed Engl 19:929

Katsuki T (1999) Epoxidation of allylic alcohols. In: Jacobsen E, Pfaltz A, Yamamoto H (eds) Comprehensive asymmetric catalysis I–III. Springer, Berlin, Heidelberg, New York, p 621f

Maruoka K (2006) Chiral phase transfer catalysis for practical asymmetric synthesis. 45th Tutzing Symposiom Organocatalysis, Tutzing, Germany, October 8–11

Maruoka K, Ooi T (2003) Enantioselective amino acid synthesis by chiral phase-transfer catalysis. Chem Rev 103:3013–3028

Mulholland GK, O'Donnell MJ, Chin FT, Delgado F (2002) Nucleophilic approach for preparing radiolabeled imaging agents and associated compounds. WO0244144

O'Donnell MJ (2001) The preparation of optically active α-amino acids from the benzophenone imines of glycine derivatives. Aldrichim Acta 34:3

O'Donnell MJ, Bennett WD, Wu S (1989) The stereoselective synthesis of α-amino acids by phase-transfer catalysis. J Am Chem Soc 111:2353

Ooi T, Kameda M, Maruoka K (1999) Molecular design of a C_2-symmetric chiral phase-transfer catalyst for practical synthesis of α-amino acids. J Am Chem Soc 121:6519

Ooi T, Kameda M, Maruoka K (2003) Design of N-spiro C_2-symmetric chiral quaternary ammonium bromides as novel chiral phase-transfer catalysts: synthesis and application to practical asymmetric synthesis of α-amino acids. J Am Chem Soc 125:5139–5151

Park HG, Jeong BS, Yoo MS, Lee JH, Park MK, Lee YJ, Kim MJ, Jew SS (2002) Highly enantioselective and practical chinchona-derived phase-transfer catalysts for the synthesis of α-amino acids. Angew Chem Int Ed 41:3036

Patterson DE, Xie S, Jones LA, Osterhout MH, Henry CG, Roper TD (2007) Synthesis of 4-fluoro-β-(4-fluorophenyl)-L-phenylalanine by an asymmetric phase-transfer catalyzed alkylation: synthesis on scale and catalyst stability. Org Process Res Dev 11:624–627

Shi Y (2004) Organocatalytic asymmetric epoxidation of olefins by chiral ketones. Acc Chem Res 32:488

Sigman MS, Jacobsen EN (1998a) Schiff base catalysts for the asymmetric Strecker reaction identified and optimized from parallel synthetic libraries. J Am Chem Soc 120:4901

Sigman MS, Jacobsen EN (1998b) Schiff base catalysts for the asymmetric Strecker reaction identified and optimized from parallel synthetic libraries. Book of Abstracts, 216th ACS Meeting, Boston, 23–27 August 1998

Sigman MS, Vachal P, Jacobsen EN (2000) A general catalyst for the asymmetric Strecker reaction. Angew Chem Int Ed Engl 39:1279

Tsogoeva SB, Wöltinger J, Jost C, Reichert D, Kühnle A, Krimmer HP, Drauz K (2002) Juliá–Colonna asymmetric epoxidation in a continuously operated chemzyme membrane reactor. Synlett 707

Tu Y, Wang ZX, Shi Y (1996) An efficient asymmetric epoxidation method for *trans*-olefins mediated by a fructose-derived ketone. J Am Chem Soc 118:9806

Vachal P, Jacobsen EN (2000) Enantioselective catalytic addition of HCN to ketimines. Catalytic synthesis of quaternary amino acids. Org Lett 2:867–870

Wandrey C, Flaschel E (1979) Process development and economic aspects in enzyme engineering. Acylase L-methionine system. Adv Biochem Eng Biotechnol 12:147–218

Wandrey C, Leuchtenberger W, Kula MR (1981) Process for the continuous enzymatic change of water soluble α-ketocarboxylic acids into the corresponding amino acids. US 4304858

Wang ZX, Tu F, Frohn M, R Zhang J, Shi Y (1997) An efficient catalytic asymmetric epoxidation method. J Am Chem Soc 119:11224

Ernst Schering Foundation Symposium Proceedings, Vol. 2, pp. 159–181
DOI 10.1007/2789_2008_081
© Springer-Verlag Berlin Heidelberg
Published Online: 30 April 2008

Nucleophilic Carbenes as Organocatalysts

F. Glorius[✉], K. Hirano

Organisch Chemisches Institut, Westfälische Wilhelms-Universität Münster,
Corrensstraße 40, 48149 Münster, Germany
email: *glorius@uni-muenster.de*

1	Introduction	159
2	Conjugate Umpolung	163
3	Mechanistic Proposal	166
4	Ketones and Imines as Electrophiles	167
5	Conjugate Umpolung of Crotonaldehyde Derivatives	169
6	Conjugate Umpolung of α-Substituted Cinnamaldehyde Derivatives	170
7	Intramolecular Variants	172
8	Formation of β-Lactones	174
References		177

Abstract. N-Heterocyclic carbenes (NHC) have become an important class of organocatalysts and class of ligands for transition-metal catalysis. In organocatalyzed umpolung reactions, thiazolium salt-derived NHC have been used successfully for decades. Even so, during recent years there has been an increased interest in NHC-catalyzed transformations and many new reactions have been developed. This article focuses on the use of NHC in the conjugate umpolung of α,β-unsaturated aldehydes.

1 Introduction

For many years, carbenes seemed to be elusive compounds and their isolation appeared impossible. Therefore, the first isolation and characterization of a free N-heterocyclic carbene (NHC) in 1991 marked a true

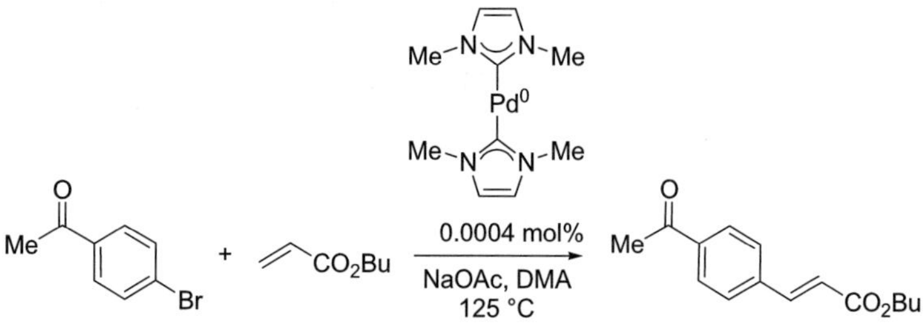

Scheme 1. First application of NHC as ligands in catalysis

breakthrough and initiated extensive research programs (Arduengo et al. 1991; see also Igau et al. 1988; for reviews on NHC, see Nolan 2006; Glorius 2007; Herrmann 2002; Herrmann and Köcher 1997; Cavell and McGuinness 2004; Peris and Crabtree 2004; Hillier et al. 2002; Bourissou et al. 2000; Arduengo 1999; Arduengo and Krafczyk 1998). Consequently, one of the first applications of NHC as ligands in transition-metal catalysis—in a palladium-catalyzed Heck reaction—was reported (Herrmann et al. 1995; Scheme 1). Since then, many other applications in transition-metal catalysis have been reported and NHC have become an important ligand class for many transformations such as cross coupling and metathesis reactions. This is because NHC exhibit a number of attractive ligand properties:

1. NHC are strong σ-donor ligands and are even more electron-rich than the already electron-rich trialkylphosphines. Whereas in the early 1990s π-acceptor properties of NHC were disputed, it became obvious that NHC can act as π-acceptors and that the degree of π-acceptance depends on the nature of the metal bound and the NHC itself (Sanderson et al. 2006).
2. In many cases, very robust complexes are formed, being insensitive to heat, air and moisture.
3. The shape of the carbene ligands can be varied over a wide range, allowing the design and adjustment of the NHC for a variety of different applications.

In organocatalysis, NHC were used long before Arduengo's finding. For decades, thiazoliums salt-derived carbenes were well-known cata-

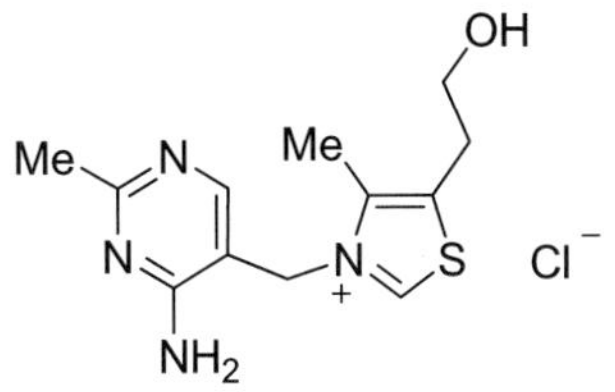

Fig. 1. Thiamine (vitamin B_1)

lysts for umpolung reactions such as the Benzoin condensation (Wöhler and Liebig 1832; Lapworth 1903; Breslow 1958) or the Stetter reaction (Stetter and Kuhlmann 1976). Moreover, in the biochemistry of the naturally occurring thiazolium salt coenzyme thiamine (vitamin B_1) the catalytically active species is most probably an NHC (Fig. 1). In recent years, an increased interest in NHC as organocatalysts has resulted in many powerful transformations such as (asymmetric) benzoin condensations (Enders et al. 2006; Enders and Niemeier 2004; Enders and Kallfass 2002; Hachisu et al. 2003, 2004; Dudding and Houk 2004; Pesch et al. 2004), transesterifications, polymerizations, redox reactions and several others. (For excellent reviews on NHC in organocatalysis see Enders and Balensiefer 2004; Johnson 2004; Christmann 2005; Zeitler 2005).

Again, a number of attractive properties of NHC as organocatalysts have become apparent:

1. In organocatalysis, NHC can have several different modes of action, the most prominent ones of which are shown in Scheme 2. First of all, the pronounced nucleophilicity of NHC allows the addition to electrophiles such as aldehydes (a). In addition, the catalyst in the form of a positively charged azolium salt is strongly electron- withdrawing and results in a significant acidification of the α-position (b). Furthermore, an enamine motif incorporating the NHC can result and react as a nucleophile (c). Finally, the catalyst can act as a leaving group (d), depart from the molecule and thus, enter the catalytic cycle once again.
2. The electronic and steric properties of the NHC can be tuned over a wide range by choosing different nitrogen heterocycles

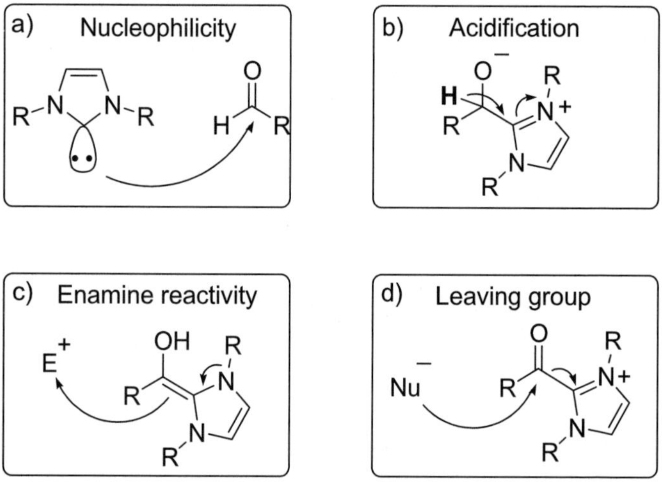

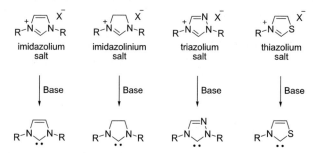

Scheme 2. Modes of action of NHCs in organocatalysis

Scheme 3. Different types of NHC

(Scheme 3). Subtle structural differences can have a pronounced effect on catalytic activity and selectivity.

3. Based on the R substituents in the 1- and 3-positions of the NHC, the area around the carbene carbon can be greatly influenced, allowing the design and adjustment for a variety of different applications (Fig. 2).

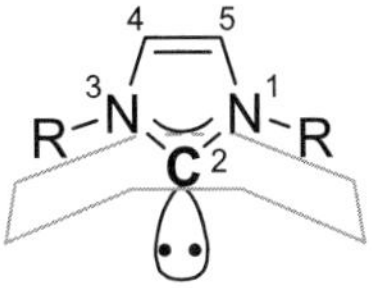

Fig. 2. Shape of NHC

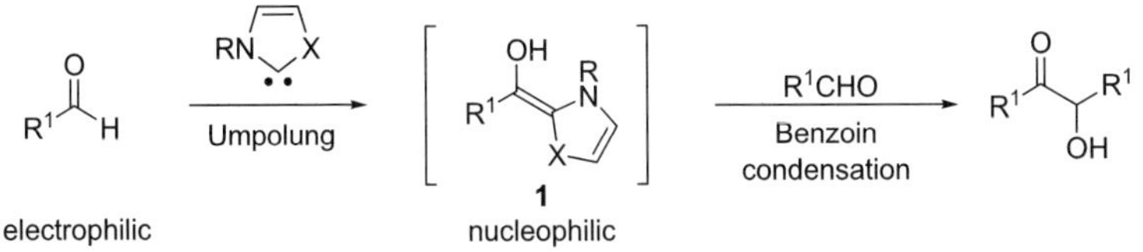

Scheme 4. Benzoin-type umpolung

Thiazolium and triazolium salt-derived NHC, in particular, are well known catalysts for benzoin- and Stetter-type umpolung reactions. In the course of these reactions, the NHC catalyst adds to the electrophilic aldehyde, resulting in the formation of a nucleophilic enamine species. Subsequently, this enamine can react with a series of different electrophiles such as aldehydes (benzoin condensation) or α,β-unsaturated substrates (Stetter reaction) (Scheme 4).

2 Conjugate Umpolung

Our work in the area of NHC as ligands in transition-metal catalysis (Glorius et al. 2002; Altenhoff et al. 2003, 2004, 2006; Burstein et al. 2005; Tewes et al. 2007) inspired us to think of applications of NHC in the area of organocatalysis (for excellent reviews on modern organocatalysis see Seayad and List 2005; Dalko and Moisan 2001, 2004; Berkessel and Gröger 2004). We envisioned the umpolung of α,β-unsaturated aldehydes. Attack by an NHC would result in a conjugate enamine and its reaction with an aldehyde could result in the formation of a γ-butyrolactone as outlined in Scheme 5. γ-Butyrolactones are important substructures of many biologically active molecules and it is

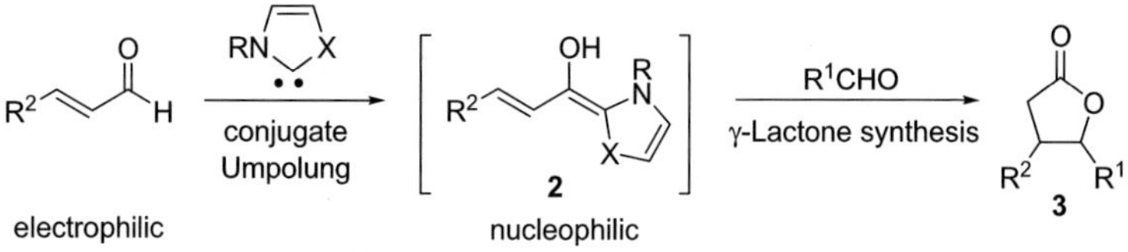

Scheme 5. Conjugate umpolung

estimated that around 10% of all naturally occurring compounds contain a γ-butyrolactone ring (Seitz and Reiser 2005).

Initially, a solution of cinnamaldehyde and 4-chlorobenzaldehyde in tetrahydrofuran (THF) was treated with different azolium salts under basic conditions (Scheme 6). The use of thiazolium salt **4** resulted in no formation of the desired γ-butyrolactone, only benzoin products were formed. In contrast, using the NHC IMes [1,3-di(2,4,6-trimethylphenyl)imidazol-2-ylidene; generated in situ from the salt IMesHCl by deprotonation], γ-butyrolactone **3a** was isolated in 53% yield and a 80:20 *cis/trans* ratio. This different outcome might be explained by the increased steric demand of IMes compared to **4** (Scheme 7). Most likely, IMes reversibly adds to the aldehyde groups of both substrates resulting in the intermediates **1a** and **2a**. Whereas the mesityl groups shield the former aldehyde carbon in both intermediates, the conjugate position of **2a** is still accessible and can add to the electrophilic aldehyde.

We (and in parallel Bode et al.) were pleased to find that the IMes-catalyzed conjugate umpolung of α,β-unsaturated aldehydes has a broad scope (Burstein and Glorius 2004; Burstein et al. 2006; Schrader et al. 2007; Sohn et al. 2004; He and Bode 2005; Sohn and Bode 2005) (For related applications of NHC in organocatalysis, see Chow and Bode 2004; Reynolds et al. 2004; Chan and Scheidt 2005; Reynolds and Rovis 2005; Zeitler 2006; Nair et al. 2006a,b; He et al. 2006; Fischer et al. 2006; Chiang et al. 2007; Philips et al. 2007; Maki et al. 2007). Under optimized conditions, a 1:1 mixture of cinnamaldehyde and the benzaldehyde derivative in THF was treated with IMes (prepared in situ from IMes·HCl and an excess of KOtBu) and stirred at ambient temperature for 16 h (Table 1). A variety of differently substituted aromatic

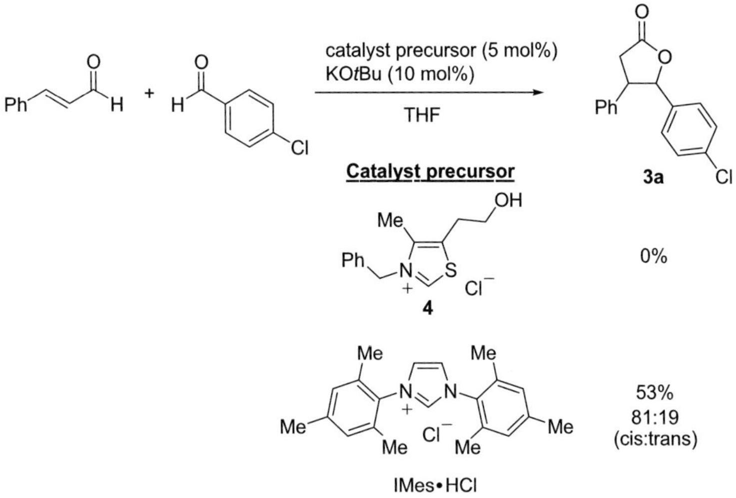

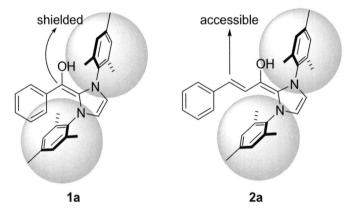

Scheme 6. Initial screening of different azolium salts

Scheme 7. A certain steric demand of the NHC seems to be key to success

aldehydes could be used and resulted in yields of γ-lactone products ranging from 30% to 70% with generally hardly any benzoin or Stetter product being produced. It should be noted that using a 2:1 or a 1:2

Table 1 IMes-catalyzed reaction of cinnamaldehyde with aromatic aldehydes[a]

Entry	Ar	3	Yield (%)	cis/trans[b]
1	4-ClC$_6$H$_4$	a	53	81:19
2	4-BrC$_6$H$_4$	b	49	80:20
3	4-MeO(CO)C$_6$H$_4$	c	70	79:21
4	4-F$_3$CC$_6$H$_4$	d	44	77:23
5	3-FC$_6$H$_4$	e	52	78:22
6	3-ClC$_6$H$_4$	f	61	79:21
7	3-BrC$_6$H$_4$	g	66	79:21
8[c]	2-ClC$_6$H$_4$	h	32	23:77

[a] General reaction conditions: IMes·HCl (0.05 mmol), KOtBu (0.1 mmol), THF (6 ml); cinnamaldehyde (1.0 mmol), ArCHO (1.0 mmol), 16 h at rt. Isolated combined yield of separately isolated diastereomers
[b] Determined by GC-MS
[c] Isolated as mixture of diastereomers

ratio of the two substrates allows a significant increase in yield. The *cis*-diastereomer was formed predominantly in all cases, with the exception of 2-chlorobenzaldehyde (Table 1, entry 8). The typical *cis/trans*-ratio was found to be 80:20 and the diastereomers were separated by column chromatography.

3 Mechanistic Proposal

The formation of the observed products can be explained by the following catalytic cycle (Scheme 8). Addition of the nucleophilic carbene leads to adduct **I**, followed by proton transfer to give conjugate enamine **IIa**. Homoenolate equivalent **IIa** (see also resonance structure **IIb**) can add to the aldehyde reaction partner providing zwitterion **III** and after

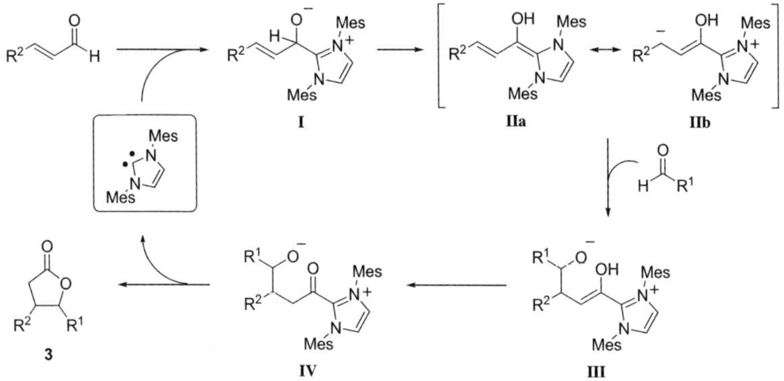

Scheme 8. Proposed catalytic cycle of the conjugate umpolung

tautomerism zwitterion **IV**. In this compound the imidazolium moiety represents a good leaving group. Thus, attack of the alkoxide to the carbonyl group results in the liberation of the NHC catalyst and the formation of the γ-butyrolactone product. This mechanistic proposal was backed by an investigation of the intermediates of the catalytic cycle using ESI-MS (Schrader et al. 2007).

4 Ketones and Imines as Electrophiles

As many natural products contain γ-butyrolactone fragments (Seitz and Reiser 2005) the scope of this conjugate umpolung reaction was investigated in detail.

Gratifyingly, ketones can also be used as the electrophilic component in this reaction, resulting in the formation of γ-butyrolactones bearing a valuable quaternary stereocenter. The reaction of cinnamaldehyde with one equivalent of α,α,α-trifluoroacetophenone catalyzed by IMes provided the corresponding γ-butyrolactone **5a** in 70% yield. The yield can be increased to 84% by using a twofold excess of the ketone (Table 2, entry 1). The reaction could be scaled up to 30 mmol without a reduction of the yield (entry 2). Furthermore, the use of DBU (1,8-diaza-bicyclo[5.4.0]undec-7-ene) as a base instead of KOtBu often resulted in superior results, slightly improving the yield and sim-

6
single diastereomer
7 examples, 60-78%

7
1:1 mixture of diastereomers
8 examples, 85-98%

Scheme 9. Spirocyclic products as reported by Nair et al. (2006a)

plifying the reaction set up (entry 3). Methyl benzoylformate and 1-phenyl-1,2-propanedione were also found to be suitable ketone electrophiles in this reaction resulting in the formation of the corresponding γ-butyrolactones in good yields (Table 2, entries 6–9). In addition, instead of cinnamaldehyde itself, electron-rich methoxy or *N,N*-dimethylamino substituted cinnamaldehyde derivatives were successfully used (entries 4, 5, 8, 9). In addition, 1,2-diketones can also be used, resulting in interesting spirocyclic products **6** and **7** (Scheme 9) (Nair et al. 2006a). In conclusion it can be summarized that activated ketones are good electrophiles in this conjugate umpolung.

In parallel, the use of imines as electrophiles in this reaction was described (He and Bode 2005). This insightful investigation started off with the application of diaryl substituted aldimines **8** (Scheme 10). However, due to the low level of electrophilicity of **8**, the conjugate enamine (**IIa**) selected another molecule of cinnamaldehyde as the reaction partner, giving access to a γ-butyrolactone. Consequently, more electrophilic sulfinimine **9** was selected as a potential coupling partner. In this case, however, the imine was too reactive and inhibited the reaction by the formation of a stable adduct with the carbene catalyst. In accordance with the fairy tale Goldilocks and the bears (too hard, too soft, just right), using slightly more electron-rich sulfonimines **10** resulted in the formation of the desired lactame products. Even though the scope of this transformation still seems to be rather narrow, these results beautifully show the power of a rational optimization.

Table 2 Reaction of cinnamaldehyde and derivatives with activated ketones[a]

Entry	Ar	R	5	Yield (%)	l/u[b]
1[c]	Ph	CF_3	a	84	66:34
2[c,d]	Ph	CF_3	a	84	64:36
3[e]	Ph	CF_3	a	92	68:32
4[c]	4-(MeO)C_6H_4	CF_3	b	92	66:34
5[c]	4-(Me$_2$N)C_6H_4	CF_3	c	74	70:30
6[f]	Ph	C(O)Me	d	55	58:42
7	Ph	CO_2Me	e	78	50:50
8[c]	4-(MeO)C_6H_4	CO_2Me	f	94	47:53
9	4-(Me$_2$N)C_6H_4	CO_2Me	g	98	44:56

[a] General reaction conditions: IMes·HCl (0.05 mmol), DBU (0.05 mmol), THF (2.5 ml), cinnamaldehyde derivative (0.5 mmol), ketone (1.0 mmol), 16 h at rt. Yield given for the isolated mixture of diastereomers
[b] Determined by GC-MS
[c] Reaction conditions: IMes·HCl (0.05 mmol), KOtBu (0.1 mmol), THF (3 ml); cinnamaldehyde derivative (1 mmol), ketone (2.0 mmol), 16 h at rt
[d] 30-mmol scale
[e] 10-mmol scale
[f] Run at 60°C

5 Conjugate Umpolung of Crotonaldehyde Derivatives

Crotonaldehyde derivates, aliphatically substituted α,β-unsaturated aldehydes were also successfully used in the NHC-catalyzed lactone formation (Scheme 11). Good yields up to 90% and good stereoselectivities up to 93:7 were obtained in these transformations. In these cases, DBU was found to give better results than KOtBu.

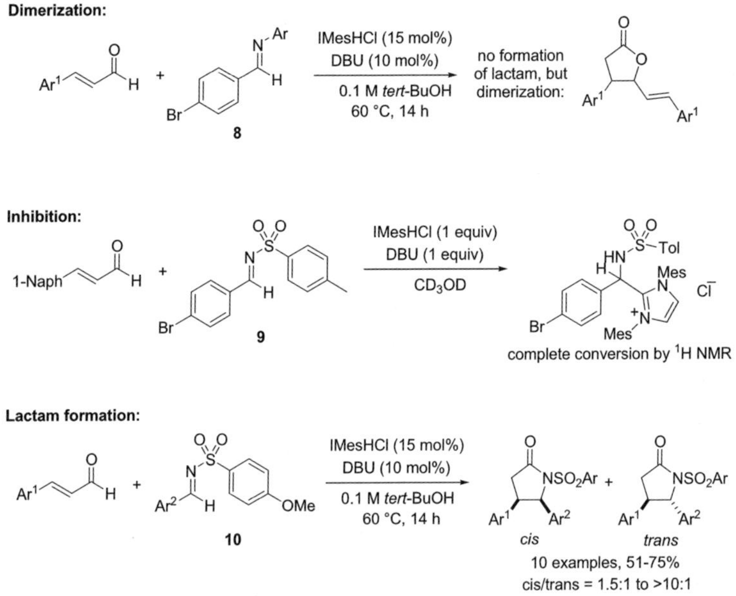

Scheme 10. Conjugate umpolung using different imine substrates (Sohn et al. 2005)

6 Conjugate Umpolung of α-Substituted Cinnamaldehyde Derivatives

A particularly challenging class of substrates are α-substituted cinnamaldehyde derivatives. Under conditions optimized for the previously mentioned reactions using IMes as the catalyst, the use of α-methyl cinnamaldehyde and trifluoroacetophenone did not give any of the desired product. This can easily be understood when analyzing the structure of the conjugate enamine of α-methyl cinnamaldehyde in the conjugated planar conformation. This planar arrangement is disfavored, due to the steric demand of the mesityl groups that results in an unfavorable steric interaction with the α-methyl group. Consequently, the size of the imidazolium substituents was reduced, and thus the dimethyl substituted imidazolylidene IMe provided 10% of the desired lactone product.

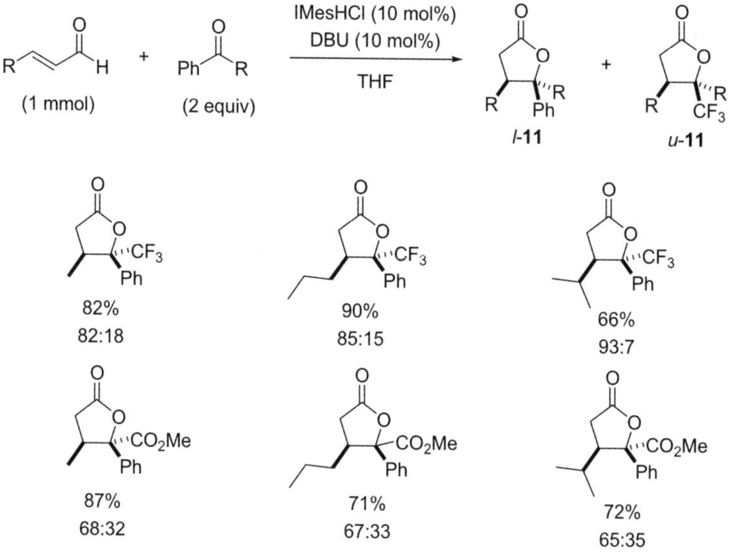

Nucleophilic Carbenes as Organocatalysts

Scheme 11. Transformations with crotonaldehyde derivatives

Whereas this limited success was based on a rational analysis of this problem, the breakthrough using the dimethyl substituted benzimidazolylidene was completely unexpected. Using this catalyst and DMF as the optimal solvent, 83% of the desired γ-butyrolactone **12** was formed in the reaction of α-methylcinnamaldehyde and trifluoroacetophenone (Scheme 12).

This protocol was successfully applied for the synthesis of a number of γ-butyrolactones (Scheme 13). Of the four possible diastereomers, mainly **12-I** and **12-II** were obtained. In these two major diastereomers the methyl-group at C3 is oriented trans relative to the aromatic group at C4. In most cases, isomer **12-I** was predominantly formed. However, in the case of 2-methyl-5-phenyl-2,4-pentadienal as the unsaturated substrate, diastereomer **12c-II** was formed in excess. Stereochemistry of these new compounds was assigned by X-ray structural analysis of **12c-II** and NMR correlation.

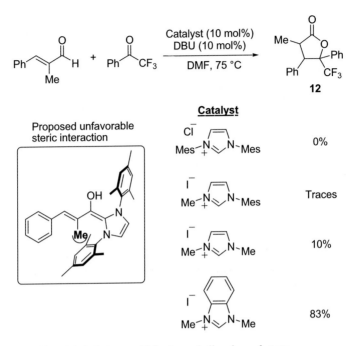

Scheme 12. α-Methyl cinnamaldehyde as challenging substrate

7 Intramolecular Variants

The aforementioned *inter*molecular reactions generate a γ-butyrolactone with up to three contiguous stereocenters. An *intra*molecular variant of this reaction would be attractive, because more complex systems form, higher stereoselectivities are expected and fewer reactive electrophiles could potentially be used, thereby significantly expanding the scope of this transformation. However, an often complex, multistep substrate synthesis decreases the attractivity of *intra*molecular reactions. Consequently, our investigation commenced with the design of readily accessible cyclization precursors.

2-Butenediol **13** was envisioned to be an ideally suited building block, allowing the synthesis of substrates for the conjugate umpolung cyclization reaction in only two steps. A highly regioselective epox-

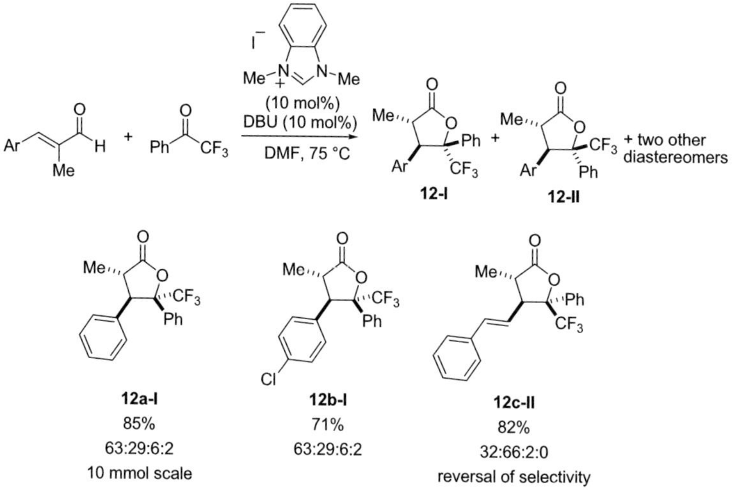

Scheme 13. Use of α-methyl cinnamaldehyde derivatives (major product isomer shown in each case)

ides opening was followed by the parallel oxidation of the resulting hydroxy groups with Dess–Martin-periodinane in good yield of 53% in both cases (Scheme 14). Using IMes as the catalyst in THF at 60 °C resulted in the cyclization of **14** and **16** to the bi- and tricyclic γ-butyrolactones **15** and **17** (Scheme 14). Besides the γ-butyrolactone ring, a tetrahydrofuran ring also forms. In both cases, only a single diastereomer was obtained. Intriguingly, this represents the first successful application of nonactivated, enolizable ketones as electrophiles in the conjugate umpolung of cinnamaldehyde derivatives.

Another class of substrates for an intramolecular homoenolate addition, leading to the formation of six-membered rings (Scheme 15), was easily synthesized in a few steps. For these substrates, the IMes-catalyzed conjugate umpolung cyclization results in the formation of the γ-butyrolactone ring and, in addition, of a six-membered ring. Again, in two cases, only a single diastereomer was obtained, interestingly, the depicted *trans*-stereoisomer.

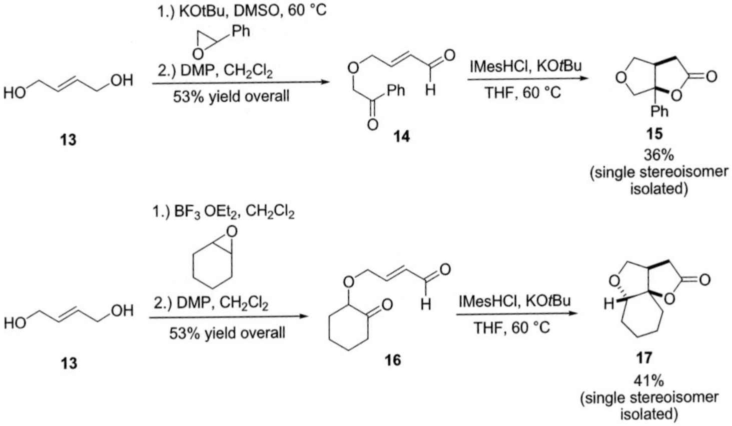

Scheme 14. Intramolecular reactions using an ether linkage

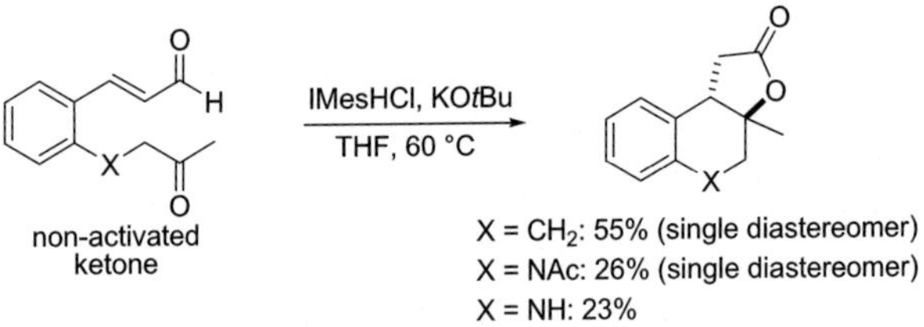

Scheme 15. Intramolecular reactions

8   Formation of β-Lactones

Not only can this umpolung reaction be used to form 5-membered γ-butyrolactones, but 4-membered β-lactones can be formed also. Interestingly, this change does not rely on a change of catalyst, but rather the reaction conditions are crucial for the reaction outcome. Using the same substrates and the same catalyst, but changing the base, the solvent and the reaction temperature allowed a change of the outcome of this reaction. Under optimized reaction conditions, β-lactones **18** formed with

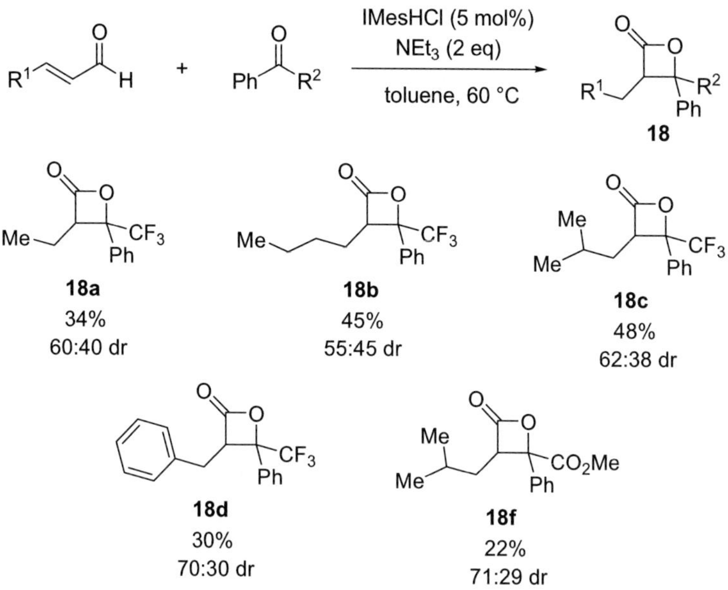

Scheme 16. β-Lactone formation

IMes as the catalyst, two equivalents of triethylamine as the base in toluene at 60 °C (Scheme 16).

The mechanistic proposal for the formation of these β-lactone products is related to that for the formation of γ-lactones (Scheme 17). Initial formation of the conjugate enamine **IIa** is followed by a proton transfer from oxygen to carbon thereby forming the enolate **V**. In an aldol-type reaction this enolate attacks the electrophilic ketone providing zwitterions **VI**. The subsequent cyclization to the β-lactone **18** then liberates the NHC catalyst.

This formation of β-lactones is strongly related to a serendipitous finding made by Nair et al. (Nair et al. 2006b; Chiang et al. 2007; Phillips et al. 2007). Interestingly, they found that the IMes-catalyzed coupling of α,β-unsaturated aldehydes with α,β-unsaturated ketones led to the stereoselective formation of *trans*-substituted cyclopentenes

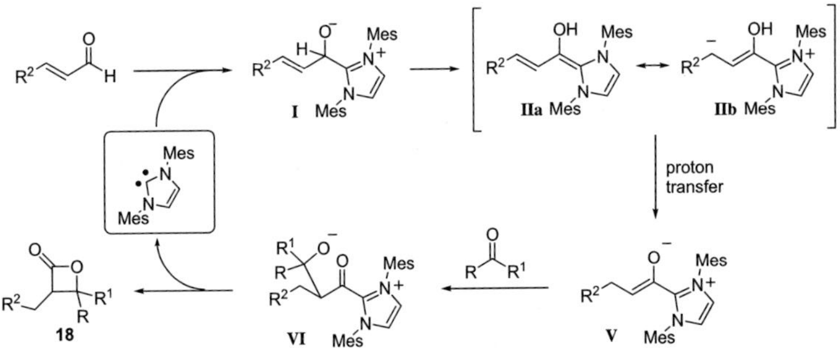

Scheme 17. Proposed catalytic cycle for the formation of β-lactones

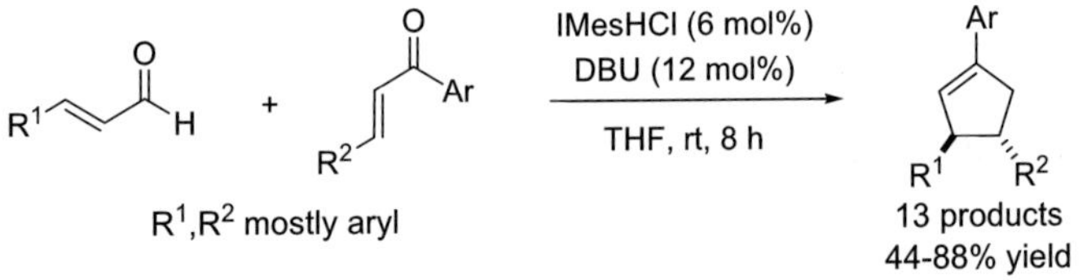

Scheme 18. Formation of cyclopentenes

(Scheme 18). The formation can be explained by the initial conjugate umpolung of the aldehyde and subsequent 1,4-addition to the unsaturated ketone. After proton transfer, an intramolecular aldol-type addition results in the formation of the aforementioned zwitterions. Nucleophilic displacement of the imidazolium moiety by the alkoxide provides the β-lactone, which exhibits increased strain, since it is annulated to a cyclopentane ring. Consequently, the β-lactone breaks apart and liberates CO_2 and the observed cyclopentene products (Scheme 19).

In conclusion, the conjugate umpolung of α,β-unsaturated aldehydes represents a versatile and powerful method to synthesize different cyclic products such as β- and γ-lactones and cyclopentenes. More valuable applications based on the NHC-catalyzed umpolung are expected to be discovered in due course.

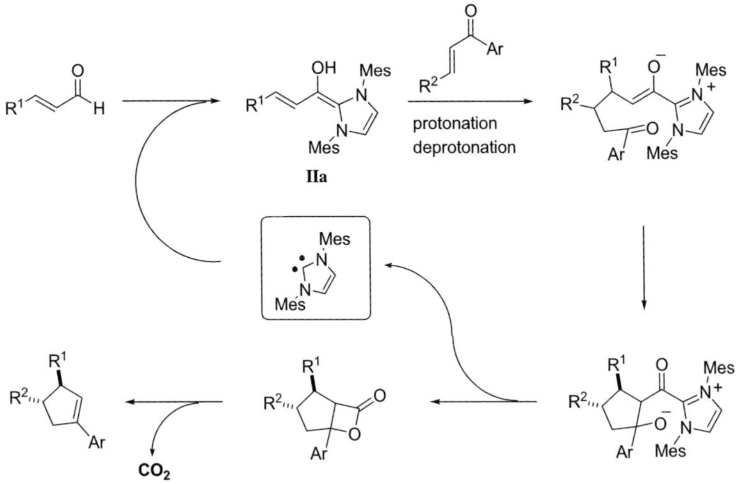

Scheme 19. Mechanistic proposal

Acknowledgements. Generous financial support by the Deutsche Forschungsgemeinschaft (Priority program organocatalysis), the Fonds der Chemischen Industrie (Dozentenstipendium for F.G.), the Deutsche Akademische Austauschdienst (fellowship for K.H.) and the BASF AG (BASF Catalysis Award to F.G.) as well as donations by Bayer AG are gratefully acknowledged. In addition, the research of F.G. was also generously supported by the Alfried Krupp Prize for Young University Teachers of the Alfried Krupp von Bohlen und Halbach Foundation.

References

Altenhoff G, Goddard R, Lehmann CW, Glorius F (2003) Ein N-heterocyclischer Carbenligand mit flexiblem sterischem Anspruch ermöglicht die Suzuki-Kreuzkupplung sterisch gehinderter Arylchloride bei Raumtemperatur. Angew Chem 115:3818–3821

Altenhoff G, Goddard R, Lehmann CW, Glorius F (2004) Sterically demanding, bioxazoline-derived N-heterocyclic carbene ligands with restricted flexibility for catalysis. J Am Chem Soc 126:15195–15201

Altenhoff G, Würtz S, Glorius F (2006) The first palladium-catalyzed Sonogashira coupling of unactivated secondary alkyl bromides. Tetrahedron Lett 47:2925–2928

Arduengo AJ 3rd (1999) Looking for stable carbenes: the difficulty in starting anew. Acc Chem Res 32:913–921

Arduengo AJ 3rd, Harlow RL, Kline M (1991) A stable crystalline carbene. J Am Chem Soc 113:361–363

Arduengo AJ 3rd, Krafczyk R (1998) The quest for stable carbenes. Chem Unserer Zeit 32:6–14

Berkessel A, Gröger H (2004) Asymmetric Organocatalysis. VCH, Weinheim

Bourissou D, Guerret O, Gabbaï FP, Bertrand G (2000) Stable carbenes. Chem Rev 100:39–92

Breslow RJ (1958) On the mechanism of thiamine action. IV. Evidence from studies on model systems. J Am Chem Soc 80:3719–3726

Burstein C, Glorius F (2004) Organocatalyzed conjugate umpolung of α,β-unsaturated aldehydes for the synthesis of γ-butyrolactones. Angew Chem Int Ed 43:6205–6208

Burstein C, Lehmann CW, Glorius F (2005) Imidazo[1,5-a]pyridine-3-ylidenes-pyridine derived N-heterocyclic carbene ligands. Tetrahedron 61:6207–6217

Burstein C, Tschan S, Xie X, Glorius F (2006) N-Heterocyclic carbene-catalyzed conjugate umpolung for the synthesis of γ-butyrolactones. Synthesis 2006:2418–2439

Cavell KJ, McGuiness DS (2004) Redox processes involving hydrocarbylmetal (N-heterocyclic carbene) complexes and associated imidazolium salts: ramifications for catalysis. Coord Chem Rev 248:671–681

Chan A, Scheidt KA (2005) Conversion of α,β-unsaturated aldehydes into saturated esters: an umpolung reaction catalyzed by nucleophilic carbenes. Org Lett 7:905–908

Chiang OC, Kaeobamrung J, Bode JW (2007) Enantioselective, cyclopentene-forming annulations via NHC-catalyzed benzoin-oxy-cope reactions. J Am Chem Soc 129:3520–3521

Chow KYK, Bode JW (2004) Catalytic generation of activated carboxylates: direct, stereoselective synthesis of β-hydroxyesters from. J Am Chem Soc 126:8126–8127

Christmann M (2005) New developments in the asymmetric stetter reaction. Angew Chem Int Ed 44:2632–2634

Dalko PI, Moisan L (2001) Enantioselective organocatalysis. Angew Chem Int Ed 40:3726–3748

Dalko PI, Moisan L (2004) In the golden age of organocatalysis. Angew Chem Int Ed 43:5138–5175

Dudding T, Houk KN (2004) Computational predictions of stereochemistry in asymmetric thiazolium- and triazolium-catalyzed benzoin condensations. Proc Natl Acad Sci USA 101:5770–5775

Enders D, Balensiefer T (2004) Nucleophilic carbenes in asymmetric organocatalysis. Acc Chem Res 37:534–541

Enders D, Kallfass U (2002) An efficient nucleophilic carbene catalyst for the asymmetric benzoin condensation. Angew Chem Int Ed 41:1743–1745

Enders D, Niemeier O (2004) Thiazol-2-ylidene catalysis in intramolecular crossed aldehyde-ketone benzoin reactions. Synlett 2004:2111–2114

Enders D, Niemeier O, Balensiefer T (2006) Asymmetric intramolecular crossed-benzoin reactions by N-heterocyclic carbene catalysis. Angew Chem Int Ed 45:1463–1467

Fischer C, Smith SW, Powell DA, Fu GC (2006) Umpolung of Michael acceptors catalyzed by N-heterocyclic carbenes. J Am Chem Soc 128:1472–4173

Glorius F (2007) (ed) N-Heterocyclic carbenes in transition metal catalysis. (Topics in Organometalic Chemistry, vol 28) Springer, Berlin Heidelberg New York

Glorius F, Altenhoff G, Goddard R, Lehmann C (2002) Oxazolines as chiral building blocks for imidazolium salts and N-heterocyclic carbene ligands. Chem Comm 2002:2704–2705

Hachisu Y, Bode JW, Suzuki K (2003) Catalytic intramolecular crossed aldehyde-ketone benzoin reactions: a novel synthesis of functionalized preanthraquinones. J Am Chem Soc 125:8432–8433

Hachisu Y, Bode JW, Suzuki K (2004) Thiazolium ylide-catalyzed intramolecular aldehyde-ketone benzoin-forming reactions: substrate scope. Adv Synth Catal 346:1097–1100

He M, Bode JW (2005) Catalytic synthesis of γ-lactams via direct annulations of enals and N-sulfonylimines. Org Lett 7:3131–3134

He M, Struble JR, Bode JW (2006) Highly enantioselective azadiene Diels–Alder reactions catalyzed by chiral N-heterocyclic carbenes. J Am Chem Soc 128:8418–8420

Herrmann WA (2002) N-Heterocyclic carbenes: a new concept in organometallic catalysis. Angew Chem Int Ed 41:1290–1309

Herrmann WA, Elison M, Fischer J, Köcher C, Artus GRJ (1995) Metal complexes of N-heterocyclic carbenes—a new structural principle for catalysts in homogeneous catalysis. Angew Chem Int Ed 34:2371–2374

Herrmann WA, Köcher C (1997) N-Heterocyclic carbenes. Angew Chem Int Ed 36:2162–2187

Hillier AC, Grasa GA, Viciu MS, Lee HM, Yang CL, Nolan SP (2002) Catalytic cross-coupling reactions mediated by palladium/nucleophilic carbene systems. J Organomet Chem 653:69–82

Igau A, Grützmacher H, Baceiredo A, Bertrand G (1988) Analogous α,α'-biscarbenoid triply bonded species: synthesis of a stable λ^3-phosphinocarbene-λ^5-phosphaacetylene. J Am Chem Soc 110:6463–6466

Johnson JS (2004) Catalyzed reactions of acyl anion equivalents. Angew Chem Int Ed 43:1326–1328

Lapworth A (1903) XCVI—Reactions involving the addition of hydrogen cyanide to carbon compounds. J Chem Soc 83:995–1005

Maki BE, Chan A, Phillips EM, Scheidt KA (2007) Tandem oxidation of allylic and benzylic alcohols to esters catalyzed by N-heterocyclic carbenes. Org Lett 9:371–374

Nair V, Vellalath S, Poonoth M, Suresh E (2006a) N-Heterocyclic carbene catalyzed reaction of enals and 1,2-dicarbonyl compounds: stereoselective synthesis of spiro γ-butyrolactones. Org Lett 8:507–509

Nair V, Vellalath S, Poonoth M, Suresh E (2006b) N-heterocyclic carbene-catalyzed reaction of chalcones and enals via homoenolate: an efficient synthesis of 1,3,4-trisubstituted cyclopentenes. J Am Chem Soc 128:8736–8737

Nolan SP (ed) (2006) N-Heterocyclic carbenes in synthesis. Wiley-VCH, Weinheim

Peris E, Crabtree RH (2004) Recent homogeneous catalytic applications of chelate and pincer *N*-heterocyclic carbenes. Coord Chem Rev 248:2239–2246

Pesch J, Harms K, Bach T (2004) Preparation of axially chiral N,N'-diarylimidazolium and N-arylthiazolium salts and evaluation of their catalytic potential in the benzoin and in the intramolecular stetter reactions. Eur J Chem 9:2025–2035

Phillips EM, Wadamoto M, Chan A, Scheidt KA (2007) A highly enantioselective intramolecular michael reaction catalyzed by N-heterocyclic carbenes. Angew Chem Int Ed 46:3107–3110

Reynolds NT, Read de Alaniz J, Rovis T (2004) Conversion of α-haloaldehydes into acylating agents by an internal redox reaction catalyzed by nucleophilic carbenes. J Am Chem Soc 126:9518–9519

Reynoldt NT, Rovis T (2005) Enantioselective protonation of catalytically generated chiral enolates as an approach to the synthesis of α-chloroesters. J Am Chem Soc 127:16406–16407

Sanderson MD, Kamplain JW, Bielawski CW (2006) Quinone-annulated N-heterocyclic carbene-transition-metal complexes: observation of π-backbonding using FT-IR spectroscopy and cyclic voltammetry. J Am Chem Soc 128:16514–16515

Schrader W, Handayani PP, Burstein C, Glorius F (2007) Investigating organocatalytic reactions: mass spectrometric studies of a conjugate umpolung reaction. Chem Comm 2007:716–718

Seayad J, List B (2005) Asymmetric organocatalysis. Org Biomol Chem 3: 719–724

Seitz M, Reiser O (2005) Synthetic approaches towards structurally diverse γ-butyrolactone natural-product-like compounds. Curr Opin Chem Biol 9: 285–292

Sohn SS, Bode JW (2005) Catalytic generation of activated carboxylates from enals: a product-determining role for the base. Org Lett 7:3873–3876

Sohn SS, Rosen EL, Bode JW (2004) N-Heterocyclic carbene-catalyzed generation of homoenolates: γ-butyrolactones by direct annulations of enals and aldehydes. J Am Chem Soc 126:14370–14371

Stetter H, Kuhlmann H (1976) Addition von aliphatischen, heterocyclischen und aromatischen Aldehyden an α,β-ungesättigte Ketone, Nitrile und Ester. Chem Ber 109:2890–2896

Tewes F, Schlecker A, Harms K, Glorius F (2007) Carbohydrate-containing N-heterocyclic carbene complexes. J Organomet Chem 692:4593–4602

Wöhler F, Liebig J (1832) Untersuchungen über das Radikal der Benzoesäure. Ann Pharm 3:249–282

Zeitler K (2005) Extending mechanistic routes in heterazolium catalysis-promising concepts for versatile synthetic methods. Angew Chemie Int Ed 44: 7506–7510

Zeitler K (2006) Stereoselective synthesis of (E)-α,β-unsaturated esters via carbene-catalyzed redox esterification. Org Lett 8:637–640

N-Heterocyclic Carbenes: Organocatalysts Displaying Diverse Modes of Action

K. Zeitler(✉)

Institut für Organische Chemie, Universität Regensburg, Universitätsstr. 31, 93053 Regensburg, Germany
email: kirsten.zeitler@chemie.uni-regensburg.de

1 Introduction 183
2 Catalyst Structures and Preparation 185
3 Classification of NHC-Mediated Reactions 190
References 199

Abstract. Within the context of Lewis base catalysis N-heterocyclic carbenes represent an extremely versatile class of organocatalyst that allows for a great variety of different transformations. Starting from the early investigations on benzoin, and later Stetter reactions, the mechanistic diversity of N-heterocyclic carbenes, depending on their properties, has led to the development of several unprecedented catalytic reactions. This article will provide an overview of the versatile reactivity of N-heterocyclic carbenes.

1 Introduction

Chemists have been inspired by Nature for hundreds of years, not only trying to understand the chemistry that occurs in living systems, but also trying to extend Nature based on the learned facts. Although already pointed out by Langenbeck in the late 1920s (Langenbeck 1928) that, unlike frequent remarks regarding analogies of enzymes to inorganic catalysts, "strange to say the investigation of organic compounds concerning their enzyme-like properties has been neglected", only in recent years have organocatalysts received widespread attention (Dalko and Moisan 2004; Pellisier 2007). During the last decade it has been demonstrated that such small (purely) organic molecules can function as efficient and highly selective catalysts, which are generally non-toxic, inexpensive to prepare, can easily be linked to solid supports, and allow novel modes of substrate activation (Lelais and MacMillan 2007; Seayad and List 2005). Hence, asymmetric organocatalysis complements the established fields of (transition)-metal catalysis and biocatalysis (List and Yang 2006).

Referring to a mechanistic classification of organocatalysts (Seayad and List 2005), currently the two most prominent classes are Brønsted acid catalysts and Lewis base catalysts. Within the latter class chiral secondary amines (enamine, iminium, dienamine activation; for a short review please refer to List 2006) play an important role and can be considered as—by now—already widely extended mimetics of type I aldolases, whereas acylation catalysts, for example, refer to hydrolases or peptidases (Spivey and McDaid 2007). Thiamine-dependent enzymes, a versatile class of C–C bond forming and destructing biocatalysts (Pohl et al. 2002) with their common catalytically active coenzyme thiamine (vitamin B_1), are understood to be the biomimetic roots of carbene catalysis, a further class of nucleophilic, Lewis base catalysis with increasing importance in the last 5 years.

This rapidly growing interest in N-heterocyclic (NHC) carbenes might be partly due to their important role as ligands for transition metal complexes (Glorius 2007; Nolan 2006), but is also attributed to their highly versatile character as organocatalysts (Enders et al. 2007a,b; Marion et al. 2007; Zeitler 2005). Based on this functional duality a comparison to phosphines can be drawn. Although some similarities can be found, NHC compounds have already proven to be not merely 'phosphine mimics', but to be important in their own right.[1] This is especially true as carbene catalysis offers the opportunity to swap tradi-

---

[1] Some aspects of the different electronic and steric properties of phosphines and carbenes have been summarized in short comparative overviews (Glorius 2007; Kantchev et al. 2007).

tional reactivity patterns by pursuing the concept of umpolung (polarity reversal) (Seebach 1979) and thus exemplifies the power of organocatalysis in terms of the development and application of novel retrosynthetic bond disconnections.

In this chapter, an overview of the different modes of action of NHC carbenes and their impact on the catalytic availability of certain classes of reactive intermediates will be provided, prefaced by a short discussion of the current state of the art concerning the preparation of (chiral) heterazolium catalysts including methods for the immobilization of this versatile class of organocatalyst.

2   Catalyst Structures and Preparation

Since Breslow's seminal paper (Breslow 1958) proposing a mechanistic model for the catalytic activity of coenzyme thiamine *via* an *in situ* formed carbene species by deprotonation of the thiazolium salt, and its subsequent reaction with an aldehyde to generate an 'active aldehyde' (the so-called 'Breslow-intermediate'), a large number of different achiral, but also chiral heterazolium precatalysts, have been prepared (Fig. 1). Besides sufficient catalytic activity, the introduction of chirality to mimic the chiral environment naturally provided by the enzyme have been the main goals. Thus, with regard to the optimization of catalysts for the different types of reaction (see Sect. 3), catalyst development and tuning has been the key to the significant progress in the field of enantioselective carbene catalysis and has been essential for the recent impact of this class of nucleophilic catalysis (Enders et al. 2007a,b).

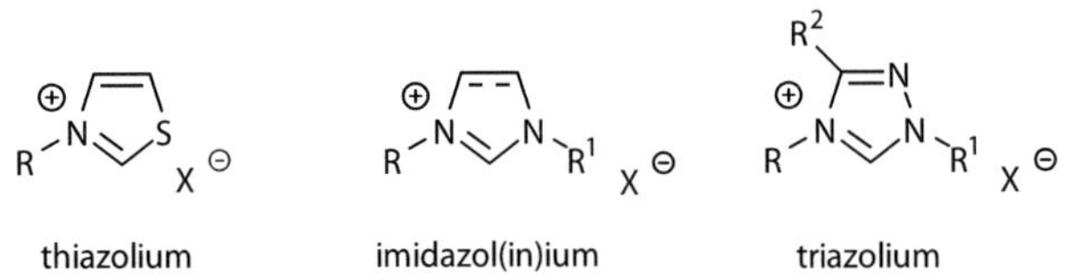

Fig. 1. General structures of heterazolium precatalysts

In general, the three most common, major classes of NHC carbenes can be accessed by deprotonation of their corresponding heterazolium precursors[2], i.e. thiazolium, imidazol(in)ium and triazolium salts, which comprise differences in both the electronic and steric nature for the respective carbenes.

With respect to the application of asymmetric carbene catalysis as a tool for enantioselective synthesis, the last decade's major success is based on substantial improvements in catalyst development. Early reports dealt with implementing chirality in thiazolium scaffolds (Sheehan and Hunneman 1966; Sheehan and Hara 1974; Dvorak and Rawal 1998), but their catalytic performance suffered from either low yields or low *ee*-values. In this regard, the investigation of triazole heterocycles as an alternative core structure (Enders et al. 1995) has played a crucial role to provide heterazolium precatalysts improving both asymmetric benzoin and Stetter reactions. An intramolecular Stetter reaction yielding chromanones upon cyclization of salicylaldehyde-derived substrates is commonly used as a benchmark reaction to compare catalyst efficiency (Scheme 1; Ciganek 1995; Enders et al. 1996; Kerr et al. 2002; Kerr and Rovis 2004).

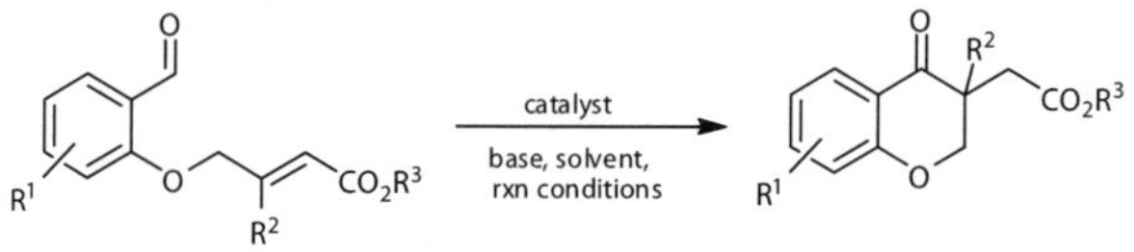

Scheme 1. Intramolecular Stetter reaction as benchmark reaction for catalyst efficiency

Apart from the higher reactivity over their thiazolium counterparts, these triazolium salts allow not only the introduction of a second group of greater steric demand at the former 'unfunctionalized' position of sulfur, but also the integration of the triazolium core within further stabilized bi- or polycyclic scaffolds of enhanced rigidity. Only by application of these types of catalysts *ee*-values greater than 90% can be

---

[2] An alternative method for *in situ* carbene generation starts from methanol, chloroform or pentafluorobenzene adducts that allow the release of the free carbenes at elevated temperature, etc. (Csihony 2005; Coulembier et al. 2005, 2006; Enders et al. 1995).

achieved indicating the importance of this additional conformational fixation.

Whereas the first chiral triazolium precatalysts relied on amines derived from the chiral pool as enantio-differentiating building blocks, most of the current successfully used triazolium ions stem from chiral 1,2-amino alcohols (Scheme 2; Knight and Leeper 1998; Kerr et al. 2005). Alternatively, enantiopure γ-amino acids can be used to target bicyclic structures, thus avoiding stability problems of the former, first-generation catalysts that are prone to destructive ring opening reactions at the non-substituted carbon (Teles et al. 1996). Natural amino acids provide a common source for both approaches offering diverse steric arrangements as well as ready availability.

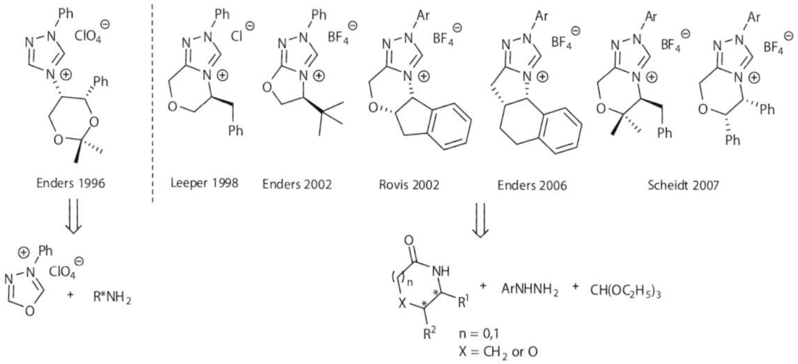

Scheme 2. Chiral triazolium precatalysts

To build up the desired cyclic key precursors, containing an amide bond as an 'anchor' to connect the triazole moiety, different strategies can be pursued. Cyclization of γ-amino acids directly yields pyrrolidones, whereas β-amino alcohols allow access to two different chiral bicyclic frameworks. They can form either oxazolidin-2-ones (Enders and Kallfass 2001) or morpholinones upon treatment with phosgene or chloro acetylchloride, respectively (Scheme 3).

Methylation with Meerwein's reagent affords the imino ethers which are treated *in situ* with aryl hydrazines and consequently cyclized with triethyl orthoformate to yield both 5/5 and 6/5 bicyclic chiral triazolium scaffolds (Kerr et al. 2005).

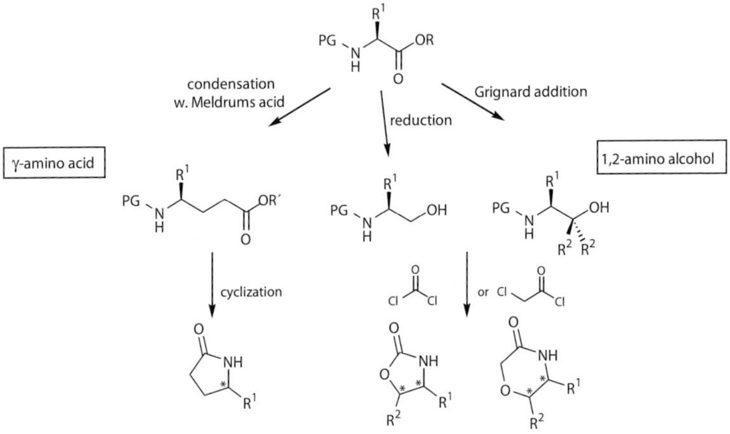

Scheme 3. Amino acids as precursors for various triazolium ions *via* 'lactam–route'

Triazolylidene carbenes show great versatility as they not only catalyze classical umpolung reactions, but have also been shown to be successful catalysts for enantioselective extended umpolung reactions (for details, please see the following sections).

Only restricted activity towards classical umpolung reactions, such as benzoin, acyloin or Stetter reactions is known for imidazolylidene carbenes (Marion et al. 2007; Enders et al. 2007a,b; Matsumoto and Tomioka 2006). Imidazolium salts, which are typically prepared from their bisimine precursors (Nolan 2006),[3] as well as benzimidazolium salts can provide twofold steric shielding at both nitrogen atoms. Chirality can be integrated in the sterically demanding side chains that stem from the corresponding chiral amines (Kano et al. 2005), whereas members of the saturated imidazolinylidene carbenes might also implement chiral bisamines in their backbone (Fig. 2; Matsumoto and Tomioka 2006).

---

[3] Only recently an additional, elegant access to unsymmetrically substituted imidazolium salts has been disclosed (Fürstner et al. 2006). Unsymmetric bisaryl-substituted *N*-heterocyclic carbenes (imidazolidinium derived) can be prepared from ethyl chlorooxoacetate (Waltman and Grubbs 2004).

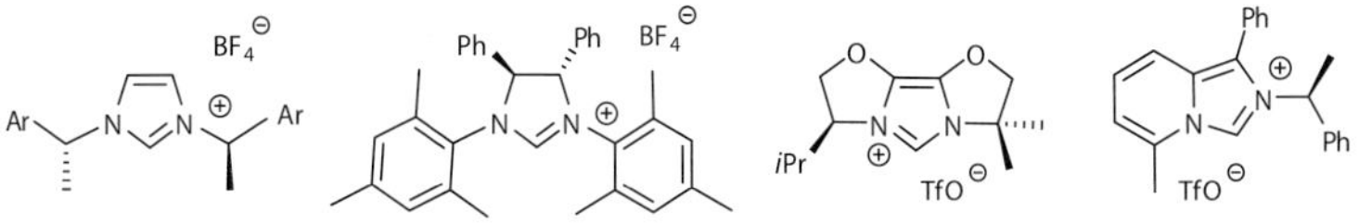

Fig. 2. Chiral imidazolium precatalysts

Flexible steric bulk is a characteristic of a new class of imidazolium salts derived from bisoxazolines (IBiox) (Glorius et al. 2002; Altenhoff et al. 2004), which were tested for the organocatalytic synthesis of butyrolactones (Burstein and Glorius 2004).

Only recently, based on earlier reports on achiral examples (Alcarazo et al. 2005; Burstein et al. 2005), have optically active imidazopyridinium salts been prepared (Schmidt and Movassaghi 2007).

So far, chiral imidazolium precatalysts have been used successfully for kinetic resolutions of racemic secondary alcohols *via* enantioselective acylation (Kano et al. 2005; Suzuki et al. 2004).

Apart form a great number of chiral NHC carbenes that have been used as ligands in enantioselective transition-metal catalysis (Gade and Bellemin-Laponnanz 2007), some less usual heterazolium salts have been tested in organocatalytic transformations. A planar-chiral thiazolium salt (Pesch et al. 2004) and a rotaxane-derived precatalyst were reported (Tachibana et al. 2004), as well as catalytically active peptides containing an unnatural thiazolium-substituted alanine amino acid (Fig. 3; Mennen et al. 2005a,b).

Because of the growing importance of sustainable synthetic methods on the one hand and the need for simple reaction procedures that allow

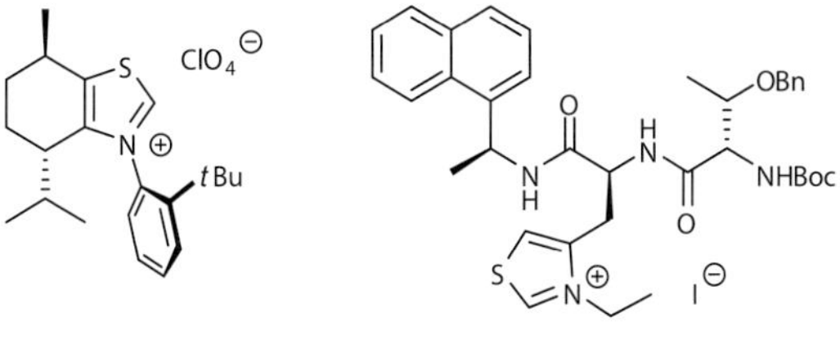

Fig. 3. Structurally unique chiral heterazolium salts

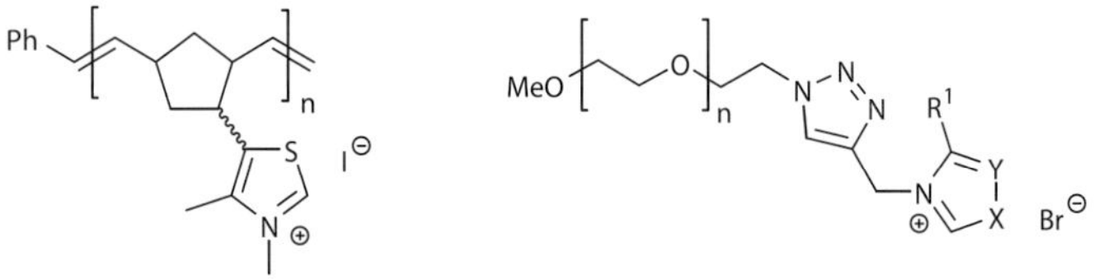

Fig. 4. Immobilized organocatalytic carbene precursors

for parallel synthesis on the other, immobilization of organocatalysts (Cozzi 2006; Benaglia 2006) is attracting increasing attention. Apart from some combinations of heterazolium salts with 'non-innocent' imidazolium-derived ionic liquids (Zhou et al. 2006) there are only very few examples of successful use of immobilized organocatalytic carbenes (Fig. 4). Barrett describes the application of a ROMP gel-supported thiazolium iodide for intermolecular Stetter reactions (Barrett et al. 2004). Only recently, an efficient and modular approach for the immobilization of different classes of heterazolium precatalysts *via* copper-catalyzed [3+2]-cycloaddition—and their use for both classical and extended umpolung reactions—has been reported (Zeitler and Mager 2007).

3 Classification of NHC-Mediated Reactions

Not only due to the different types of heterazolium precatalysts and their corresponding structural and electronic properties N-heterocyclic carbenes display, based on the common principle of nucleophilic Lewis base catalysis, various modes of action. They range from simple transesterifications (Grasa et al. 2002) to the catalysis of highly enantioselective and efficient hetero Diels–Alder reactions (He et al. 2006a,b). In particular, their unique capability to enable processes that proceed by inversion of the reactivity of one substrate (i.e. umpolung or polarity reversal), represents the special nature of carbene organocatalysts and points to their growing importance, as they allow new strategic retrosynthetic bond disconnections and hence offer new routes for organic synthesis. But apart from such umpolung reactions NHCs can also mediate

NHCs – Highly Versatile Organocatalysts

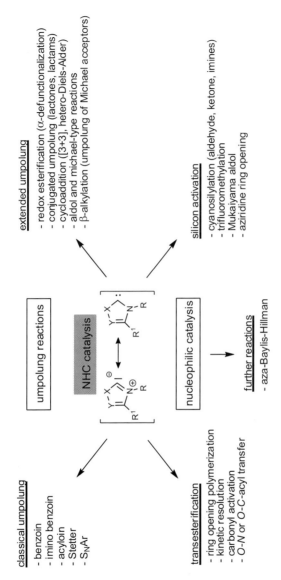

Scheme 4. Classification of NHC-catalyzed transformations

a large number of other transformations by different types of nucleophilic catalysis (Enders et al. 2007; Marion et al. 2007; Zeitler 2005).

With regard to a mechanistic classification of the known carbene-catalyzed processes, a further subdivison of these two major classes might be useful. Scheme 4 provides an attempt at a mechanism-based classification.

Within the field of nucleophilic carbene catalysis, with the exception of umpolung-directed transformations (see below), one of the most important classes in terms of applications already realized are transesterifications (Scheme 5; Grasa et al. 2003; Nyce et al. 2002; Singh et al. 2004) and related reactions (Dove et al. 2006).

Transesterifications start with nucleophilic attack on the ester carbonyl carbon by the NHC, and are widely used for living ring opening polymerizations (ROP) of cyclic ester monomers, such as lactides (Nyce et al. 2003) and lactones or cyclic siloxanes (Rodriguez et al. 2007), to generate well-defined metal-free polyesters or siloxanes. Additionally, N-heterocyclic carbenes also catalyze the ring opening of cyclic carbosiloxanes providing an organocatalytic access to narrowly dispersed polycarbosiloxanes (PCS) (Lohmeijer et al. 2006). Moreover, transesterification activity of NHCs has been shown not to be restricted to carboxylate esters, but could be extended to the important class of phosphorus esters (Singh and Nolan 2005). Chiral carbenes have been successfully applied to the kinetic resolution of racemic secondary alcohols using vinylesters as acyl donors (Suzuki et al. 2006a; Kano et al. 2005; Suzuki et al. 2004).

The amidation of unactivated esters by amino alcohols (Movassaghi and Schmidt 2005) can be considered as a related transformation; here,

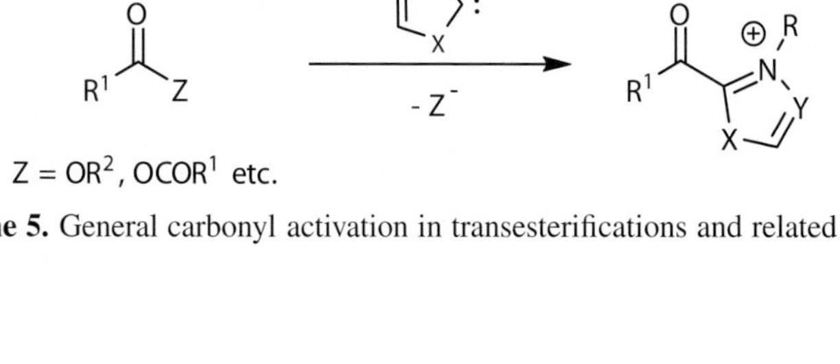

$Z = OR^2, OCOR^1$ etc.

Scheme 5. General carbonyl activation in transesterifications and related reactions

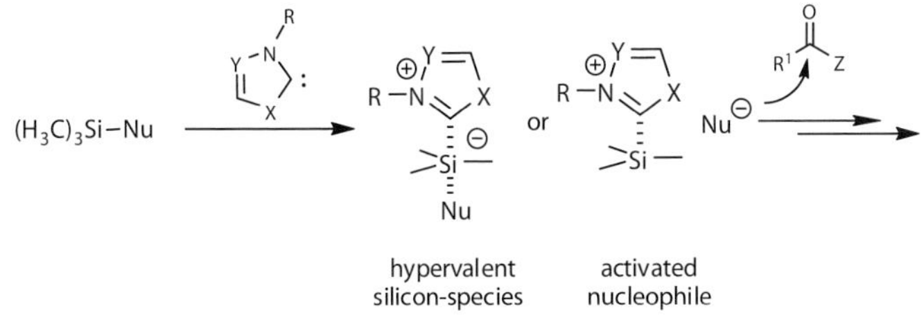

Scheme 6. Proposed mechanism for the carbene-mediated activation of silicon compounds

an initial transesterification step is followed by a rapid $O \rightarrow N$ acyl transfer to yield the desired amide. Furthermore, a carbene-promoted efficient $O \rightarrow C$ acyl transfer has recently been reported in the context of the rearrangement of α-amino acid-derived *O*-acyl carbonates to their corresponding *C*-acylated isomers (Thomson et al. 2006). Less common carbonyl compounds that can undergo related NHC-mediated activation include isocyanates. Their catalyzed trimerization affording isocyanurates has proven to be strongly dependent on the catalyst's nature (Duong et al. 2004). In addition, nucleophilic carbene attack to activate acid anhydrides is also proposed as an initiating step in the ring opening reaction of aziridines (Sun et al. 2006).

The second major class of 'non-umpolung' nucleophilic carbene catalysis comprises reactions by initial NHC-activation of various silicon compounds. Their proposed common pathway is thought to lead to a hypervalent silicon complex[4] and thus provide carbene-catalyzed activation of the corresponding nucleophiles such as TMSCN, TMSCF$_3$ etc. (Kano et al. 2006; Song et al. 2005; 2006). It is not only certain carbon–silicon bonds that can be effectively activated, but a comparable activation of Si–O bonds, e.g. of trimethylsily enol ethers etc., allows for mild, NHC-promoted Mukaiyama aldol reactions (Scheme 6; Song et al. 2007).

---

[4] Although there is some experimental evidence for the formation of hypervalent silicon species (Fukada et al. 2006) an alternative reaction pathway *via* preliminary carbene-mediated activation of the carbonyl or imine species respectively (1,2-addition), and subsequent reaction of the anionic species with TMSCN etc. (electrophilic trapping of the alkoxide) has been proposed (Suzuki et al. 2006b; Marion et al. 2007).

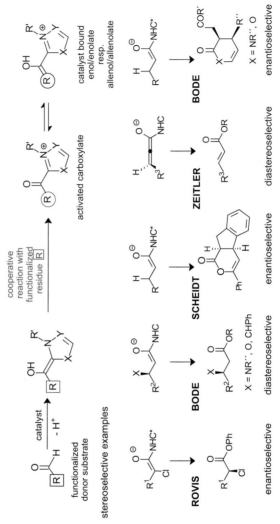

Scheme 7. Examples for stereoselective extended umpolung reactions (Reynolds and Rovis 2005; Chow and Bode 2004; Sohn and Bode 2006; Phillips et al. 2007; Zeitler 2006; He et al. 2006a,b)

The ring opening reaction of N-tosylaziridines with certain silylated nucleophiles *via* a similar activation has been reported recently (Wu et al. 2006).

A NHC-catalyzed aza–Morita–Baylis–Hillman reaction (aza–MBH) following a standard nucleophile-mediated MBH mechanism has been disclosed very recently (He et al. 2007). Although combined with a preceding equilibrium for the reversible formation of imine–carbene adducts this reaction has similarities with phosphines and their organocatalytic reactivity (Methot and Roush 2004).

Classical umpolung reactions such as Stetter reactions (Christmann 2005), benzoin reactions and nucleophilic aromatic substitution (S_NAr) (Suzuki et al. 2003) have already become important tools in organic synthesis as exemplified by their application in the syntheses of several natural product (Nicolaou et al. 2007; Harrington and Tius 2001). Significant progress has been reported for enantioselective crossed acyloin formations (Takikawa et al. 2006; Enders et al. 2006), but also new applications, such as an elegant desymmetrization reaction (Liu and Rovis 2006), have been described recently. An interesting domino reaction involving a combination of amine and carbene catalysis allows simplified one-pot access to β-substituted ester derivatives with high enantioselectivity (Zhao et Córdova 2007). The aerobic ring opening of N-tosylaziridines with aldehydes is proposed to proceed *via* an alkylated 'Breslow intermediate' which is subsequently attacked by dioxygene as the electrophile (Liu et al. 2006).[5]

Moreover, the growing group of *extended umpolung reactions* is attracting considerable attention. This integration of functionalized aldehyde substrates in umpolung reactions (Reynolds et al. 2004; Chow and Bode 2004), and hence their cooperative interaction to yield new, defunctionalized intermediates has already been applied in a remarkable number of stereoselective variations (Scheme 7).

In general, all extended umpolung reactions deal with aldehyde substrates that bear somehow reducible side chains prone to subsequent interaction at the stage of the common so-called 'Breslow intermediate' (Zeitler 2005). Their subsequent participation leads—depending on the

---

[5]For two further aziridine opening reactions please refer to transesterification and silicon activation section above.

original substrate structure—to the formation of two major classes of catalyst-bound reactive intermediates, i.e., homoenolates and enolates. In the case of propargylic aldehyde substrates allenolates are proposed to be the reactive species allowing for the stereoselective formation of α,β-unsaturated esters (Scheme 8; Zeitler 2006).

Based on this covalent catalyst integration the use of enantiopure catalysts potentially allows the generation of the corresponding chiral homoenolate and chiral enolate intermediates. In fact, both reactive species are strongly related as a simple β-protonation (or any other electrophilic attack) of the homoenolate affords the corresponding enolate, as shown by Bode and coworkers in the context of the enantioselective hetero Diels–Alder reactions with either aza- or oxodienes (He et al. 2006a,b). Until now, however, in most cases of catalytically generated homoenolates, after the electrophilic attack the subsequent tautomerization of the initial enolate intermediate affords an activated ester which is then trapped by suitable nucleophiles. Depending on the nature of the electrophile acyclic, i.e. simple esters (Chan and Scheidt 2005), or cyclic products such as γ-lactones or γ-lactams can be formed (Burstein and Glorius 2004; Sohn et al. 2004; He and Bode 2005). The reaction of 1,2-diketones yields spiro products (Nair et al. 2006a) whereas a formal [3+3] cycloaddition with azomethines generates bicyclic pyridazinones (Chan and Scheidt 2007).

Apart form the aforementioned highly enantioselective hetero-Diels–Alder reactions, that proceed with very low catalyst loadings, the catalytically accessible enolates have also been used for related intramolecular Michael reactions (Philips et al. 2007) and for the desymmetrization of 1,3-diketones yielding cyclopentenes *via* an intramolecular aldol reaction (Wadamoto et al. 2007). The formation of cyclopentenes, however, presents a special case, so—depending on the stereochemical nature of the enone substrates (s-*cis* or s-*trans*) and the stereochemistry of the final products—two different mechanisms are discussed in the literature. Whereas *trans*-cyclopentenes are proposed to be available upon conjugate addition of a homoenolate to chalcones,

NHCs – Highly Versatile Organocatalysts

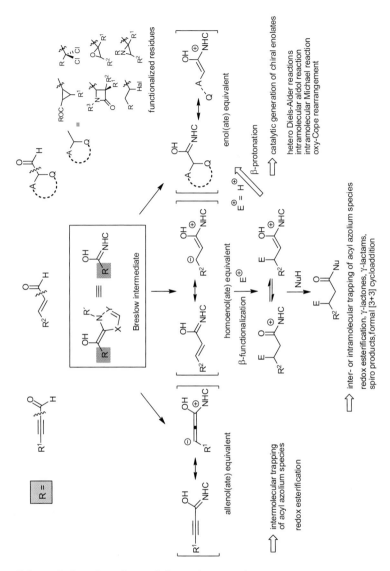

Scheme 8. Overview of extended umpolung reactions

followed by β-lacton formation and subsequent CO_2 extrusion[6] (Nair et al. 2006b), the *cis* products have been demonstrated to stem from a crossed benzoin-oxy-Cope cascade that leads to the identical intermediate as proposed earlier from conjugate addition (Chiang et al. 2007). With respect to the differences in the equilibria of s-*cis* and s-*trans* conformers of different enones this mechanistic reaction pathway provides an explanation for the varying stereochemical outcome (Chiang et al. 2007).

In addition, an umpolung reaction without participation of an aldehyde moiety has been disclosed recently (Fischer et al. 2006). Michael acceptors (unsaturated esters, amides and nitriles) that bear pendant leaving groups such as bromides or tosylates can undergo an organocatalytic, intramolecular β-alkylation. Carbene addition at the β-position of the Michael acceptor and subsequent tautomerization also promotes the generation of a homoenolate equivalent which is trapped intramolecularly under ring closure.[7]

Finally, there are also some special NHC-mediated transformations that do not completely fit into the classification, such as triazolylidene-catalyzed hydroacylations (Chan and Scheidt 2006). Aldehydes can serve as hydride donors for activated ketones partly following a standard 1,2-addition of the NHC to the aldehyde, but instead of the usual carbonyl umpolung a hydride ('H-umpolung') transfer is initiated. A related 'Cannizzaro-type' transformation has been described for indazole-derived carbene catalysts (Schmidt et al. 2007).

In conclusion, it is obvious that *N*-heterocyclic carbenes and their unique, but concurrently versatile reactivity have already proven wide applicability. The offered catalytic (and enantioselective) access to important reactive intermediates such as homoenolates or enolates as well as their additional catalytic properties will smooth the way for mild and sustainable synthesis of multifunctionalized compounds.

---

[6]Conjugate or vinylogous umpolung can also lead to the formation of β-lactones; interestingly their stability is highly dependent on the substrates. For stable β-lactones please see: Burstein et al. 2006).

[7]The first step of this unprecedented umpolung reaction is identical to the standard MBH mechanism. Only the tautomerization event differentiates between α- (enolate) and β-functionalization (homoenolate).

Acknowledgements. Generous financial support by the Deutsche Forschungsgemeinschaft (DFG priority program SPP1179 'organocatalysis') and by the Fonds der Chemischen Industrie (Liebig fellowship) is gratefully acknowledged.

References

Alcarazo M, Roseblade SJ, Cowley AR, Fernandez R, Brown JM, Lassaletta JM (2005) Imidazo[1,5-a]pyridine: a versatile architecture for stable N-heterocyclic carbenes. J Am Chem Soc 127:3290–3291

Altenhoff G, Goddard R, Lehmann CW, Glorius F (2004) Sterically demanding, bisoxazoline–derived N–heterocyclic carbene ligands with restricted flexibility for catalysis. J Am Chem Soc 126:15195–15201

Barrett AGM, Love AC, Tedeschi L (2004) ROMPgel-supported thiazolium iodide: an efficient supported organic catalyst for parallel Stetter reactions. Org Lett 6:3377–3380

Benaglia M (2006) Recoverable and recyclable chiral organic catalysts. New J Chem 30:1525–1533

Breslow R (1958) Mechanism of thiamine action. IV. Evidence from studies on model systems. J Am Chem Soc 80:3719–3726

Burstein C, Glorius F (2004) Organocatalyzed conjugate umpolung of α,β-unsaturated aldehydes for the synthesis of γ-butyrolactones. Angew Chem Int Ed 43:6205–6208

Burstein C, Lehmann CW, Glorius F (2005) Imidazo[1,5-a]pyridine-3-ylidenes-pyridine derived N-heterocyclic carbene ligands. Tetrahedron 61:6207–6217

Burstein C, Tschan S, Xie X, Glorius F (2006) N-Heterocyclic carbene-catalyzed conjugate umpolung for the synthesis of γ–butyrolactones. Synthesis 14:2418–2439

Chan A, Scheidt KA (2005) Conversion of α,β-unsaturated aldehydes into saturated esters: an umpolung reaction catalyzed by nucleophilic carbenes. Org Lett 7:905–908

Chan A, Scheidt KA (2006) Hydroacylation of activated ketones catalyzed by N-heterocyclic carbenes. J Am Chem Soc 128:4558–4559

Chan A, Scheidt KA (2007) Highly stereoselective formal [3+3] cycloaddition of enals and azomethine imines catalyzed by N-heterocyclic carbenes. J Am Chem Soc 129:5334–5335

Chiang PC, Kaeobamrung J, Bode JW (2007) Enantioselective, cyclopentene-forming annulations via NHC-catalyzed benzoin-oxy-Cope reactions. J Am Chem Soc 129:3520–3521

Chow KYK, Bode JW (2004) Catalytic generation of activated carboxylates: Direct, stereoselective synthesis of β–hydroxyesters from epoxyaldehydes. J Am Chem Soc 126:8126–8127

Christmann M (2005) New developments in the asymmetric Stetter reaction. Angew Chem Int Ed 44:2632–2634

Ciganek E (1995) Esters of 2,3-dihydro-3-oxobenzofuran-2-acetic acid and 3,4-dihydro-4-oxo-2*H*-1-benzopyran-3-acetic acid by intramolecular Stetter reactions. Synthesis 10:1311–1314

Coulembier O, Dove AP, Pratt RC, Sentman AC, Culkin DA, Mespouille L, Dubois P, Waymouth RM, Hedrick JL (2005) Latent, thermally activated organic catalysts for the on-demand living polymerization of lactide. Angew Chem Int Ed 44:4964–4968

Coulembier O, Lohmeijer BG, Dove AP, Pratt RC, Mespouille L, Culkin DA, Benight SJ, Dubois P, Waymouth RM, Hedrick JL (2006) Alcohol adducts of *N*-heterocyclic carbenes: latent catalysts for the thermally-controlled living polymerization of cyclic esters. Macromolecules 39:5671–5628

Cozzi F (2006) Immobilization of organic catalysts. When, why, and how. Adv Synth Cat 348:1367–1390

Csihony S, Culkin DA, Sentman AC, Dove AP, Waymouth RM, Hedrick JL (2005) Single-component catalyst/initiators for the organocatalytic ring-opening polymerization of lactide. J Am Chem Soc 127:9079–9084

Dalko PI, Moisan L (2004) In the golden age of organocatalysis. Angew Chem Int Ed 43:5138–5175

Dove AP, Pratt RC, Lohmeijer BGG, Li H, Hagberg EC, Waymouth RM, Hedrick JL (2006) *N*-Heterocyclic carbenes as organic catalysts. In: Nolan SP (ed) *N*-Heterocyclic carbenes in synthesis. Wiley-VCH, Weinheim, pp 275–296

Duong HA, Cross MJ, Louie J (2004) *N*-Heterocyclic carbenes as highly efficient catalysts for the cyclotrimerization of isocyanates. Org Lett 6:4679–4681

Dvorak CA, Rawal VH (1998) Catalysis of benzoin condensation by conformationally-restricted chiral bicyclic thiazolium salts. Tetrahedron Lett 39:2925–2928

Enders D, Balensiefer T, Niemeier O, Christmann M (2007a) Nucleophilc *N*–Heterocyclic carbenes in asymmetric organocatalysis. In: Dalko PI (ed) Enantioselective Organocatalysis: reactions and experimental procedures. Wiley-VCH, Weinheim, pp 331–355

Enders D, Breuer K, Raabe G, Runsink J, Teles JH, Melder JP, Ebel K, Brode S (1995) Preparation, structure, and reactivity of 1,3,4-triphenyl-4,5-dihydro-1*H*-1,2,4-triazol-5-ylidene, a new stable carbene. Angew Chem Int Ed 34:1021–1023

Enders D, Breuer K, Runsink J, Teles JH (1996) The first asymmetric intramolecular Stetter reaction. Helv Chim Acta 79:1899–1902

Enders D, Kallfass U (2002) An efficient nucleophilic carbene catalyst for the asymmetric benzoin condensation. Angew Chem Int Ed 41:1743–1745

Enders D, Niemeier O, Balensiefer T (2006) Asymmetric intramolecular crossed-benzoin reactions by N-heterocyclic carbene catalysis. Angew Chem Int Ed 45:1463–1467

Enders D, Niemeier O, Henseler A (2007b) Organocatalysis by N-heterocyclic carbenes. Chem Rev 107:5606–5655

Fischer C, Smith SW, Powell DA, Fu GC (2006) Umpolung of Michael acceptors catalyzed by N-heterocyclic carbenes. J Am Chem Soc 128:1472–1473

Fürstner A, Alcarazo M, César V, Lehmann CW (2006) Convenient, scalable and flexible method for the preparation of imidazolium salts with previously inaccessible substitution patterns. Chem Comm 2006(20):2176–2178

Fukuda Y, Maeda Y, Kondo K, Aoyama T (2006) Strecker reaction of aldimines catalyzed by a nucleophilic N-heterocyclic carbene. Synthesis 12:1937–1939

Gade LH, Bellemin-Laponnaz S (2007) Chiral N-heterocyclic carbenes as stereodirecting ligands in asymmetric catalysis. Top Organomet Chem 21:117–157

Glorius F (2007) N–Heterocyclic carbenes in catalysis–an introduction. Top Organomet Chem 21:1–20

Glorius F, Altenhoff G, Goddard R, Lehmann C (2002) Oxazolines as chiral building blocks for imidazolium salts and N-heterocyclic carbene ligands. Chem. Comm 2704–2705

Grasa GA, Gueveli T, Singh R, Nolan SP (2003) Efficient transesterification/acylation reactions mediated by N-heterocyclic carbene catalysts. J Org Chem 68:2812–2819

Grasa GA, Kissling RM, Nolan SP (2002) N-Heterocyclic carbenes as versatile nucleophilic catalysts for transesterification/acylation reactions. Org Lett 4:3583–3586

Harrington PE, Tius MA (2001) Synthesis and absolute stereochemistry of roseophilin. J Am Chem Soc 123:8509–8514

He M, Bode JW (2005) Catalytic synthesis of γ-lactams *via* direct annulations of enals and N-sulfonylimines. Org Lett 7:3131–3134

He M, Struble JR, Bode JW (2006b) Highly enantioselective azadiene Diels–Alder reactions catalyzed by chiral N-heterocyclic carbenes. J Am Chem Soc 128:8418–8420

He M, Uc GJ, Bode JW (2006a) Chiral N–heterocyclic carbene catalyzed, enantioselective oxodiene Diels–Alder reactions with low catalyst loadings. J Am Chem Soc 128:15088–15089

He L, Jian TY, Ye S (2007) N-Heterocyclic carbene catalyzed Aza-Morita-Baylis-Hillman reaction of cyclic enones with N-tosylarylimines. J Org Chem 72:7466–7468

Kano T, Sasaki K, Konishi T, Mii H, Maruoka K (2006) Highly efficient trialkylsilylcyanation of aldehydes, ketones, and imines catalyzed by a nucleophilic N-heterocyclic carbene. Tetrahedron Lett 47:4615–4618

Kano T, Sasaki K, Maruoka K (2005) Enantioselective acylation of secondary alcohols catalyzed by chiral N-heterocyclic carbenes. Org Lett 7:1347–1349

Kantchev EAB, O'Brien CJ, Organ MG (2007) Palladium complexes of N-heterocyclic carbenes as catalysts for cross-coupling reactions – a synthetic chemist's perspective. Angew Chem Int Ed 46:2768–2813

Kerr MS, Rovis T (2004) Enantioselective synthesis of quaternary stereocenters *via* a catalytic asymmetric Stetter reaction. J Am Chem Soc 126:8876–8877

Kerr MS, Read de Alaniz J, Rovis T (2002) A highly enantioselective catalytic intramolecular Stetter reaction. J Am Chem Soc 124:10298–10299

Kerr MS, Read de Alaniz J, Rovis T (2005) An efficient synthesis of achiral and chiral 1,2,4-triazolium salts: bench stable precursors for N-heterocyclic carbenes. J Org Chem 70:5725–5728

Knight RL, Leeper FJ (1998) Comparison of chiral thiazolium and triazolium salts as asymmetric catalysts for the benzoin condensation. J Chem Soc Perkin Trans 1:1891–1894

Langenbeck W (1928) Über Ähnlichkeiten in der katalytischen Wirkung von Fermenten und definierten organischen Stoffen. Angew Chem 41:740–745

Lelais G, MacMillan DWC (2007) History and perspective of chiral organic catalysts. In: Mikami K, Lautens M (eds) New frontiers in asymmetric catalysis. Wiley-VCH, Weinheim, pp 313–358

List B (2006) The ying and yang of asymmetric aminocatalysis. Chem Comm 2006(8):819–824

List B, Yang JW (2006) The organic approach to asymmetric catalysis. Science 313:1584–1586

Liu Q, Rovis T (2006) Asymmetric synthesis of hydrobenzofuranones *via* desymmetrization of cyclohexadienones using the intramolecular Stetter reaction. J Am Chem Soc 128:2552–2553

Liu YK, Li R, Yue L, Li BJ, Chen YC, Wu Y, Ding LS (2006) Unexpected ring-opening reactions of aziridines with aldehydes catalyzed by nucleophilic carbenes under aerobic conditions. Org Lett 8:1521–1524

Lohmeijer BGG, Dubois G, Leibfarth F, Pratt RC, Nederberg F, Nelson A, Waymouth RM, Wade C, Hedrick JL (2006) Organocatalytic living ring-opening polymerization of cyclic carbosiloxanes. Org Lett 8:4683–4686

Marion N, Diez-Gonzalez S, Nolan SP (2007) *N*-Heterocyclic carbenes as organocatalysts. Angew Chem Int Ed 46:2988–3000

Matsumoto Y, Tomioka K (2006) C2 symmetric chiral *N*-heterocyclic carbene catalyst for asymmetric intramolecular Stetter reaction. Tetrahedron Lett 47:5843–5848

Mennen SM, Blank JT, Tran-Dube MB, Imbriglio JE, Miller SJ (2005a) A peptide-catalyzed asymmetric Stetter reaction. Chem Comm 2005(2): 195–197

Mennen SM, Gipson JD, Kim YR, Miller SJ (2005b) Thiazolylalanine-derived catalysts for enantioselective intermolecular aldehyde-imine cross-couplings. J Am Chem Soc 127:1654–1655

Methot JL, Roush WR (2004) Nucleophilic phosphine organocatalysis. Adv Synth Cat 346:1035–1050

Movassaghi M, Schmidt MA (2005) *N*-Heterocyclic carbene-catalyzed amidation of unactivated esters with amino alcohols. Org Lett 7:2453–2456

Nair V, Vellalath S, Poonoth M, Mohan R, Suresh E (2006a) *N*-Heterocyclic carbene catalyzed reaction of enals and 1,2-dicarbonyl compounds: stereoselective synthesis of spiro γ-butyrolactones. Org Lett 8:507–509

Nair V, Vellalath S, Poonoth M, Suresh E (2006b) *N*-Heterocyclic carbene-catalyzed reaction of chalcones and enals *via* homoenolate: an efficient synthesis of 1,3,4-trisubstituted cyclopentenes. J Am Chem Soc 128: 8736–8737

Nicolaou KC, Tang Y, Wang J (2007) Formal synthesis of (±)–platensimycin. Chem Comm 2007(19):1922–1923

Nolan SP (2006) *N*-Heterocyclic carbenes in synthesis. Wiley-VCH, Weinheim

Nyce GW, Glauser T, Connor EF, Moeck A, Waymouth RM, Hedrick JL (2003) In situ generation of carbenes: a general and versatile platform for organocatalytic living polymerization. J Am Chem Soc 125:3046–3056

Nyce GW, Lamboy JA, Connor EF, Waymouth RM, Hedrick JL (2002) Expanding the catalytic activity of nucleophilic *N*-heterocyclic carbenes for transesterification reactions. Org Lett 4:3587–3590

Pellissier H (2007) Asymmetric organocatalysis. Tetrahedron 63:9267–9331

Pesch J, Harms K, Bach T (2004) Preparation of axially chiral *N,N'*-diarylimidazolium and *N*-arylthiazolium salts and evaluation of their catalytic potential in the benzoin and in the intramolecular Stetter reactions. Eur J Org Chem 2004(9):2025–2035

Phillips EM, Wadamoto M, Chan A, Scheidt KA (2007) A highly enantioselective intramolecular Michael reaction catalyzed by *N*-heterocyclic carbenes. Angew Chem Int Ed 46:3107–3110

Pohl M, Lingen B, Müller M (2002) Thiamin-diphosphate-dependent enzymes: new aspects of asymmetric C–C bond formation. Chem Eur J 8:5288–5295

Reynolds NT, Read de Alaniz J, Rovis T (2004) Conversion of α-haloaldehydes into acylating agents by an internal redox reaction catalyzed by nucleophilic carbenes. J Am Chem Soc 126:9518–9519

Reynolds NT, Rovis T (2005) Enantioselective protonation of catalytically generated chiral enolates as an approach to the synthesis of α-chloroesters. J Am Chem Soc 127:16406–16407

Rodriguez M, Marrot S, Kato T, Sterin S, Fleury E, Baceiredo A (2007) Catalytic activity of N–heterocyclic carbenes in ring opening polymerization of cyclic siloxanes. J Organomet Chem 692:705–708

Schmidt A, Habeck T, Snovydovych B, Eisfeld W (2007) Addition reaction and redox esterifications of carbonyl compounds by N–heterocyclic carbenes of indazole. Org Lett 9:3515–3518

Schmidt MA, Movassaghi M (2007) Synthesis of optically active imidazopyridinium salts and their corresponding NHCs. Tetrahedron Lett 48:101–104

Seayad J, List B (2005) Asymmetric organocatalysis. Org Biomol Chem 3:719–724

Seebach D (1979) Methods of reactivity umpolung. Angew Chem Int Ed 18:239–258

Sheehan JC, Hara T (1974) Asymmetric thiazolium salt catalysis of the benzoin condensation. J Org Chem 39:1196–1199

Sheehan JC, Hunneman DH (1966) Homogeneous asymmetric catalysis. J Am Chem Soc 88:3666–3667

Singh R, Kissling RM, Letellier MA, Nolan SP (2004) Transesterification/acylation of secondary alcohols mediated by N-heterocyclic carbene catalysts. J Org Chem 69:209–212

Singh R, Nolan SP (2005) Synthesis of phosphorus esters by transesterification mediated by N-heterocyclic carbenes (NHCs). Chem Comm 2005(43):5456–5458

Sohn SS, Bode JW (2006) N-Heterocyclic carbene catalyzed C–C bond cleavage in redox esterifications of chiral formylcyclopropanes. Angew Chem Int Ed 45:6021–6024

Sohn SS, Rosen EL, Bode JW (2004) N-Heterocyclic carbene–catalyzed generation of homoenolates γ–butyrolactones by direct annulations of enals and aldehydes. J Am Chem Soc 126:14370–14371

Song J J, Gallou F, Reeves JT, Tan Z, Yee NK, Senanayake CH (2006) Activation of TMSCN by N–heterocyclic carbenes for facile cyanosilylation of carbonyl compounds. J Org Chem 71:1273–1276

Song JJ, Tan Z, Reeves JT, Gallou F, Yee NK, Senanayake CH (2005) N-Heterocyclic carbene catalyzed trifluoromethylation of carbonyl compounds. Org Lett 7:2193–2196

Song JJ, Tan Z, Reeves JT, Yee NK, Senanayake CH (2007) N-Heterocyclic carbene-catalyzed Mukaiyama aldol reactions. Org Lett 9:1013–1016

Spivey AC, McDaid P (2007) Asymmetric acyl transfer reactions. In: Dalko PI (ed) Enantioselective Organocatalysis: reactions and experimental procedures. Wiley-VCH, Weinheim, pp 287–329

Sun X, Ye S, Wu J (2006) N-Heterocyclic carbene: an efficient catalyst for the ring-opening reaction of aziridine with acid anhydride. Eur J Org Chem 2006(21):4787–4790

Suzuki Y, Muramatsu K, Yamauchi K, Kaorie Y, Sato M (2006a) Chiral N-heterocyclic carbenes as asymmetric acylation catalysts. Tetrahedron 62:302–310

Suzuki Y, Abu Bakar MD, Muramatsu K, Sato M (2006b) Cyanosilylation of aldehydes catalyzed by N–heterocyclic carbenes. Tetrahedron 62:4227–4231

Suzuki Y, Toyota T, Imada F, Sato M, Miyashita A (2003) Nucleophilic acylation of arylfluorides catalyzed by imidazolidenyl carbene. Chem Comm 2003(11):1314–1315

Suzuki Y, Yamauchi K, Muramatsu K, Sato M (2004) First example of chiral N-heterocyclic carbenes as catalysts for kinetic resolution. Chem Comm 2004(23):2770–2771

Tachibana Y, Kihara N, Takata T (2004) Asymmetric Benzoin Condensation Catalyzed by Chiral Rotaxanes Tethering a Thiazolium Salt Moiety via the Cooperation of the Component: Can Rotaxane Be an Effective Reaction Field? J Am Chem Soc 126:3438–3439

Takikawa H, Hachisu Y, Bode JW, Suzuki K (2006) Catalytic enantioselective crossed aldehyde–ketone benzoin cyclization. Angew Chem Int Ed 45:3492–3494

Teles JH, Melder JP, Ebel K, Schneider R, Gehrer E, Harder W, Brode S, Enders D, Breuer K, Raabe G (1996) The chemistry of stable carbenes. Part 2. Benzoin-type condensations of formaldehyde catalyzed by stable carbenes. Helv Chim Acta 79:61–83

Thomson JE, Rix K, Smith AD (2006) Efficient N–heterocyclic carbene-catalyzed O- to C-acyl transfer. Org Lett 8:3785–3788

Wadamoto M, Phillips EM, Reynolds TE, Scheidt KA (2007) Enantioselective Synthesis of α,α-disubstituted cyclopentenes by an N-heterocyclic carbene-catalyzed desymmetrization of 1,3-diketones. J Am Chem Soc 129:10098–10099

Waltmann AW, Grubbs RH (2004) A new class of chelating N-heterocyclic carbene ligands and their complexes with palladium. Organometallics 23:3105–3107

Wu J, Sun X, Ye S, Sun W (2006) N-Heterocyclic carbene. A highly efficient catalyst in the reactions of aziridines with silylated nucleophiles. Tetrahedron Lett 47:4813–4816

Zeitler K (2005) Extending mechanistic routes in heterazolium catalysis–promising concepts for versatile synthetic methods. Angew Chem Int Ed 44:7506–7510

Zeitler K (2006) Stereoselective synthesis of (E)-α,β—unsaturated esters via carbene-catalyzed redox esterification. Org Lett 8:637–640

Zeitler K, Mager I (2007) An efficient and versatile approach for the immobilization of carbene precursors via copper–catalyzed [3+2]-cycloaddition and their catalytic application. Adv Synth Cat 349:1851–1857

Zhao GL, Córdova A (2007) A one-pot combination of amine and heterocyclic carbene catalysis: direct asymmetric synthesis of β-hydroxy and β-malonate esters from α,β-unsaturated aldehydes. Tetrahedron Lett 48:5976–5980

Zhou ZZ, Ji FQ, Cao M, Yang GF (2006) An efficient intramolecular Stetter reaction in room temperature ionic liquids promoted by microwave irradiation. Adv Synth Cat 348:1826–1830

Ernst Schering Foundation Symposium Proceedings, Vol. 2, pp. 207–253
DOI 10.1007/2789_2008_085
© Springer-Verlag Berlin Heidelberg
Published Online: 30 April 2008

New Developments in Enantioselective Brønsted Acid Catalysis: Chiral Ion Pair Catalysis and Beyond

M. Rueping[✉], E. Sugiono

Degussa Endowed Professorship, Institute of Organic Chemistry and Chemical Biology,
Johann Wolfgang Goethe-Universität Frankfurt am Main, Max-von-Laue Strasse 7,
60438 Frankfurt am Main, Germany
email: *M.rueping@chemie.uni-frankfurt.de*

1	Development of the First Highly Enantioselective Organocatalytic Reduction of Ketimines 209
2	Biomimetic Reductions: Amino Acid Dehydrogenases as the Role Model for the Brønsted Acid Catalyzed Transfer Hydrogenation of Ketimines 209
3	The First Enantioselective Brønsted Acid Catalyzed Transfer Hydrogenation of Ketimines: Chiral Phosphates Are Efficient Metal-Free Catalysts 213
4	Highly Enantioselective Brønsted Acid Catalyzed Transfer Hydrogenation of Quinolines: Development of a New Organocatalytic Cascade Reaction 216
5	Enantioselective Reduction of Benzoxazines, Benzthiazines, and Benzoxazinones: Remarkably Low Catalyst Loading in the Asymmetric Brønsted Acid Catalyzed Transferhydrogenation 221
6	First Organocatalytic Enantioselective Reduction of Pyridines . . . 225
7	First Highly Enantioselective Brønsted Acid Catalyzed Strecker Reaction: Use of C-Nucleophiles in Chiral Ion Pair Catalysis 230
8	Asymmetric Brønsted Acid Catalyzed Imino-Azaenamine Reaction 233

9	Development of the First Brønsted Acid Assisted Enantioselective Brønsted Acid Catalyzed Direct Mannich Reaction 236
10	Cooperative Co-Catalysis: The Effective Interplay of Two Brønsted Acids in the Enantioselective Synthesis of Isoquinuclidines 238
11	Asymmetric Brønsted Acid Catalyzed Carbonyl Activation: The First Organocatalytic Electrocyclic Reaction 241
References...................................245	

Abstract. The design of catalytic reactions that proceed with high enantioselectivity is an important goal in organic synthesis. Increased interest in this research area has resulted in substantial progress, particularly in the field of metal catalyzed transformations. In recent years small organic molecules have been used as organocatalysts for a variety of enantioselective reactions. Among these, secondary amine catalysts are the most widely applied and can be used in the activation of the nucleophilic component through enamine formation (enamine catalysis), or by formation of an iminum intermediate to activate the electrophile (iminium catalysis). Additionally, chiral diols and thioureas, as well as carbene- and DMAP-derivatives (hydrogen bonding-, nucleophilic catalysis), have been shown to be versatile catalysts for enantioselective transformations. An alternative to these strategies is the activation of an electrophile or nucleophile by use of a chiral Brønsted acid. Compared to amino-, carbene-, pyridine- and hydrogen-bonding catalyzed transformations, enantioselective Brønsted acid catalysis has only recently emerged as important and promising area of research. In the course of our research program we were able to contribute significantly to the field of enantioselective Brønsted acid catalysis over the last 2 years, and could demonstrate for the first time that in various enantioselective transformations chiral Brønsted acid catalysts can give better or at least comparable results to metal-catalyzed processes. In this chapter we will highlight some of our most recent results and will, additionally, describe how we initially entered the field of asymmetric Brønsted acid catalysis by starting of from a biomimetic approach using nature as a role model.

1 Development of the First Highly Enantioselective Organocatalytic Reduction of Ketimines

The hydrogenation of unsaturated organic compounds, such as olefins, carbonyls and imines is one of the most important and utilized transformation in both academia and in the chemical industry. With the ever increasing number of biologically active substances with hydrogen as part of the stereocenter, it is not surprising that the development of efficient asymmetric reductions has become a central research area in enantioselective catalysis. So far, most of these enantioselective reductions rely on chiral transition metal catalysts, and highly enantioselective hydrogenations of ketones and alkenes are known. However, the asymmetric hydrogenation of ketimines to obtain chiral amines still represents a challenge and only less effective and selective reductions are available. Current methods include transition metal catalyzed high pressure hydrogenations (Blaser et al. 2003; Tang and Zhang 2003), hydrosilylations (Riant et al. 2004; Carpentier and Bette 2002; organocatalytic hydrosilylation: Malkov et al. 2004), or transfer hydrogenations (Kadyrov and Riermeier 2003), using a variety of chiral transition metals, such Pd, Ti, Rh, Ru, and Ir-complexes. While almost all of these metal catalyzed transformations show high reactivities and selectivities none proved to have a broad substrate scope. Hence, so far exclusively enzymatic resolutions are applied for the large-scale production of chiral amines, with the typical disadvantage of biocatalytic resolutions, limited substrate scope and recycling of the undesired byproducts. Given the above limitations we decided to develop a new biomimetic enantioselective reductive amination or transfer hydrogenation of ketimines using nature as the role model.

2 Biomimetic Reductions: Amino Acid Dehydrogenases as the Role Model for the Brønsted Acid Catalyzed Transfer Hydrogenation of Ketimines

Dehydrogenases are important enzymes involved in the assimilation or dissimilation of ammonia. For instance, glutamate dehydrogenase (GDH) catalyzes the reductive amination of 2-ketogluterate to glutamate with the use of nicotineamide dinucleotide as hydride donor (Eq. 1).

$$2\text{-ketogluarate} + NH_3 + NAD(P)H \rightleftharpoons \text{glutamate} + NAD(P)^+ \quad (1)$$

The reaction is reversible, although the equilibrium lies in the direction of glutamate formation. Likewise, other amino acid dehydrogenases assimilate ammonia analogous to the GDH reaction, but are specific for other keto acid/amino acid substrate pairs, e.g. alanine dehydrogenase catalyzes the reductive amination of pyruvate to alanine. Typically, a high activity of these dehydrogenases is observed if a high concentration of ammonia is in the environment or under stress conditions such as high temperature, salinity, senescence, or environmental pollution.

A mechanism for the catalytic activity of dehydrogenases has been proposed following detailed kinetic studies and crystallographic structures. One important feature in the reductive amination of the 2-ketoglutarate is the protonation of the α-iminoglutarate by aspartate D165, resulting in the formation of an imium-ion. Subsequent hydride transfer from the nicotineamide dinucleotide leads to the formation of the amino acid glutamate (Fig. 1).

To support this mechanism the putative catalytic aspartyl residue D165 in the active site of GDH has been replaced with other amino acids, such as alanine, serine or histidine by site directed mutagenesis. Although the modified enzymes still appeared to be correctly folded the high catalytic activity was lost, indicating the vital role of the amino acid D165 for its catalytic activity.

Based on the above activation mechanism we wondered whether it would be possible to develop a biomimetic, organocatalytic reductive amination or transfer hydrogenation of ketimines. We reasoned that the activation of the imine by catalytic protonation through the Brønsted acid should enable the hydrogen transfer from a suitable NADH mimic to yield the corresponding amine (Fig. 2). Hence, initial experiments focused on the examination of various Brønsted acids in combination with different hydride sources (Rueping et al. 2005a).

From this survey we found that several proton acids catalyze the reduction of ketimines using Hantzsch dihydropyridine **2** as the hydride donor. With regard to the different acids tested diphenyl phosphate

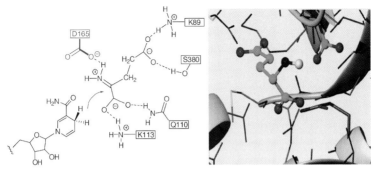

Fig. 1. Representation of the proposed catalytic activation. Aspartate D165 protonates the α-iminoglutarate to form an iminium ion which after hydride transfer from NADH results in the formation of the amino acid glutamate (*left*). Active site of a GDH crystal structure: the glutamate is located in a hydrogen bond distance from the catalytically essential D165 (*right*)

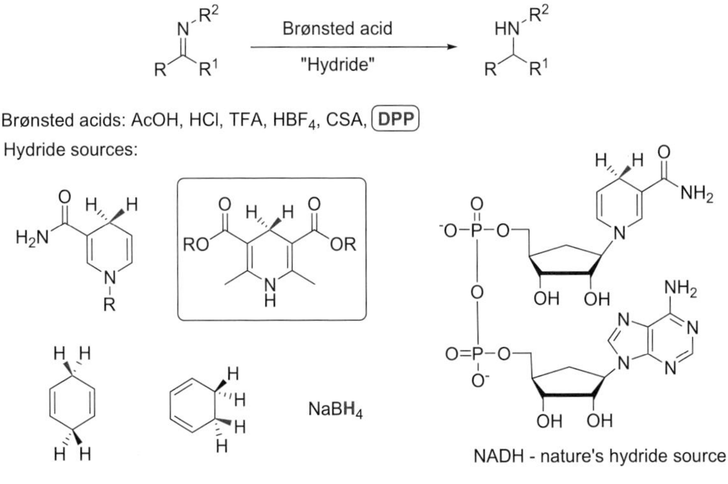

Fig. 2. Development of a new Brønsted acid catalyzed transfer hydrogenation

(DPP) appeared to be the best catalyst for this transformation, showing the highest reaction rate and best yields of the amine formed. Further

Table 1 Scope of the new Brønsted acid catalyzed transfer hydrogenation

Entry	1	R	R^1	R^2	Yield of 3 [%][a]
1	a	Ph	Et	Ph	70
2	b	Ph	Et	PMP	92
3	c	Napht	Me	Ph	75
4	d	Biphe	Me	Ph	87
5	e	Biphe	Me	PMP	67
6	f	Ph	Bu	PMP	92
7	g	Ph	i-Pr	PMP	70
8	h	2-CH_3-C_6H_4	Me	PMP	72
9	i	2-F-C_6H_4	Me	PMP	67
10	j	4-CF^3-C_6H_4	Me	PMP	70
11	k	2,4-CH_3-C_6H_3	Me	PMP	66
12	l	3-CH_3-C_6H_4	Ph	PMP	86
13	m	4-OMe-C_6H_4	Me	PMP	76
14	n	Ph	CO_2Me	PMP	77

[a] Isolated yields of **3** after chromatography

exploration of the reaction conditions revealed that non-polar aromatic or halogenated solvents proved to be superior to polar ones.

In general, differently substituted alkyl and aryl ketimines, including α-imino esters, can be reduced in good to excellent yields (Table 1).

Starting from a biomimetic approach we developed an unprecedented organocatalytic Brønsted acid catalyzed transfer hydrogenation of ketimines using Hantzsch dihydropyridine as reducing agent. The operational simplicity and practicability, as well as the mild reaction conditions, render this transformation an attractive approach to various amines. More importantly, however, the use of catalytic amounts of DPP should allow the extension of this procedure to an asymmetric organocatalytic reduction by employing chiral phosphoric acids as catalysts.

3 The First Enantioselective Brønsted Acid Catalyzed Transfer Hydrogenation of Ketimines: Chiral Phosphates Are Efficient Metal-Free Catalysts

The development of the new DPP catalyzed transfer hydrogenation encouraged us to explore a catalytic enantioselective variant of this transformation, as it would not only be the first example of an enantioselective organocatalytic hydrogenation of ketimines but would, additionally, provide a new approach and access to chiral amines. Initial experiments focused therefore on the application of various Brønsted acids (Schreiner 2003; Pihko 2004; Bolm et al. 2005), including chiral thiourea-, diol-, amidinium-catalysts as well as chiral sulfonic acids. However, none of these catalysts afforded satisfactory yields and selectivities. Therefore, we decided to prepare chiral phosphoric acid catalysts **5** and **6** which are based on the BINOL core structure (Rueping et al. 2006a). These catalysts had previously been used for resolutions of racemic amines, as ligand in metal catalyzed transformations and, more recently, in metal-free catalysis. Starting from the protected (*R*)-3,3'-dibromo-1,1'-binaphthyl-2,2'-diol or the octahydro derivative **4** the corresponding 3,3-aryl substituted catalysts were readily prepared by Suzuki coupling, followed by deprotection and phosphorylation (Scheme 1). Alternatively, the silylated BINOL-derivatives **4b** were subjected to a Brook-type rearrangement using *t*-BuLi, and subsequent phosphorylation resulted in the chiral Brønsted acids **6**.

With the chiral BINOL-phosphates in hand we started to examine the enantioselective transferhydrogenation of ketimines **1**. After reaction optimization, including a survey of different solvents, temperatures, BINOL-phosphates, and Hantzsch dihydropyridines, we found that indeed enantioselectivities are observed and the best selectivities are obtained with Brønsted acid **5a** and Hantzsch ester **2a** (Table 2). In general, for the first time, high enantioselectivities and good yields are observed in this newly developed metal-free reduction procedure (Rueping et al. 2005b; Hofmann et al. 2005; Storer et al. 2006).

With regard to the mechanism we assume that, similar to the dehydrogenase (Fig. 1), the ketimine **1** will be activated by protonation through Brønsted acid **5** which results in the formation of a chiral ion-pair, an iminium ion **A**. Subsequent hydrogen transfer from the dihy-

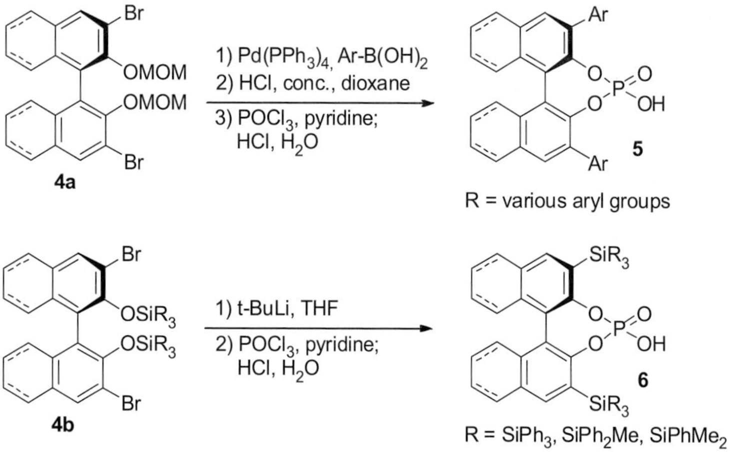

Scheme 1. Synthesis of differently substituted BINOL-phosphates **5** and **6**

dropyridine **2a** yields the chiral amine **7** and pyridinium salt **B**, which undergoes proton transfer to regenerate the Brønsted acid catalyst **5** (Fig. 3).

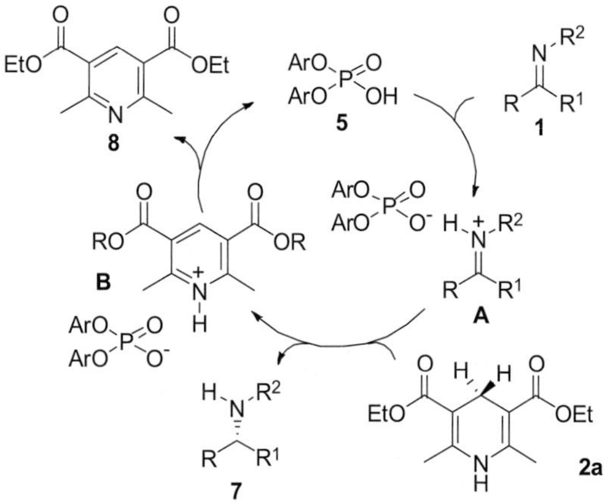

Fig. 3. Proposed mechanism for the transfer hydrogenation

Table 2 Scope of the first enantioselective Brønsted acid catalyzed transfer hydrogenation

Entry	7	R	R′	Yield [%]^a	ee [%]^b
1	a	PMP	2-naphthyl	82	70 (94)^c
2	b	Ph	2-naphthyl	69	68
3	c	PMP	4-CF$_3$-Phenyl	71	72
4	d	Ph	4-CF$_3$-Phenyl	58	70
5	e	PMP	Phenyl	76	74
6	f	Ph	Phenyl	71	72
7	g	PMP	2-F-Phenyl	82	84
8	h	PMP	2-CH$_3$-Phenyl	74	78
9	i	PMP	2,4-(CH$_3$)-Phenyl	91	78
10	j	PMP	4-Biphenyl	71	74 (98)^c
11	k	PMP	2-CH$_3$O-Phenyl	76	72
12	l	PMP	3-Br-Phenyl	62	72
13	m	PMP	2-CF$_3$-Phenyl	46	82

^{a} Isolated yields of **7** after chromatography
^{b} Enantioselectivities were determined by HPLC analysis
^{c} After one recrystallization from methanol

The absolute configuration of the amine **7** may be explained by a stereochemical model based on the X-ray crystal structure of the chiral BINOL-phosphate (Fig. 4). In the transition state the ketimine is activated by the Brønsted acid in such a way, that the nucleophile has to approach from the less hindered *si* face as the *re* face is effectively shielded by the large aryl substituent of the catalyst (Fig. 4, left). Furthermore, a bifunctional activation seems to be plausible, where next to the ketimine protonation, the dihydropyridine is activated through a hydrogen bond from the Lewis basic oxygen of the phosphoryl group.

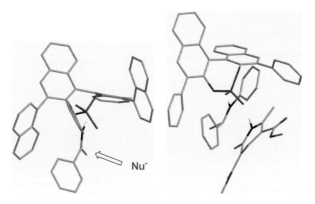

Fig. 4. Transition state model based on the X-ray crystal structure of **5**

4 Highly Enantioselective Brønsted Acid Catalyzed Transfer Hydrogenation of Quinolines: Development of a New Organocatalytic Cascade Reaction

Although numerous metal-catalyzed enantioselective hydrogenations using various chiral metal-complexes have been reported most of these catalysts failed to give satisfactory results in the asymmetric hydrogenation of aromatic and heteroaromatic compounds and examples of efficient transformations are rare (Glorius 2005). This is also true for the partial reduction of readily available quinoline derivatives (Wang et al. 2003; Lu et al. 2004; Yang and Zhou 2004; Xu et al. 2005; Reetz and Li 2006); the most convenient route to 1,2,3,4-tetrahydroquinolines, which are of great synthetic importance in the preparation of pharmaceuticals, agrochemicals, and in material sciences. Furthermore, many alkaloid natural products consist of this structural key element. Given the importance of this class of molecules, together with our recently developed biomimetic, enantioselective Brønsted acid catalyzed transfer hydrogenation of ketimines (Table 2; Rueping et al. 2005b), we wondered whether it would be possible to extend our procedure to the enantioselective hydrogenation of quinolines **9**. This would not only represent the first example of a metal-free reduction of heteroaromatic compounds but would additionally give straightforward access to optically pure

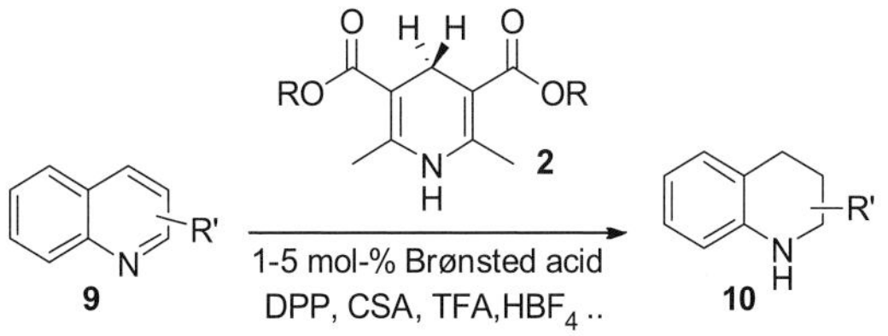

Scheme 2. First Brønsted acid catalyzed transfer hydrogenation of quinolines

tetrahydroquinolines **10**. We reasoned that activation of the quinoline by catalytic protonation through a Brønsted acid would enable a cascade hydrogenation, which involves a 1,4-hydride addition, protonation and 1,2-hydride addition sequence to generate the desired tetrahydroquinolines **10** (Scheme 2; Rueping et al. 2006b).

Following up on our idea, we initially focused on the exploration of appropriate Brønsted acids and the examination of reaction parameters, such as catalyst loading, hydride source, temperature, solvent and concentration. While the use of various Brønsted acids with different acidities resulted in the desired products, the application of DPP again showed the best reactivities and yielded a diverse set of tetrahydroquinolines with different substituents in 2-, 3- or 4-position.

Having developed this first metal-free reduction of quinolines we wanted to extend this procedure to an asymmetric variant by employing a chiral BINOL-phosphate **5** (Rueping et al. 2006c). After optimization of the reaction conditions we explored the scope of the Brønsted acid catalyzed cascade transfer hydrogenation of quinolines. In general, high enantioselectivities and good isolated yields of several tetrahydroquinolines with aromatic and heteroaromatic residues, as well as aliphatic substituents are observed (Table 3). Interestingly, this metal-free hydrogenation procedure is even compatible with halogenated aromatic and aliphatic residues, such as 3-bromophenyl or chloromethyl substituents.

Having established a general and highly enantioselective cascade transfer hydrogenation of 2-substituted quinolines, we decided to apply this new methodology to the synthesis of biologically active tetrahydroquinoline natural products, such as Galipinine (Rakotoson et al. 1998; Jacquemond-Collet et al. 1999), Cuspareine (Houghton et al. 1999;

Table 3 Scope of the first enantioselective Brønsted acid catalyzed transfer hydrogenation of quinolines

Ar: Phenanthryl

Entry	10	R	t[h]	Yield [%][a]	ee [%][b]
1	a	phenyl	12	92	97
2	b	2-F-Phenyl	30	93	98
3	c	2-CH$_3$-Phenyl	48	54[c]	91
4	d	2,4-(CH$_3$)-Phenyl	60	65[d]	97
5	e	2-naphthyl	12	93	>99
6	f	3-Br-Phenyl	18	92	98
7	g	4-(CF$_3$)-Phenyl	30	91	>99
8	h	1,1′-biphenyl-4-yl	12	91	>99
9	i	4-CH$_3$O-phenyl	12	90[e]	98
10	j	2-furyl	12	93	91
11	k	choromethyl	12	91[e]	88
12	I	n-butyl	12	91	87
13	m	n-pentyl	12	88	90
14	n	2-phenylethyl	12	90	90
15	o	(benzodioxol-methyl)	12	94[e]	91
16	p	(3,4-dimethoxyphenyl-methyl)	12	95[e]	90

[a] Isolated yields of **10** after chromatography
[b] Enantioselectivities were determined by HPLC analysis
[c] 45% recovered starting material
[d] 5mol% catalyst
[e] 1mol% catalyst

Jacquemond-Collet et al. 1999) and Angustureine (Jacquemond-Collet et al. 1999).

The Brønsted acid catalyzed enantioselective hydrogenation of the corresponding readily available 2-substituted quinolines (for an interesting approach to 2-alkyl tetrahydroquinolines by an aza-xylene Diels–Alder reaction, see: Steinhagen and Corey 1999; Avemaria et al. 2003), which we prepared by simple alkylation of 2-methylquinoline, generated the tetrahydroquinoline derivatives in excellent enantioselectivities and subsequent N-methylation gave the desired natural products in good overall yields (Fig. 5).

Mechanistically, we assume that the first step of the enantioselective cascade hydrogenation is the protonation of the quinoline **9** through the chiral Brønsted acid catalyst **5** to generate the iminium ion **A** (Fig. 6). Subsequent first hydrogen transfer from the dihydropyridine **4** in 4-position of the quinoline generates the enamine **11** and pyridinium salt **B**, which undergoes proton transfer to regenerate the Brønsted acid **5** and Hantzsch pyridine **8**. The enamine **11** undergoes a Brønsted acid catalyzed protonation to form the iminium ion **C**, which will again be subjected to hydrogen transfer to give the desired tetrahydroquinoline **10** and subsequent proton transfer will recycle the Brønsted acid **5**.

This so far unprecedented Brønsted acid catalyzed cascade transfer hydrogenation provides direct access to a variety of 2-aryl- and alkyl-

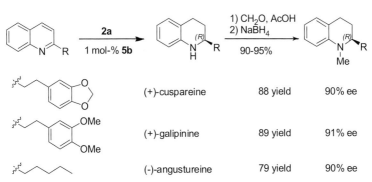

Fig. 5. Synthesis of natural product alkaloids applying the enantioselective Brønsted acid catalyzed transfer hydrogenation of quinolines

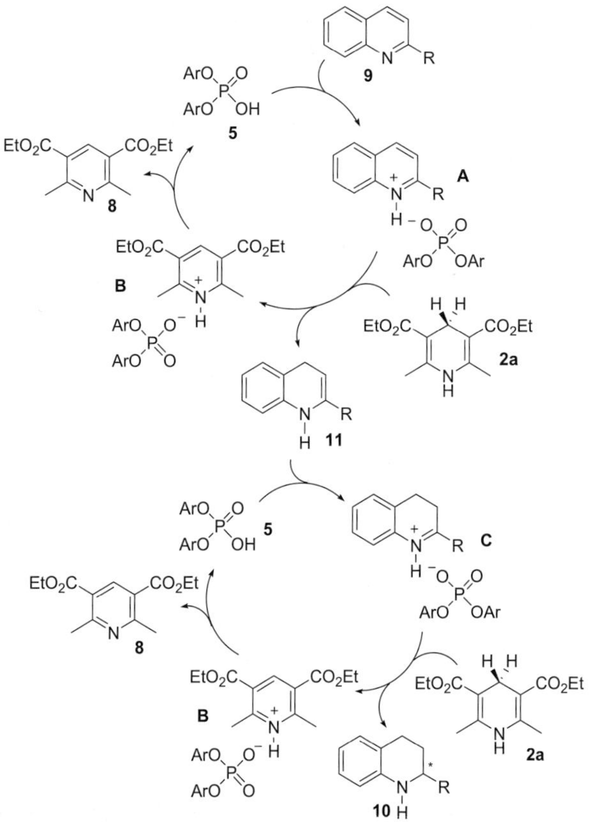

Fig. 6. Proposed mechanism of the Brønsted acid catalyzed transfer hydrogenation of quinolines

substituted tetrahydroquinolines with excellent enantioselectivities and good isolated yields. The mild reaction conditions of this metal-free process, the operational simplicity and practicability, as well as the low catalyst loading render this transformation an attractive approach to optically active tetrahydroquinolines and their derivatives. This we have demonstrated in the enantioselective synthesis of the naturally occurring alkaloids Galipinine, Cuspareine and Angustureine.

5 Enantioselective Reduction of Benzoxazines, Benzthiazines, and Benzoxazinones: Remarkably Low Catalyst Loading in the Asymmetric Brønsted Acid Catalyzed Transferhydrogenation

To date various highly enantioselective processes based on chiral metal-complexes have been described. However, so far only few catalysts have been reported to catalyze the reduction of cyclic imines, such as benzoxazine, benzthiazine and benzoxazinone derivatives (Glorius 2005; Wang et al. 2003; Lu et al. 2004; Yang and Zhou 2004; Xu et al. 2005; Reetz and Li 2006; Noyori 1996; Satoh et al. 1998; Zhou et al. 2005). This is surprising as the corresponding dihydro-$2H$-benzoxazines, dihydro-$2H$-benzothiazines and dihydro-$2H$-benzoxazinones represent the structural motif of various natural products with interesting biological activities. These compounds have also been employed as important chiral building blocks in the synthesis of many pharmaceuticals, such as promising anti-depressants, calcium antagonists as well as anti-inflammatory, anti-nociceptive, anti-bacterial and anti-microbial agents (Brown and Djerassi 1964; Belattar and Saxton 1992; Krohn et al. 1993; Kleemann et al. 2001; Achari et al. 2004; Ilas et al. 2005).

Based on our newly developed transfer-hydrogenation of ketoimines (Rueping et al. 2005b) and quinolines (Rueping et al. 2006c) we believed it might be possible to extend our chiral ion pair catalysis procedure to the enantioselective transfer hydrogenation of benzoxazines and benzthiazines (Rueping et al. 2006d). We reasoned that, similar to enzymatic transfer hydrogenations, activation of benzoxazines **12** and benzthiazines **14** by catalytic protonation through Brønsted acid **5** would generate a chiral ion pair **A**, which would subsequently undergo a hydride transfer addition from Hantzsch dihydropyridine **2** to give the desired dihydro-$2H$-benzoxazine **13** and dihydro-$2H$-benzothiazines **15** (Fig. 7).

Hence, our initial investigations concentrated on finding the appropriate catalyst and reaction conditions. Again, catalyst **5b** showed the best enantioselectivities and remarkable reactivities. Therefore, we decided to perform a detailed investigation of the catalyst loading (Table 4).

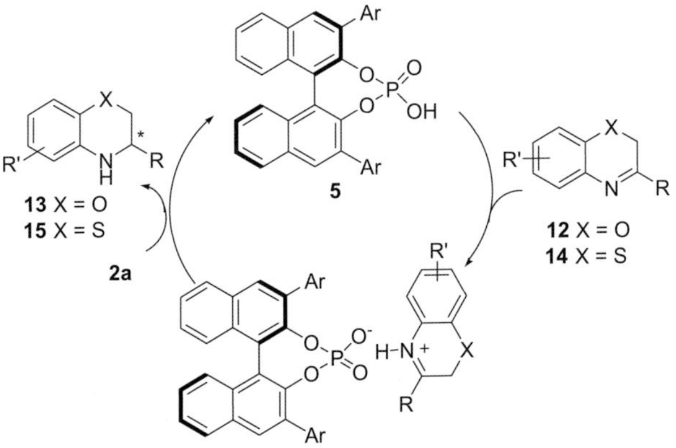

Fig. 7. Brønsted acid catalyzed transfer hydrogenation of benzoxazines and benzthiazines

Table 4 Evaluation of the catalyst loading for the enantioselective transfer hydrogenation

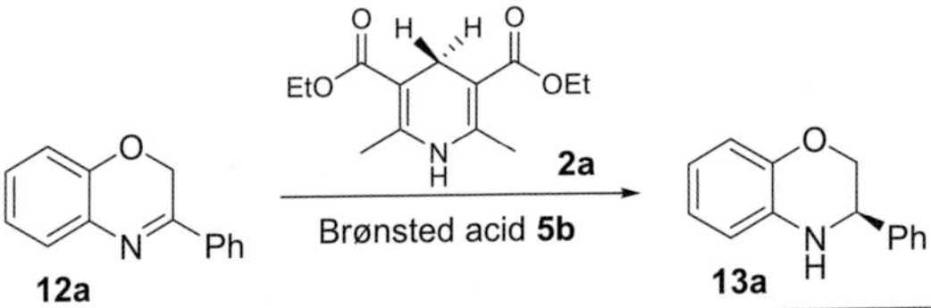

Entry	Loading of **5b** [%]	Yield [%][a]	ee [%][b]
1	10	91	96
2	5	95	96
3	2	93	96
4	1	94	96
5	0.1	95	96
6	0.01	90	93

[a] Isolated yields of **13a** after chromatography
[b] Enantioselectivities were determined by HPLC analysis

Beginning with 10 mol% of the chiral Brønsted acid catalyst **5b** we were able to decrease the catalyst loading to 0.01 mol% (substrate/catalyst ratio = 10,000:1) without a considerable loss in reactivity and selectivity. With just 0.01 mol% of Brønsted acid **5b** the corresponding 2*H*-dihydro-benzoxazine **13a** could be isolated in a remarkable 90% yield and 93% *ee* and with a TON of 9000 and a TOF 500 h^{-1}.

It should be noted that this is, to date, the lowest catalyst loading reported for an organocatalytic enantioselective transformation. Furthermore, if compared to enantioselective metal-catalyzed reductions of heterocyclic systems, which employ 0.5–5 mol% (Glorius 2005; Wang et al. 2003; Lu et al. 2004; Yang and Zhou 2004; Xu et al. 2005; Reetz and Li 2006; Noyori 1996; Satoh et al. 1998; Zhou et al. 2005), our metal-free transformation requires the lowest amount of catalyst, demonstrating the high potential of BINOL-phosphates, and Brønsted acid catalysts in general. With the optimized conditions in hand we started to explore the scope of the Brønsted acid catalyzed transfer hydrogenation of benzoxazines **12a–e** and benzthiazines **14a–e** (Table 5). In general, differently substituted 3-aryl-dihydro-2*H*-benzoxazines **13a–e** and -benzthiazines **15a–e** bearing either electron withdrawing or electron donating groups were obtained in good yields and with excellent enantioselectivities (94%–99% *ee*). Furthermore, the new metal-free procedure shows advantages over the application of most metal catalysts, which are known to be poisoned by sulfur containing substrates.

Having established a general and highly enantioselective transfer hydrogenation of **12** and benzthiazines **14**, we also decided to apply this efficient methodology to the reduction of benzoxazinones **16** as this would lead to the valuable cyclic *aryl*-substituted amino acid derivatives **17**. Hence, we applied our organocatalytic enantioselective hydrogenation procedure to 3-aryl-substituted benzoxazinones **16** using the same reaction conditions (Table 6). Once again, we were able to isolate the corresponding differently substituted, electron-rich and electron-deficient aryl- and heteroaryl-glycine derivatives **17** in good yields and with excellent enantioselectivities (90%–99% *ee*).

The newly developed procedure described represents the first metal-free reduction of benzoxazines and benzthiazines in general, and provides the corresponding dihydro-2*H*-benzoxazine and dihydro-2*H*-

Table 5 Scope of the enantioselective transfer hydrogenation of benzoxazines and benzthiazines

Entry	Product		Yield [%][a]	ee [%][b]
1	13a		95	98
2	13b		93	>99
3	13c		94	98
4	13d		92	98
5	13e		95	>99
6	15a		87	>99
7	15b		50	96

Table 5 (continued)

Entry	Product		Yield [%][a]	ee [%][b]
8	15c	(benzothiazine with Ph-substituted aryl)	78	94
9	15d	(benzothiazine with m-Br aryl)	51	94
10	15e	(benzothiazine with p-F aryl)	70	>99

[a] Isolated yields after chromatography
[b] Enantioselectivities were determined by HPLC analysis

benzthiazines in good isolated yields and with excellent enantioselectivities (93%–99% ee). Additionally, we were able to extend this valuable methodology to the enantioselective synthesis of cyclic and linear aryl- and heteroaryl-glycine derivatives. Attractive features of our ion-pair catalysis hydrogenation process include the mild reaction conditions, operational simplicity and practicability as well as a broad substrate scope, including sulfur containing substrates. It should be highlighted that this organocatalytic procedure allows the reduction of catalyst loading to 0.01 mol% of Brønsted acid **5b** without considerable loss in reactivity and selectivity.

6 First Organocatalytic Enantioselective Reduction of Pyridines

The asymmetric reduction of aromatic and heteroaromatic compounds still represents a great challenge (Glorius 2005; Noyori 2002; Knowles 2002; Ohkuma et al. 2000; Ohkuma and Noyori 2004; Nishiyama and

Table 6 Scope of the enantioselective transfer hydrogenation of benzoxazinones

Entry	Product	Yield [%][a]	ee [%][b]
1	17a	85	98
2	17b	92	>99
3	17c	91	>99
4	17d	90	>99
5	17e	55	96
6	17f	90	98
7	17g	81	90

[a] Isolated yields of **17** after chromatography
[b] Enantioselectivities were determined by HPLC analysis

Itoh 2000; Blaser et al. 2003; Tang and Zhang 2003). This is particularly true with regard to the enantioselective hydrogenation of substituted pyridine derivatives (metal-catalyzed enantioselective reductions of pyridines: Legault and Charette 2005; Lei et al. 2006; an efficient diastereoselective pyridine reduction: Glorius et al. 2004), which can readily be prepared using various methods. The corresponding chiral piperidines are not only important starting materials for numerous biologically active compounds but also important structural building blocks which occur in many alkaloid natural products including pumiliotoxins, gephyrotoxins, and in over 50 further members of the 2,5-disubstituted decahydroquinoline class of compounds (reviews: Daly 1998; O'Hagan 2000; Daly et al. 2005; Michael 2005).

Gephyrotoxin

Pumiliotoxin-C

Due to the extensive biological activity, as well as the lack of enantioselective catalytic methods for the preparation of these products, we considered it most important to investigate a Brønsted acid catalyzed enantioselective reduction of pyridines. In this context, and building on our previous chiral ion-pair catalysis results (Rueping et al. 2005b, 2006c, d), we decided to examine a BINOL-phosphate (other binol phosphate catalyzed reactions: (Akiyama et al. 2004, 2005a,b, 2006a,b; Uraguchi and Terada 2004; Uraguchi et al. 2004; Rowland et al. 2005; Terada et al. 2006a,b; Seayad et al. 2006; Mayer and List 2006; Itoh et al. 2006; Nakashima and Yamamoto 2006; Hasegawa et al. 2006; Hoffmann et al. 2006; Chen et al. 2006; Liu et al. 2006; Terada and Sorimachi 2007; Martin and List 2006; Kang et al. 2007; review: Akiyama et al. 2006c) catalyzed reduction of pyridines. We assumed that activation of pyridines by catalytic protonation using a Brønsted acid would allow a cascade hydrogenation in which the Hantzsch dihydropyridine would function as the hydride source. Hence, we began our study with the BINOL-phosphate catalyzed reduction of the tri-substituted pyridines **18**, whereby the reaction parameters including the solvent, temperature, catalyst loading, substrate concentration and the hydride source were varied. The results showed that our new enantioselective, Brønsted-acid catalyzed cascade transfer hydrogenation of pyridines can indeed be performed and that the partially reduced enantiomerically enriched pyridines **19** can be isolated (Table 7).

In general this organocatalytic pyridine reduction procedure provides azadecalinones **19a–f** as well as the tetrahydropyridines **19g–j** in good isolated yields and with excellent enantioselectivities (up to 92% *ee*) (Rueping and Antonchick 2007).

The newly developed enantioselective organocatalytic reduction of pyridines can, for example, be used as a key step in the synthesis of decahydroquinolines from the pumiliotoxin family (Scheme 3). Hence, the BINOL-phosphate catalyzed reduction of the corresponding pyridine, which can readily be prepared according to a procedure described by Bohlmann and Rahtz (Bohlmann and Rahtz 1957; Bagley et al. 2002), gives the corresponding 2-propyl-hexahydroquinolinone. This can subsequently be transformed, according to a reported sequence into *diepi*-pumiliotoxin C (Sklenicka et al. 2002).

The first enantioselective Brønsted acid catalyzed cascade reduction of pyridines provides the corresponding products in good yields and with excellent enantioselectivities (up to 92% *ee*). The hexahydroquinolinones and tetrahydropyridines isolated are not only known starting products for various natural products such as pumiliotoxin C but also for numerous interesting biologically active compounds. As previously only metal catalyzed hydrogenations of pyridines have been described, and these did not result in the valuable products described herein and gave lower enantioselectivities, our newly developed metal-free Brønsted acid catalyzed procedure represents an important contribution to the enantioselective reduction of pyridines.

New Developments in Enantioselective Brønsted Acid Catalysis

Table 7 Scope of the enantioselective transfer hydrogenation of pyridines

Compound	ee / yield	Compound	ee / yield
19a	91% ee[a], 84% yield[b]	19f	92% ee, 73% yield
19b	91% ee, 72% yield	19g	90% ee, 73% yield
19c	89% ee, 69% yield	19h	84% ee, 55% yield
19d	92% ee, 66% yield	19i	86% ee, 47% yield
19e	87% ee, 83% yield	19j	89% ee, 68% yield

[a] Isolated yields of chromatography
[b] Enantioselectivities were determined by HPLC analysis

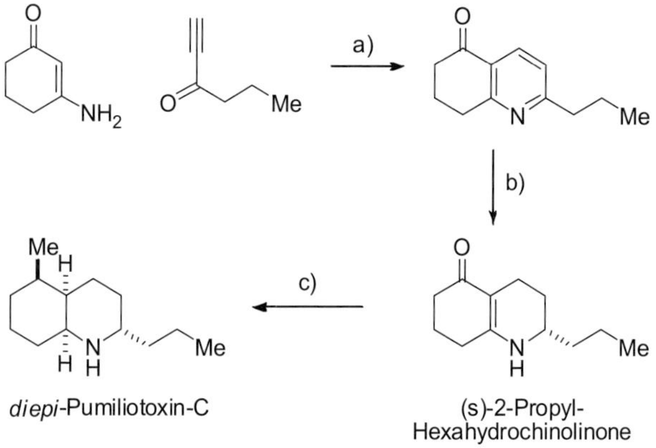

Scheme 3. Enantioselective synthesis of *diepi*-pumiliotoxin-C; (a) EtOH, 50 °C, 12 h, then 140 °C, 2 h; (Bohlmann and Rahtz 1957; Bagley et al. 2002) (b) 5 Mol % (*S*)-**3f**, **2** (4 equiv.) at 50 °C in benzene; (c) (Sklenicka et al. 2002)

7 First Highly Enantioselective Brønsted Acid Catalyzed Strecker Reaction: Use of C-Nucleophiles in Chiral Ion Pair Catalysis

The hydrocyanation of imines, the Strecker reaction, is considered the most practical and direct route to α-amino acids (Strecker 1850). Consequently, various attempts to develop asymmetric Strecker reactions have been made (for reviews, see: Gröger 2003; Yet 2001; Spino 2004). In addition to metal-catalyzed hydrocyanations using chiral metal catalysts (Al catalysts: Sigman and Jacobsen 1998a; Takamura et al. 2000; Krueger et al. 1999; Byrne et al. 2000; Josephsohn et al. 2001; Ishitani et al. 1998; Kobayashi and Ishitani 2000; Chavarot et al. 2001; Masumoto et al. 2003), promising metal-free, enantioselective variants of this reaction have recently been disclosed. These processes are based on chiral guanidines (Corey and Grogan 1999), ureas and thioureas (Sigman and Jacobsen 1998b; Sigman et al. 2000; Vachal and Jacobsen 2000; Vachal and Jacobsen 2002; Wenzel et al. 2003; Tsogoeva et al. 2005), bis-*N*-oxides [for the application of stoichiometric amounts of bis(N-oxides), see: Liu et al. 2001; Jiao et al. 2003] and ammonium

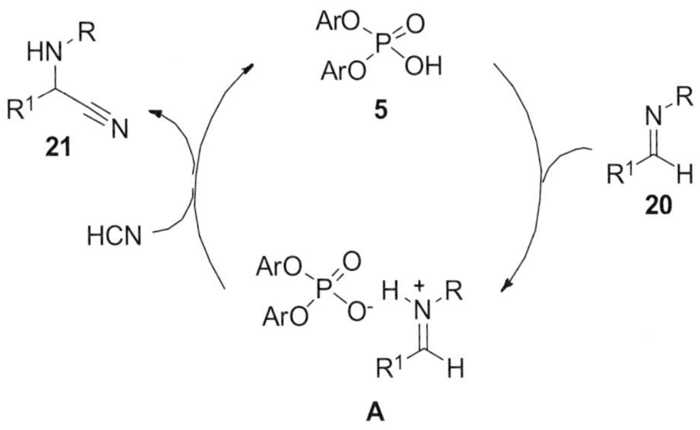

Fig. 8. Proposed catalytic cycle for the asymmetric hydrocyanation

salts (Huang and Corey 2004). The importance of the Strecker reaction and the resulting products prompted us to examine a new chiral Brønsted acid catalyst for this important transformation (Rueping et al. 2006e). Based on the above described Brønsted acid catalyzed transfer hydrogenations using BINOL-phosphate catalysts **5**, we reasoned that activation of imine **20** by catalytic protonation would generate the iminium ion **A**, a chiral ion pair which would subsequently undergo addition of HCN to give the desired amino nitrile **21** and the regenerated Brønsted acid **5** (Fig. 8).

Hence, initial explorations concentrated on varying the chiral BINOL-phosphate as well as reaction parameters including different protected imines, cyanide sources, catalyst loadings, temperatures, and concentrations. From these experiments the best results, with respect to yield and selectivity, were obtained with benzyl-protected aldimines and HCN at –40 °C using 10 mol% of catalyst **5b**.

With the optimized conditions in hand we explored the scope of the Brønsted acid catalyzed hydrocyanation of various imines (Table 8). In general, high enantioselectivities and good isolated yields of several aromatic and heteroaromatic, *N*-benzyl- and *N*-paramethoxy-benzyl protected amino nitriles, bearing either electron-withdrawing or electron-donating groups are obtained. These products are important precursors for the synthesis of amino acids and diamines. Hence, in order

Table 8 Scope of the asymmetric BINOL-phosphate catalyzed Strecker reaction

$$\underset{20}{R'\underset{H}{\overset{N^{-R}}{\diagdown}}} \xrightarrow[\text{HCN, toluene, -40 °C}]{10 \text{ mol\% catalyst } \mathbf{5b}} \underset{21}{R'\underset{CN}{\overset{HN^{-R}}{\diagdown}}}$$

5b Ar: Phenanthryl

Entry	21	R	R'	Yield [%][a]	ee [%][b]
1	a	Bn	4-CF$_3$C$_6$H$_4$	75	97
2	b	PMB	4-CF$_3$C$_6$H$_4$	53	96
3	c	Bn	3,5-(F)-C$_6$H$_3$	59	98
4	d	Bn	1-naphthyl	85	99
5	e	PMB	4-MeO-1-naphthyl	70	94
6	f	Bn	2-naphthyl	71	85
7	g	Bn	2-thienyl	77	95
8	h	Bn	2-furyl	84	89
9	i	Bn	5-CH$_3$-2-furyl	85	92
10	j	PMB	Phenyl	87	89
11	k	Bn	4-CH$_3$-C$_6$H$_4$	97	93
12	l	Bn	4-MeO-C$_6$H$_4$	55	87
13	m	Bn	4-ClC$_6$H$_4$	69	85
14	n	PMB	4-ClC$_6$H$_4$	60	86
15	o	Bn	benzo[1,3]dioxol-5-yl	88	93

[a] Isolated yields of the corresponding acetamide after chromatography
[b] Enantioselectivities were determined by HPLC analysis

to demonstrate the preparation of these compounds we used established procedures to afford the *p*-methoxyphenyl glycine (Sigman et al. 2000) and the corresponding diamine (Scheme 4; Hassan et al. 1998).

The organocatalytic hydrocyanation of imines provides direct access to a diverse range of aromatic amino nitriles and the corresponding amino acids and diamines in highest enantioselectivities. The use of

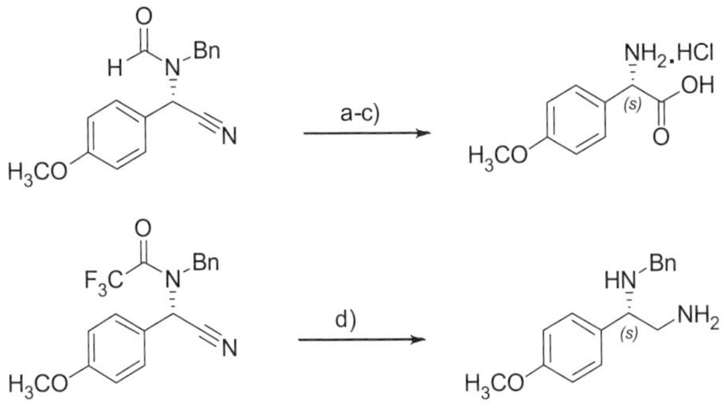

Scheme 4. Transformation of amino nitriles in amino acids and diamines: a) 65% H_2SO_4, b) HCl conc., c) H_2/Pd-C, d) $LiAlH_4$

BINOL-phosphates as efficient Brønsted acid catalysts in the enantioselective Strecker reaction shows that C-nucleophiles can be applied in the chiral ion-pair catalysis procedure. This, in turn, not only increases the diversity of possible transformations of this catalyst but also shows the great potential chiral Brønsted acids in asymmetric catalysis.

8 Asymmetric Brønsted Acid Catalyzed Imino-Azaenamine Reaction

The possibility of using C-nucleophiles in chiral ion pair catalysis encouraged us to investigate an enantioselective Brønsted acid catalyzed imino ene reaction (Rueping et al. 2007a; Scheme 5). The reaction consists of a new BINOL-phosphate catalyzed addition of methylenehydrazines **22** to *N*-Boc-protected aldimines **23** to afford chiral aminohydrazones **24**.

Hydrazones have proven to be important synthetic intermediates that can be readily derivatized to many useful chiral blocks, including amino-aldehydes, amino-nitriles, or diamines without any racemization (Scheme 6; Pareja et al. 1999; Enders et al. 1999; Enders and Schubert 1984; Diez et al. 1998, 1999; review: Job et al. 2002).

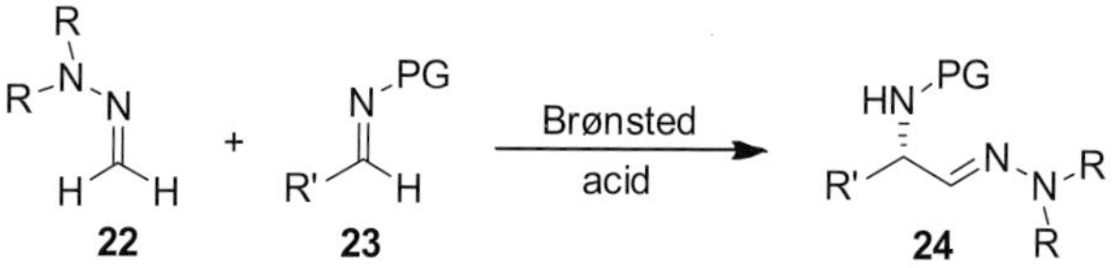

Scheme 5. Brønsted acid catalyzed imino aza-enamine reaction

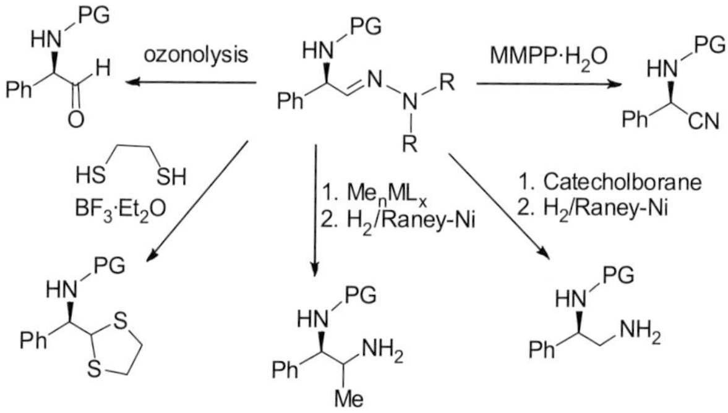

Scheme 6. Useful derivatizations of chiral amino hydrazones

Given the value of these products we decided to develop an enantioselective Brønsted acid catalyzed synthesis of amino hydrazones. Optimization of the reaction showed that *N*-Boc protected aldimines **23** in combination with the pyrrolidine-derived hydrazine **22a** gave good yields of amino-hydrazones **24**. With regard to the chiral Brønsted acid catalysts used, the use of octahydro-BINOL-phosphate **5c** resulted in the best enantioselectivities. The results are summarized in Table 9. In general, a series of *N*-Boc-protected aldimines bearing electron-withdrawing or electron-donating groups could be applied in the enantioselective aza enamine reaction resulting in the corresponding hydrazones **24a–h** in good isolated yields and with the so far highest enantioselectivities (77%–90% *ee*). The mild reaction conditions of this metal-free process, together with the operational simplicity and practicability, render this approach not only a useful procedure for the synthesis of optically active aminohydrazones but additionally, it further expands the

Table 9 Scope of the asymmetric BINOL-phosphate imino-azaenamine reaction

Entry	24	R'	Yield [%][a]	ee [%][b]
1	a	phenyl	73	77[d]
2	b	benzo[1,3]dioxol-5-yl	81	90
3	c	2-naphthyl	78	82
4	d	2-bromophenyl	82	85 (91)[c]
5	e	2-fluorophenyl	81	82
6	f	2-trifluoromethylphenyl	48	83
7	g	4-chlorophenyl	76	86
8	h	4-methoxyphenyl	71	77

[a] Isolated yields after chromatography
[b] Enantioselectivities were determined by HPLC analysis
[c] After one recrystallization from hexane-dichloromethane

repertoire of enantioselective BINOL-phosphate catalyzed transformations using C-nucleophiles.

9 Development of the First Brønsted Acid Assisted Enantioselective Brønsted Acid Catalyzed Direct Mannich Reaction

Mannich reactions represent one of the most important methods for the preparation of natural products and biologically active nitrogen-containing compounds, including β-amino acids, aldehydes and ketones (for reviews: Kobayashi and Ishitani 1999; Kleinmann 1991; Arend et al. 1998; Arend 1999; Cordova 2004). Consequently, various enantioselective variants of the Mannich reaction have been reported. However, most of the protocols focused on reactions of aldimines with preformed enolate equivalents (Fujieda et al. 1997; Ishitani et al. 1997; Kobayashi et al. 1998, 2002; Ishitani et al. 2000; Hagiwara et al. 1998; Fujii et al. 1999; Ferraris et al. 1998a,b, 1999, 2002). Hence, the development of a direct catalytic enantioselective Mannich reaction of prior unmodified carbonyl donors would be desirable, as it prevents the necessity of enolate preformation, isolation and purification (Yamasaki et al. 1999a,b; Matsunaga et al. 2003; Trost and Terrell 2003; Juhl et al. 2001; Bernardi et al. 2003; Hamashima et al. 2005).

Based on our previously developed Brønsted acid catalyzed reactions (for selected references: see Rueping et al. 2005a,b, 2006a–e, 2007a), we assumed that a direct reaction of an aromatic ketone **25** and aldimine **26** should lead to the desired β-amino ketone **27** (Fig. 9).

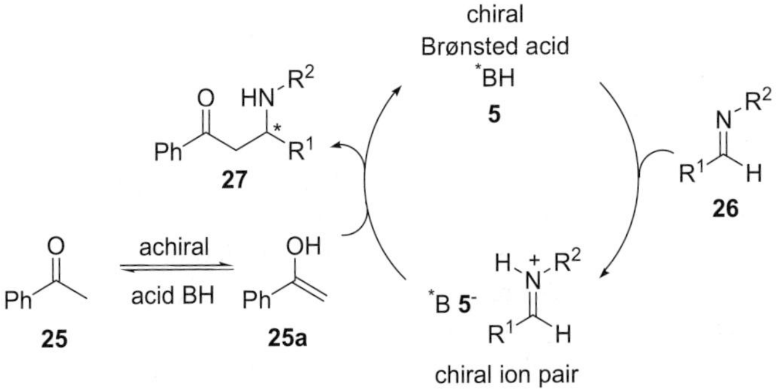

Fig. 9. Catalytic cycle for direct Brønsted acid catalyzed Mannich reaction

In the first step of this reaction a proton transfer from the chiral Brønsted acid **5** to the aldimine **26** will result in the formation of a chiral ion-pair which is now activated to react with the nucleophile **25a**. The subsequent Mannich reaction will then result in the corresponding β-amino ketone **27**. The fundamental requirement for the successful development of such a Brønsted acid assisted, asymmetric Brønsted acid catalyzed direct Mannich reaction (Rueping et al. 2007b) must be that the achiral Brønsted acid BH is *not* able to activate the aldimine **26**. Following this concept we were able to develop the first direct Mannich reaction of acetophenone and derivatives with aldimines to obtain the corresponding β-amino ketones in excellent enantioselectivities given that there is no alternative procedure which results in these products in such an efficient manner (Rueping et al. 2007b). For instance, direct Mannich reaction of **25** with **26a** resulted in the desired amino ketone **27a** with 86% *ee* (Scheme 7).

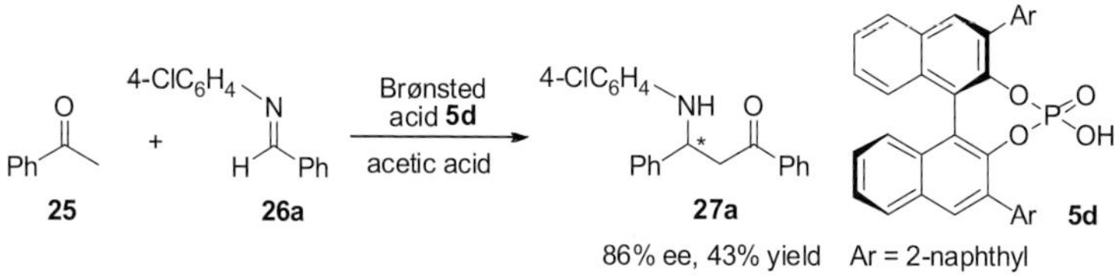

Scheme 7. Direct Brønsted acid catalyzed enantioselective Mannich reaction

A special feature of this reaction is the effective interplay of an achiral and a chiral Brønsted acid, which simultaneously—in a cooperative fashion—activate the carbonyl donor and the aldimine acceptor, thereby forming the desired enantioenriched β-amino ketones without the necessity of enolate preformation. Based on the successful application of this new concept of dual Brønsted acid catalyzed activation we decided to extend this procedure to other carbonyl donors such as cyclohexenone.

10 Cooperative Co-Catalysis: The Effective Interplay of Two Brønsted Acids in the Enantioselective Synthesis of Isoquinuclidines

Isoquinuclidines **28** (aza-bicyclo [2.2.2]octanes) consist of *N*-bicyclic structures which are the structural element of numerous natural occurring alkaloids with interesting biological properties (Sundberg and Smith 2002). Furthermore, these products can be readily converted to the biologically active pipecolic acids (Krow et al. 1982, 1999; Holmes et al. 1985). A retrosynthetic analysis shows that these isoquinuclidines **28** can be prepared from imines **29** and cyclohexenone **30** (Babu and Perumal 1998; Shi and Xu 2001; Sunden et al. 2005).

From previous work we assumed that an asymmetric Brønsted acid catalyzed reaction should enable the formation of these valuable products. Our concept, based on the direct reaction between aldimine **29** and cyclohexenone **30**, includes the simultaneous, double Brønsted acid catalyzed activation of an electrophile (by a chiral Brønsted acid *BH **5**) and a nucleophile (by an achiral Brønsted acid BH) whereby both activation processes behave co-operatively and, through effective interplay, result in the desired product (Fig. 10). However, the fundamental requirement for the successful development of such a Brønsted acid assisted, asymmetric Brønsted acid catalyzed reaction process must be that the achiral Brønsted acid BH is *not* able to activate the aldimine **29**.

Hence, the initial reactions were conducted using BINOL-phosphate **5** in combination with various achiral Brønsted acids, including protonated pyridine derivatives, alcohols and acids. Best enantioselectivities were observed with the addition of carbonic acids, phenol and hexafluoro isopropanol which provided the isoquinuclidines **28** with up to 88% *ee*. Further explorations concentrated on varying the reaction parameters including different protected imines, catalyst load-

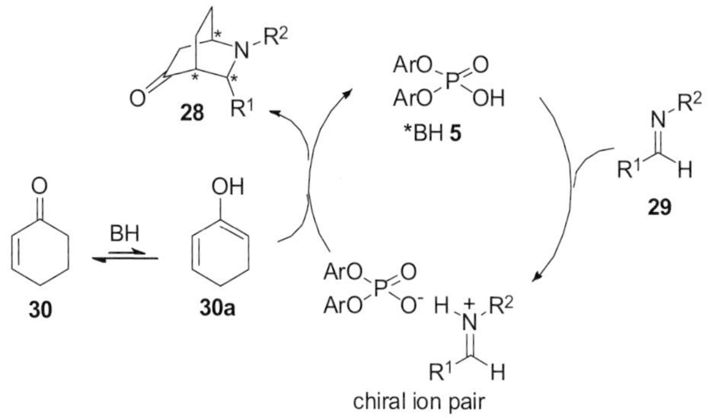

Fig. 10. Co-operative asymmetric Brønsted acid catalyzed synthesis of isoquinuclidines

ings, temperature, and concentration. From these experiments the best results were obtained when chiral BINOL-phosphates **5d or 5e** were used together with co-catalyst acetic acid in toluene. Under these optimized conditions various aldimines were applied in the double Brønsted acid catalyzed enantioselective synthesis of isoquinuclidines (Table 10). In general isoquinuclidines, with both aromatic as well as heteroaromatic residues bearing electron-withdrawing and electron-donating substituents could be isolated in good yields and with high enantiomeric ratios, whereby the *exo:endo* ratio of the products was between 1:3 and 1:9.

With regard to the reaction mechanism, we assume that our new, non-covalent, enantioselective Brønsted acid catalyzed synthesis of isoquinuclidines comprises two part reactions whereby subsequent Mannich and aza-Michael reactions are the key steps [Fig. 11; examples of such a stepwise addition–cyclization mechanism have been reported earlier for the reaction of silylenol ethers with aldimines (Birkinshaw et al. 1988; Kobayashi et al. 1995; Hermitage et al. 2004)]. In the first step, analogous to our previous reported ion pair catalysis (Rueping et al. 2005a,b, 2006a–e, 2007a), a proton transfer from the chiral Brønsted acid **5** to the aldimine occurs and the chiral ion pair **A** is formed which is now activated to react with the nucleophile. The dienol **30a**, functioning as a nucleophile, is formed from cyclohexenone **30** in the presence

Table 10 Scope of the double Brønsted acid catalyzed Mannich–Michael reaction

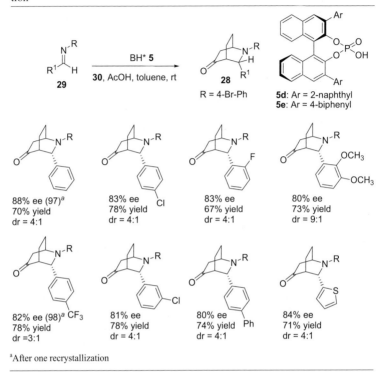

of the second achiral Bronsted acid BH, the acetic acid, *via* an accelerated adjustment of the keto-enol-equilibrium. (The rate-determining step of the reaction is presumably the dienol formation, as no considerable conversion was observed without the addition of achiral Brønsted acid.) The subsequent Mannich reaction provides the adduct **B**, an exceptionally reactive Michael acceptor, which leads directly to the corresponding isoquinuclidine **28** as well as the regenerated chiral BINOL-phosphate **5**.

The development of this so far unprecedented, double Brønsted acid catalyzed enantioselective synthesis of various aromatic and heteroaro-

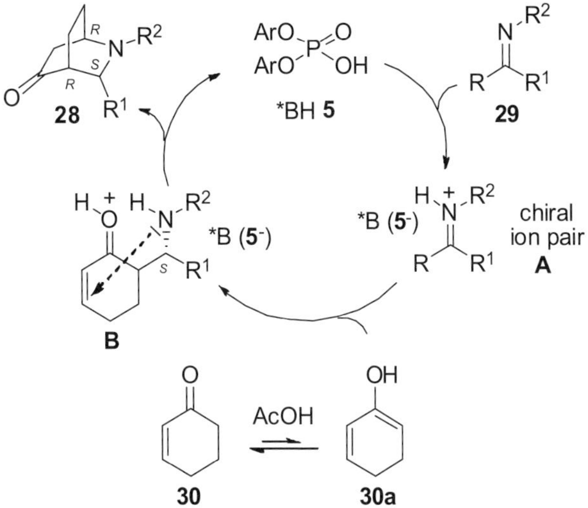

Fig. 11. Conceivable catalytic cycle of the Brønsted-acid catalyzed Mannich-aza-Michael-reaction

matic substituted isoquinuclidines demonstrates again the power of non-covalent Brønsted acid catalysis. The special feature of this Mannich-aza-Michael reaction is the effective interplay of an achiral and a chiral Brønsted acid which, simultaneously in a cooperative fashion activate the enone and aldimine. This enables an asymmetric reaction process, thereby, forming the desired isoquinuclidine products with the generation of three new stereocenters in a highly stereoselective manner (Rueping and Azap 2006).

11 Asymmetric Brønsted Acid Catalyzed Carbonyl Activation: The First Organocatalytic Electrocyclic Reaction

Within the field of chiral ion pair catalysis only aldimines and ketoimines had been activated so far. However, we have recently been successful in the activation of both the electrophile, as well as the nucleophile in a new double Brønsted acid catalyzed reaction.

The enantioselective Brønsted acid catalyzed activation of a pure carbonyl compound using a chiral BINOL-phosphate had not previously been described. Hence, we decided to develop such a reaction, a so far unprecedented Brønsted acid catalyzed enantioselective Nazarov cyclization (Rueping et al. 2007c).

The Nazarov reaction belongs to the group of electrocyclic reactions and is one of the most versatile methods for the synthesis of five membered rings which are the key structural element of numerous natural products (for reviews on the Nazarov cyclization, see: Habermas et al. 1994; Denmark 1991; Frontier and Collison 2005; Pellissier 2005; Tius 2005). Generally, the Nazarov cyclization can be catalyzed by Brønsted acids or Lewis acids. However, only a few asymmetric variations have been described, most of which require the use of large amounts of chiral metal complexes [metal-catalyzed enantioselective Nazarov reactions (Liang et al. 2003; Aggarwal and Belfield 2003); asymmetric Nazarov reactions through enantioselective protonations (Liang and Trauner 2004)].

Within the above context and building on our previous results (Rueping et al. 2005a,b, 2006a–e, 2007a; Rueping and Azap 2006), we decided to examine a metal free, BINOL-phosphate catalyzed Nazarov reaction. This would not only be the first example of a Brønsted acid catalyzed, enantioselective, electrocyclic reaction but would additionally provide a simple and direct route to optically pure cyclopentenones.

Based on our earlier work, we assumed that the catalytic protonation of a divinylketone **31** by the BINOL-phosphate **5** would result in the formation of a cyclopentadienyl cation-phosphate anion adduct **B**. Subsequent conrotatory 4π-electrocyclization would lead to oxyallyl cation **C** which, via the elimination of a proton, will form enolate **D**. Successive protonation of this enolate should then result in the formation of cyclopentenone **32** and the regenerated Brønsted acid catalyst **5** (Fig. 12).

At the outset of our experimental work we began by examining a suitable Brønsted acid catalyst for the enantioselective electrocyclization of dienone **31**. While the use of BINOL-phosphates resulted in the products **32** in good yields, better dia- and enantioselectivities were obtained with the corresponding *N*-triflyl phosphoramides **33**, which even at 0 °C gave complete conversion after 10 min. With the optimized conditions in hand, we applied various dienones to the Brønsted acid catalyzed

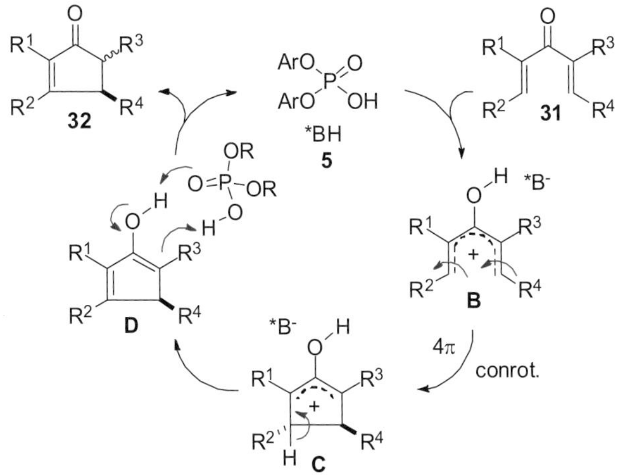

Fig. 12. Brønsted-acid catalyzed enantioselective Nazarov cyclization

enantioselective Nazarov reaction procedure (Table 11; Rueping et al. 2007c). In general it was possible to successfully transfer differently substituted dienones to the corresponding cyclopentenones **32a–h** in good yields and with excellent enantioselectivities (86%–99% *ee*). As shown in Table 11, the reaction is not only applicable to the alkyl-, aryl-substituted, but also to the dialkyl-substituted dienones.

The absolute configuration of the products was obtained from the X-ray crystal structure analysis. The *cis*-products of compound **32e** exhibits the (*S*)-configuration at both stereogenic centers. This is in agreement with a kinetic protonation through Brønsted acid **33**.

While the newly developed Brønsted acid catalyzed Nazarov reaction primarily generates the *cis*-cyclopentenones, the asymmetric metal catalyzed variations described so far often provide the *trans*-product (Liang et al. 2003; Aggarwal and Belfield 2003; Liang and Trauner 2004). To demonstrate that a route to these isomers is also possible we effectively isomerized the cyclopentenone *cis*-**32a** to the corresponding cyclopentenone *trans*-**32a** without loss of enantiomeric purity (Scheme 8).

Table 11 Scope of the first enantioselective Brønsted acid catalyzed Nazarov cyclization

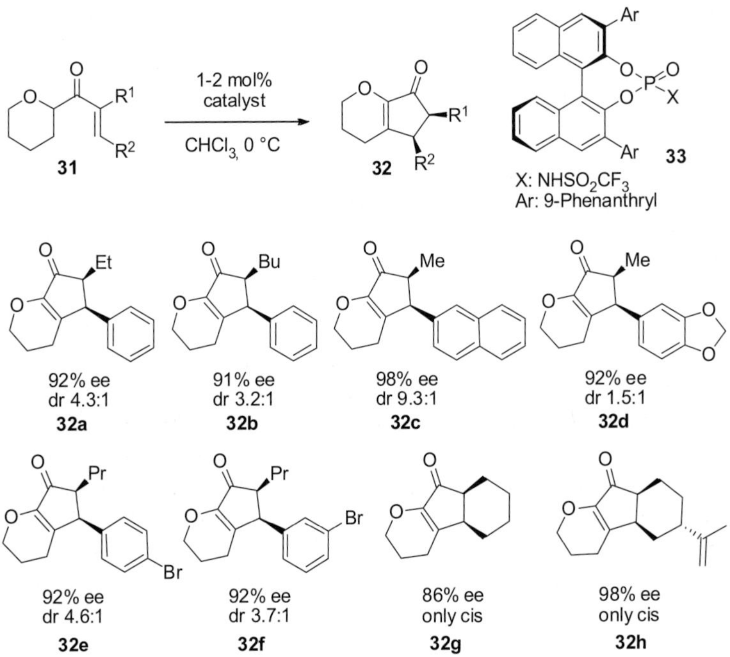
31
1-2 mol%
catalyst
CHCl3, 0 °C
32
33
X: NHSO2CF3
Ar: 9-Phenanthryl
92% ee
dr 4.3:1
32a
91% ee
dr 3.2:1
32b
98% ee
dr 9.3:1
32c
92% ee
dr 1.5:1
32d
92% ee
dr 4.6:1
32e
92% ee
dr 3.7:1
32f
86% ee
only cis
32g
98% ee
only cis
32h

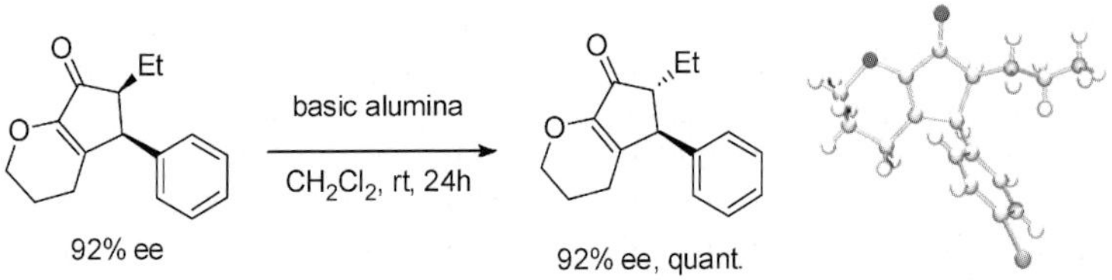
basic alumina
CH2Cl2, rt, 24h
92% ee
92% ee, quant.

Scheme 8. Isomerization without loss of enantioselectivities

We have developed the first enantioselective Brønsted acid catalyzed Nazarov reaction. This efficient method is not only the first example of an organocatalytic electrocyclic reaction but it also provides the corresponding cyclopentenones in good yields and with excellent enantioselectivities (86%–98% *ee*). The Nazarov reaction introduced represents the first BINOL-phosphate catalyzed enantio-selective carbonyl activation. Compared to the metal catalyzed reaction, special features of our new Brønsted acid catalyzed electrocyclization are the lower catalyst loadings (2 mol%), higher enantioselectivities, access to all possible stereoisomers, as well as the mild conditions and fast reaction times.

In summary, in the first 3 years of our research we have made important contributions to the field of organocatalysis. In this highly competitive and fast moving topical area of organic chemistry we have set important milestones by, for example, developing the first metal-free, highly enantioselective Brønsted catalyzed biomimetic transferhydrogenations, cascade reductions, Strecker reactions, azaenamine additions, direct Mannich reactions, pyridine reductions, as well as domino Mannich-Michael additions. The enantioselectivities observed for such transformations are impressive with most exceeding 90% enantiomeric excess. In addition to these novel chiral ion pair catalyzed transformations, we were the first to realize that chiral Brønsted acids can activate carbonyl groups which resulted in the development of the first organocatalytic electrocyclic reactions. This is not only the first example of such a method but more significantly, it opens the door for many further enantioselective carbonyl transformations.

References

Achari B, Mandal SB, Dutta PK, Chowdhury C (2004) Synlett 2004:2449
Aggarwal VK, Belfield AJ (2003) Catalytic asymmetric Nazarov reactions promoted by chiral Lewis acid complexes. Org Lett 5:5075–5078
Akiyama T, Itoh J, Fuchibe K (2006c) Adv Synth Catal 348:999
Akiyama T, Itoh J, Yokota K, Fuchibe K (2004) Enantioselective Mannich-type reaction catalyzed by a chiral Brønsted acid. Angew Chem Int Ed Engl 43:1566–1568
Akiyama T, Morita H, Itoh J, Fuchibe K (2005a) Chiral Brønsted acid catalyzed enantioselective hydrophosphonylation of imines: asymmetric synthesis of alpha-amino phosphonates. Org Lett 7:2583–2585
Akiyama T, Morita H, Fuchibe K (2006b) Chiral Brønsted acid-catalyzed inverse electron-demand aza Diels–Alder reaction. J Am Chem Soc 128: 13070–13071

Akiyama T, Saitoh Y, Morita H, Fuchibe K (2005b) Adv Synth Catal 347:1523
Akiyama T, Tamura Y, Itoh J, Morita H, Fuchibe K (2006a) Synlett 2006:141
Arend M (1999) Asymmetric catalytic aminoalkylations: new powerful methods for the enantioselective synthesis of amino acid derivatives, Mannich bases, and homoallylic amines. Angew Chem Int Ed Engl 38:2873–2874
Arend M, Westermann B, Risch N (1998) Angew Chem Int Ed Engl 37:1045
Avemaria F, Vanderheiden S, Bräse S (2003) Tetrahedron 59:6785
Babu G, Perumal PT (1998) Tetrahedron 54:1627
Bagley MC, Brace C, Dale JW, Ohnesorge M, Phillips NG, Xiong X, Bower J (2002) J Chem Soc [Perkin 1]:1663
Belattar A, Saxton JE (1992) J Chem Soc [Perkin 1]:679
Bernardi L, Gothelf AS, Hazell RG, Jørgensen KA (2003) Catalytic asymmetric Mannich reactions of glycine derivatives with imines. A new approach to optically active alpha,beta-diamino acid derivatives. J Org Chem 68:2583–2591
Birkinshaw TN, Tabor AB, Holmes AB, Raithby PR (1988) J Chem Soc Chem Commun 24:1599–1601
Blaser HU, Malan C, Pugin B, Spindler F, Steiner H, Studer M (2003) Adv Synth Catal 345:103
Bohlmann F, Rahtz D (1957) Chem Ber 90:2265
Bolm C, Rantanen T, Schiffers I, Zani L (2005) Protonated chiral catalysts: versatile tools for asymmetric synthesis. Angew Chem Int Ed Engl 44:1758–1763
Brown KS, Djerassi C (1964) J Am Chem Soc 86:2451
Byrne JJ, Chavarot M, Chavant YP, Valleé Y (2000) Tetrahedron Lett 41:873
Carpentier JF, Bette V (2002) Curr Organic Chem 6:913
Chavarot M, Byrne JJ, Chavant YP, Valleé Y (2001) Tetrahedron Asymmetry 12:1147
Chen XH, Xu XY, Liu H, Cun LF, Gong LZ (2006) Highly enantioselective organocatalytic Biginelli reaction. J Am Chem Soc 128:14802–14803
Cordova A (2004) The direct catalytic asymmetric mannich reaction. Acc Chem Res 37:102–112
Corey EJ, Grogan MJ (1999) Enantioselective synthesis of alpha-amino nitriles from N-benzhydryl imines and HCN with a chiral bicyclic guanidine as catalyst. Org Lett 1:157–160
Daly JW (1998) Thirty years of discovering arthropod alkaloids in amphibian skin. J Nat Prod 61:162–172
Daly JW, Spande TF, Garraffo HM (2005) Alkaloids from amphibian skin: a tabulation of over eight-hundred compounds. J Nat Prod 68:1556–1575
Denmark SE (1991) In: Trost BM, Flemming I (eds) Comprehensive organic synthesis, vol 5. Pergamon, Oxford, p 51

Diez E, Fernandez R, Martin-Zamora E, Pareja C, Prieto A, Lassaletta JM (1999) Tetrahedron Asymmetry 10:1145
Diez E, Lopez AM, Pareja C, Martin E, Fernandez R, Lassaletta JM (1998) Tetrahedron Lett 39:7955
Enders D, Diez E, Fernandez R, Martin-Zamora E, Munoz JM, Pappalardo RR, Lassaletta JM (1999) J Org Chem 64:6329
Enders D, Schubert H (1984) Angew Chem Int Ed Engl 23:365
Ferraris D, Young B, Cox C, Dudding T, Drury III WJ, Ryzhkov L, Taggi AE, Lectka T (2002) Catalytic, enantioselective alkylation of alpha-imino esters: the synthesis of nonnatural alpha-amino acid derivatives. J Am Chem Soc 124:67–77
Ferraris D, Young B, Cox C, Drury III WJ, Dudding T, Lectka T (1998b) Diastereo- and enantioselective alkylation of alpha-imino esters with enol silanes catalyzed by (R)-Tol-BINAP-CuClO(4).(MeCN)(2). J Org Chem 63:6090–6091
Ferraris D, Dudding T, Young B, Drury III WJ, Lectka T (1999) J Org Chem 64:2168
Ferraris D, Young B, Dudding T, Lectka T (1998a) J Am Chem Soc 120:4548
Frontier AJ, Collison C (2005) Tetrahedron 61:7577
Fujieda H, Kanai M, Kambara T, Iida A, Tomioka K (1997) J Am Chem Soc 119:2060
Fujii A, Hagiwara E, Sodeoka M (1999) J Am Chem Soc 121:5450
Glorius F (2005) Asymmetric hydrogenation of aromatic compounds. Org Biomol Chem 3:4171–4175
Glorius F, Spielkamp N, Holle S, Goddard R, Lehmann CW (2004) Efficient asymmetric hydrogenation of pyridines. Angew Chem Int Ed Engl 43: 2850–2852
Gröger H (2003) Chem Rev 103:2795
Habermas KL, Denmark SE, Jones TK (1994) Org React 45:1–158
Hagiwara E, Fujii A, Sodeoka M (1998) J Am Chem Soc 120:2474
Hamashima Y, Sasamoto N, Hotta D, Somei H, Umebayashi N, Sodeoka M (2005) Catalytic asymmetric addition of beta-ketoesters to various imines by using chiral palladium complexes. Angew Chem Int Ed Engl 44:1525–1529
Hasegawa A, Naganawa Y, Fushimi M, Ishihara K, Yamamoto H (2006) Design of Brønsted acid-assisted chiral Brønsted acid catalyst bearing a bis(triflyl)-methyl group for a Mannich-type reaction. Org Lett 8:3175–3178
Hassan NA, Bayer E, Jochims JC (1998) J Chem Soc Perkin 1:3747

Hermitage S, Howard JAK, Jay D, Pritchard RG, Probert MR, Whiting A (2004) Mechanistic studies on the formal aza-Diels–Alder reactions of N-aryl imines: evidence for the non-concertedness under Lewis-acid catalysed conditions. Org Biomol Chem 2:2451–2460

Hoffmann S, Nicoletti M, List B (2006) Catalytic asymmetric reductive amination of aldehydes via dynamic kinetic resolution. J Am Chem Soc 128: 13074–13075

Hofmann S, Seayad AM, List B (2005) Angew Chem Int Ed Engl 44:7424

Holmes AB, Thompson J, Baxter AJG, Dixon J (1985) J Chem Soc Chem Commun 1:37

Houghton PJ, Woldemariam TZ, Watanabe Y, Yates M (1999) Activity against Mycobacterium tuberculosis of alkaloid constituents of Angostura bark, Galipea officinalis. Planta Med 65:250–254

Huang J, Corey EJ (2004) A new chiral catalyst for the enantioselective Strecker synthesis of alpha-amino acids. Org Lett 6:5027–5029

Ilas J, Anderluh PS, Dolenc MS, Kikelj D (2005) Tetrahedron 61:7325

Ishitani H, Komiyama S, Kobayashi S (1998) Angew Chem Int Ed Engl 37:3186

Ishitani H, Ueno M, Kobayashi S (1997) J Am Chem Soc 119:7153

Ishitani H, Ueno M, Kobayashi S (2000) J Am Chem Soc 122:8180

Itoh J, Fuchibe K, Akiyama T (2006) Chiral Brønsted acid catalyzed enantioselective aza-Diels–Alder reaction of Brassard's diene with imines. Angew Chem Int Ed Engl 45:4796–4798

Jacquemond-Collet I, Hannedouche S, Fabre N, Fouraste I, Moulis C (1999) Phytochemistry 51:1167

Jiao Z, Feng X, Liu B, Chen F, Zhang G, Jiang Y (2003) Eur J Org Chem 2003:3818

Job A, Janeck CF, Bettray W, Peters R, Enders D (2002) Tetrahedron 58:2253

Josephsohn NS, Kuntz KW, Snapper ML, Hoveyda AH (2001) Mechanism of enantioselective Ti-catalyzed Strecker reaction: peptide-based metal complexes as bifunctional catalysts. J Am Chem Soc 123:11594–11599

Juhl K, Gathergood N, Jørgensen KA (2001) Catalytic asymmetric direct Mannich reactions of carbonyl compounds with alpha-imino esters. Angew Chem Int Ed Engl 40:2995–2997

Kadyrov R, Riermeier TH (2003) Highly enantioselective hydrogen-transfer reductive amination: catalytic asymmetric synthesis of primary amines. Angew Chem Int Ed Engl 42:5472–5474

Kang Q, Zhao ZA, You SL (2007) Highly enantioselective Friedel-Crafts reaction of indoles with imines by a chiral phosphoric acid. J Am Chem Soc 129:1484–1485

Kleemann A, Engel J, Kutscher B, Reichert D (eds) (2001) Pharmaceutical substances, 4th edn. Thieme, Stuttgart, New York

Kleinmann EF (1991) In: Trost BM, Fleming I (eds) Comprehensive organic synthesis, vol 2. Pergamon, Oxford, p 893

Knowles WS (2002) Angew Chem Int Ed Engl 41:1998

Kobayashi S, Ishitani H (1999) Catalytic enantioselective addition to imines. Chem Rev 99:1069–1094

Kobayashi S, Ishitani H (2000) Novel binuclear chiral zirconium catalysts used in enantioselective strecker reactions. Chirality 12:540–543

Kobayashi S, Ishitani H, Nagayama S (1995) Synthesis 1995:1195

Kobayashi S, Ishitani H, Ueno M (1998) J Am Chem Soc 120:431

Kobayashi S, Kobayashi J, Ishitani H, Ueno M (2002) Catalytic enantioselective addition of propionate units to imines: an efficient synthesis of anti-alpha-methyl-beta-amino acid derivatives. Chem Eur J 8:4185–4190

Krohn K, Kirst HA, Maag H (eds) (1993) Antibiotics and antiviral compounds. VCH, Weinheim

Krow GR, Johnson CA, Guare JP, Kubrak D, Henz KJ, Shaw DA, Szczepanski SW, Carey JT (1982) J Org Chem 47:5239

Krow GR, Szczepanski SW, Kim JY, Liu N, Sheikh A, Xiao Y, Yuan J (1999) J Org Chem 64:1254

Krueger CA, Kuntz KW, Dzierba CD, Wirschun WG, Gleason JD, Snapper ML, Hoveyda AH (1999) J Am Chem Soc 121:4284

Legault CY, Charette AB (2005) Catalytic asymmetric hydrogenation of N-iminopyridinium ylides: expedient approach to enantioenriched substituted piperidine derivatives. J Am Chem Soc 127:8966–8967

Lei A, Chen M, He M, Zhang X (2006) Eur J Org Chem 2006:4343

Liang G, Gradl SN, Trauner D (2003) Org Lett 5:5931

Liang G, Trauner D (2004) Enantioselective Nazarov reactions through catalytic asymmetric proton transfer. J Am Chem Soc 126:9544–9545

Liu B, Feng X, Chen F, Zhang G, Cui X, Jiang Y (2001) Synlett 2001:1551

Liu H, Cun LF, Mi AQ, Jiang YZ, Gong LZ (2006) Enantioselective direct aza hetero-Diels–Alder reaction catalyzed by chiral Brønsted acids. Org Lett 8:6023–6026

Lu SM, Han XW, G Zhou Y (2004) Adv Synth Catal 346:909

Malkov AV, Mariani A, MacDougall KN, Koèvskỳ P (2004) Role of noncovalent interactions in the enantioselective reduction of aromatic ketimines with trichlorosilane. Org Lett 6:2253–2256

Martin NJA, List B (2006) J Am Chem Soc 128:368

Masumoto S, Usuda H, Suzuki M, Kanai M, Shibasaki M (2003) Catalytic enantioselective Strecker reaction of ketoimines. J Am Chem Soc 125:5634–5635

Matsunaga S, Kumagai N, Harada S, Shibasaki M (2003) anti-Selective direct catalytic asymmetric Mannich-type reaction of hydroxyketone providing beta-amino alcohols. J Am Chem Soc 125:4712–4713

Mayer S, List B (2006) Asymmetric counteranion-directed catalysis. Angew Chem Int Ed Engl 45:4193–4195

Michael JP (2005) Indolizidine and quinolizidine alkaloids. Nat Prod Rep 22: 603–626

Nakashima D, Yamamoto H (2006) Design of chiral N-triflyl phosphoramide as a strong chiral Brønsted acid and its application to asymmetric Diels–Alder reaction. J Am Chem Soc 128:9626–9627

Nishiyama H, Itoh K (2000) In: Ojima I (ed) Catalytic asymmetric synthesis, 2nd edn. Wiley-VCH, New York, Chaps 1–2

Noyori R (1996) Acta Chem Scand 50:380

Noyori R (2002) Angew Chem Int Ed Engl 1:2008

O'Hagan D (2000) Pyrrole, pyrrolidine, pyridine, piperidine and tropane alkaloids. Nat Prod Rep 17:435–446

Ohkuma T, Noyori R (2004) In: Jacobsen EN, Pfaltz A, Yamamoto H (eds) Comprehensive asymmetric catalysis, suppl 1. Springer, Berlin Heidelberg New York, p 43

Ohkuma T, Kitamura M, Noyori R (2000) In: Ojima I (ed) Catalytic asymmetric synthesis, 2nd edn. Wiley-VCH, New York, Chap 1

Pareja C, Martin-Zamora E, Fernandez R, Lassaletta JM (1999) Stereoselective synthesis of trifluoromethylated compounds: nucleophilic addition of formaldehyde N,N-dialkylhydrazones to trifluoromethyl ketones. J Org Chem 64:8846–8854

Pellissier H (2005) Tetrahedron 61:6479

Pihko PM (2004) Activation of carbonyl compounds by double hydrogen bonding: an emerging tool in asymmetric catalysis. Angew Chem Int Ed Engl 43:2062–2064

Rakotoson JH, Fabre N, Jacquemond-Collet I, Hannedouche S, Fouraste I, Moulis C (1998) Alkaloids from Galipea officinalis. Planta Med 64:762–763

Reetz MT, Li X (2006) Asymmetric hydrogenation of quinolines catalyzed by iridium complexes of BINOL-derived diphosphonites. Chem Commun May 28:2159–2160

Riant O, Mostefai N, Courmarcel J (2004) Synthesis 2004:2943

Rowland GB, Zhang H, Rowland EB, Chennamadhavuni S, Wang Y, Antilla JC (2005) Brønsted acid-catalyzed imine amidation. J Am Chem Soc 127: 15696–15697

Rueping M, Antonchick AP (2007) Organocatalytic enantioselective reduction of pyridines. Angew Chem Int Ed Engl 46:4562–4565

Rueping M, Antonchick AP, Theissmann T (2006b) Synlett 2006:1071

Rueping M, Antonchick AP, Theissmann T (2006c) A highly enantioselective Brønsted acid catalyzed cascade reaction: organocatalytic transfer hydrogenation of quinolines and their application in the synthesis of alkaloids. Angew Chem Int Ed Engl 45:3683–3686

Rueping M, Antonchick AP, Theissmann T (2006d) Remarkably low catalyst loading in Brønsted acid catalyzed transfer hydrogenations: enantioselective reduction of benzoxazines, benzothiazines, and benzoxazinones. Angew Chem Int Ed Engl 45:6751–6755

Rueping M, Azap C (2006) Cooperative coexistence: effective interplay of two Brønsted acids in the asymmetric synthesis of isoquinuclidines. Angew Chem Int Ed Engl 45:7832–7835

Rueping M, Azap C, Sugiono E, Theissmann T (2005a) Synlett 2005:2367

Rueping M, Sugiono E, Azap C, Theissmann T, Bolte M (2005b) Enantioselective Brønsted acid catalyzed transfer hydrogenation: organocatalytic reduction of imines. Org Lett 7:3781–3783

Rueping M, Ieawsuwan W, Antonchick AP, Nachtsheim BJ (2007c) Cooperative coexistence: effective interplay of two Brønsted acids in the asymmetric synthesis of isoquinuclidines. Angew Chem Int Ed Engl 46:2097–2100

Rueping M, Sugiono E, Azap C, Theissmann T (2006a) Catalysts for fine chemical industry, vol 5. Wiley, Chichester, pp 162–176

Rueping M, Sugiono E, Azap C (2006e) A highly enantioselective Brønsted acid catalyst for the Strecker reaction. Angew Chem Int Ed Engl 45:2617

Rueping M, Sugiono E, Schoepke FR (2007b) Synlett 2007:144

Rueping M, Sugiono E, Theissmann T, Kuenkel A, Köckritz A, Pews Davtyan A, Nemati N, Beller M (2007a) An enantioselective chiral Brønsted acid catalyzed imino-azaenamine reaction. Org Lett 9:1065–1068

Satoh K, Inenaga M, Kanai K (1998) Tetrahedron Asymmetry 9:2657

Schreiner PR (2003) Metal-free organocatalysis through explicit hydrogen bonding interactions. Chem Soc Rev 32:289–296

Seayad J, Seayad AM, List B (2006) Catalytic asymmetric Pictet-Spengler reaction. J Am Chem Soc 128:1086–1087

Shi M, M Xu Y (2001) Chem Commun 2001:1876

Sigman MS, Jacobsen EN (1998a) J Am Chem Soc 120:5315

Sigman MS, Jacobsen EN (1998b) J Am Chem Soc 120:4901

Sigman MS, Vachal P, Jacobsen EN (2000) A general catalyst for the asymmetric Strecker reaction. Angew Chem Int Ed Engl 39:1279–1281

Sklenicka HM, Hsung RP, McLaughlin MJ, Wie L, Gerasyuto AI, Brennessel WB (2002) Stereoselective formal [3+3] cycloaddition approach to cis-1-azadecalins and synthesis of (–)-4a,8a-diepi-pumiliotoxin C. evidence for the first highly stereoselective 6pi-electron electrocyclic ring closures of 1-azatrienes. J Am Chem Soc 124:10435–10442

Spino C (2004) Recent developments in the catalytic asymmetric cyanation of ketimines. Angew Chem Int Ed Engl 43:1764–1766

Steinhagen H, Corey EJ (1999) Angew Chem Int Ed Engl 38:1928

Storer RI, Carrera DE, Ni Y, MacMillan DWC (2006) Enantioselective organocatalytic reductive amination. J Am Chem Soc 128:84–86

Strecker A (1850) Ann Chem Pharm 75:27

Sundberg RJ, Smith SQ (2002) The Iboga alkaloids. In: Cordell GA (ed) The alkaloids, vol 59. Academic Press, San Diego, p 281

Sunden H, Ibraheem I, Eriksson L, Cordova A (2005) Direct catalytic enantioselective aza-Diels–Alder reactions. Angew Chem Int Ed Engl 44:4877–4880

Takamura M, Hamashima Y, Usuda H, Kanai M, Shibasaki M (2000) A catalytic asymmetric Strecker-type reaction: interesting reactivity difference between TMSCN and HCN. Angew Chem Int Ed Engl 39:1650–1652

Tang W, Zhang X (2003) New chiral phosphorus ligands for enantioselective hydrogenation. Chem Rev 103:3029–3069

Terada M, Machioka K, Sorimachi K (2006b) High substrate/catalyst organocatalysis by a chiral Brønsted acid for an enantioselective aza-ene-type reaction. Angew Chem Int Ed Engl 45:2254–2257

Terada M, Sorimachi K (2007) Enantioselective friedel-crafts reaction of electron-rich alkenes catalyzed by chiral Brønsted acid. J Am Chem Soc 129:292–293

Terada M, Sorimachi K, Uraguchi D (2006a) Synlett 2006:13

Tius MA (2005) Eur J Org Chem 2005:2193

Trost BM, Terrell LR (2003) A direct catalytic asymmetric mannich-type reaction to syn-amino alcohols. J Am Chem Soc 125:338–339

Tsogoeva SB, Yalalov DA, Hateley MJ, Weckbecker C, Huthmacher K (2005) Eur J Org Chem 2005:4995

Uraguchi D, Sorimachi K, Terada M (2004) Organocatalytic asymmetric aza-Friedel-Crafts alkylation of furan. J Am Chem Soc 126:11804–11805

Uraguchi D, Terada M (2004) Chiral Brønsted acid-catalyzed direct Mannich reactions via electrophilic activation. J Am Chem Soc 126:5356–5357

Vachal P, Jacobsen EN (2000) Enantioselective catalytic addition of HCN to ketoimines. Catalytic synthesis of quaternary amino acids. Org Lett 2:867–870

Vachal P, Jacobsen EN (2002) Structure-based analysis and optimization of a highly enantioselective catalyst for the strecker reaction. J Am Chem Soc 124:10012–10014

Wang WB, Lu SM, Yang PY, Han XW, Zhou YG (2003) Highly enantioselective iridium-catalyzed hydrogenation of heteroaromatic compounds, quinolines. J Am Chem Soc 125:10536–10537

Wenzel AG, Lalonde MP, Jacobsen EN (2003) Synlett 2003:1919

Xu L, Lam KH, Ji J JWu, H Fan Q, H Lo W, Chan ASC (2005) Air-stable Ir-(P-Phos) complex for highly enantioselective hydrogenation of quinolines and their immobilization in poly(ethylene glycol) dimethyl ether (DMPEG). Chem Commun (Camb) 2005:1390–1392

Yamasaki S, Iida T, Shibasaki M (1999a) Tetrahedron Lett 40:307

Yamasaki S, Iida T, Shibasaki M (1999b) Tetrahedron 55:8857

Yang PY, Zhou YG (2004) Tetrahedron Asymmetry 15:1145

Yet L (2001) Recent developments in catalytic asymmetric Strecker-type reactions. Angew Chem Int Ed Engl 40:875–877

Zhou YG, Yang PY, Han XW (2005) Synthesis and highly enantioselective hydrogenation of exocyclic enamides: (Z)-3-arylidene-4-acetyl-3,4-dihydro-2H-1,4-benzoxazines. J Org Chem 70:1679–1683

Ernst Schering Foundation Symposium Proceedings, Vol. 2, pp. 255–279
DOI 10.1007/2789_2007_065
© Springer-Verlag Berlin Heidelberg
Published Online: 18 December 2007

Chiral Organocatalysts for Enantioselective Photochemical Reactions

S. Breitenlechner, P. Selig, T. Bach[✉]

Lehrstuhl für Organische Chemie I, Technische Universität München, Lichtenbergstr. 4, 85747 Garching, Germany
email: *thorsten.bach@ch.tum.de*

1	Introduction	256
2	Stoichiometric Template-Based Enantioselectivity	260
2.1	Properties and Synthesis of the Template	260
2.2	[2+2]-Photocycloaddition Reactions	262
2.3	Other Cycloaddition Reactions	262
2.4	Photocyclisation Reactions	264
3	Radical Reactions	265
3.1	Enantioselective Hydrogen Abstraction	266
3.2	Cyclisation Reactions	267
3.3	Norrish–Yang Cyclisation	269
4	Sensitised Reactions	271
4.1	Possible Templates and Substrates	271
4.2	PET-Catalysed Reaction	273
5	Summary	276
References		276

Abstract. In this account the development of chiral organocatalysts for enantioselective photochemical reactions is described from a personal perspective. The need for enantioselective photochemical reactions is immediately linked to the challenge of establishing the first stereogenic centre in a molecule by organic photochemistry. Hydrogen bonds were identified as ideal means to embed a prochiral substrate into a chiral environment. The combination of a binding site with a suitable chiral architecture led to the generally applicable chiral template **12**, which has been amply employed in photochemical reactions. Examples of its use are discussed in this report. It was shown that the general working mode of such a template can also be applied to radical reactions, irrespective of whether the radical chain process is initiated photolytically or chemically. While the robustness of template **12** allows for its almost quantitative recovery and its reuse, it has so far been used in a catalytic sense only in a single case. For the development of photochemical organocatalysts it was proposed to prepare modified templates, in which a light-collecting device acts simultaneously as sensitiser and as a steric shield. A prototype **40** for such a template was devised and it was shown that it can indeed induce catalytic turnover in a photoinduced electron transfer reaction.

1 Introduction

The synthesis of chiral compounds in enantiomerically pure form is one of the most important and one of the most challenging goals of modern organic synthesis. Any key step of synthetic importance, which leads to the creation of a stereogenic centre, requires to be studied with regard to its stereoselectivity. Impressive progress has been made along these lines in the last decades. The number of catalytic enantioselective reactions increases continuously (Ojima 2000; Berkessel and Gröger 2005).

In our group, organic photochemistry has been used frequently for the synthesis of chiral natural products and other biologically active compounds. A typical example is the synthesis of (+)-preussin (Basler et al. 2005), the photochemical key step of which is depicted below (Scheme 1). Starting from L-pyroglutamic acid (**1**), dihydropyrrole **2** was synthesised in eight steps, the attachment of the alkyl side chain by nucleophilic substitution with a cuprate being the decisive C–C bond formation. The subsequent photocycloaddition served to introduce the stereogenic centers at C-2 and C-3 in a diastereoselective fashion (Bach and Brummerhop 1998; Bach et al. 2000a). Oxetane **3** was hydrogenolytically cleaved and subsequently the *N*-methoxycarbonyl group was reduced to produce the desired target, (+)-preussin (**4**), which was later shown to act as cyclin-dependent kinase inhibitor (Achenbach et al. 2000).

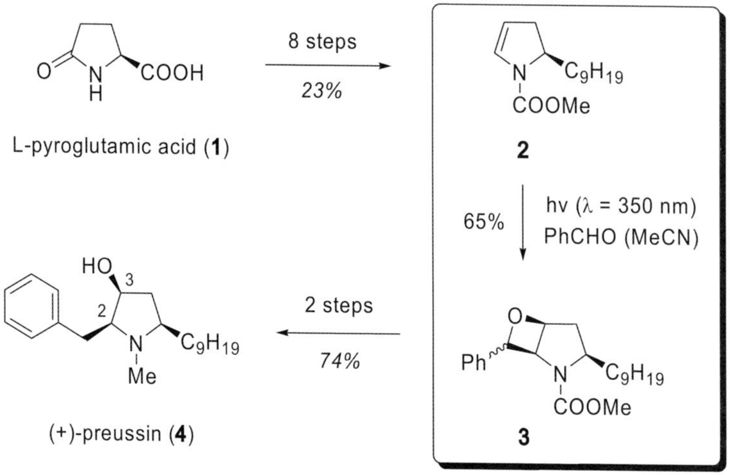

Enantioselective Photochemistry

Scheme 1.

The synthetic scheme outlined above reveals the typical approach to a chiral target molecule using organic photochemistry. It is not the *first* stereogenic centre, which is introduced photochemically, but rather an existent stereogenic centre (e.g. from natural sources) that is used to control the facial diastereoselectivity of a reaction. In the specific case, the first stereogenic centre remained part of the target molecule (substrate-induced diastereoselectivity), in many other cases the initial stereochemical information was removed (auxiliary-induced diastereoselectivity) or even destroyed. The fact that auxiliary-induced methods are still common practice in photochemistry is linked to the fact that direct enantioselective photochemical reactions have not been available until recently. Photochemistry has lagged (and still lags) behind the development in 'conventional' thermal chemistry. Although the key question of enantioselectivity is easy to ask—how enantiotopic groups or faces can be differentiated—it is more difficult to answer in photochemistry than in conventional chemistry. Scheme 2 shows a prochiral substrate represented as triangle **A** and its two possible tetrahedral product enantiomers **B** and *ent*-**B**. Below this cartoon, a typical photochemical reaction is depicted. The intramolecular [2+2]-photocyclo-

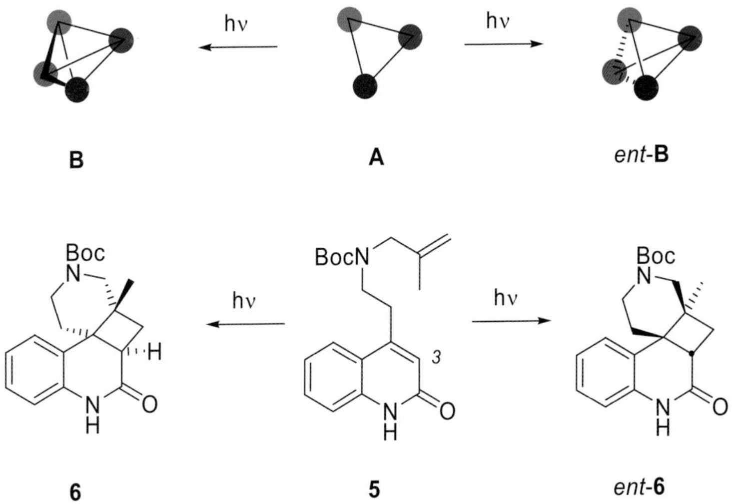

Scheme 2.

addition reaction of the prochiral quinolone **5** leads with high simple diastereoselectivity to the products **6** and *ent*-**6** (Brandes et al. 2004). According to conventional wisdom, one would use a chiral reagent or a chiral catalyst to conduct the very same reaction enantioselectively. This approach, however, is not viable in photochemistry despite the fact that the reagent 'light' is chiral, if used in circularily polarised form. The ability of circularly polarised light (CPL) to induce enantioselectivity has remained low despite intensive efforts (Rau 2004). The other solution to the quest for enantioselectivity is to allow a face differentiation—for **5** a differentiation of the *re* and *si* face relative to carbon atom C-3—by a second passive reagent, commonly called a chiral template or chiral complexing agent (Grosch and Bach 2004).

Taking into account the basic considerations outlined in the previous section, we reasoned that a chiral complexing agent, which needs to be applicable in photochemistry, requires a general binding motif. In addition, it appeared necessary to employ directed interactions in such a template rather than an unspecific complexation as encountered in chiral solvents or larger supramolecular entities. Hydrogen bonds appealed

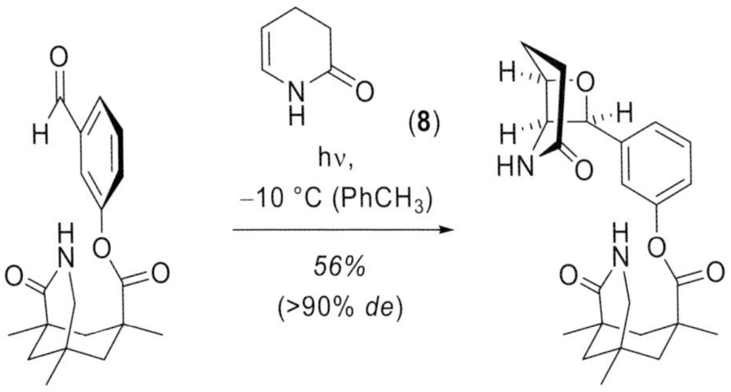

Enantioselective Photochemistry

Scheme 3.

to us as control elements in a putative template because the compatibility of photochemical conditions with Brønsted acidic or basic sites has been frequently observed. Indeed, photochemistry can be conducted—and this in an advantage to many conventional reactions—without the use of protecting groups.

In a preliminary set of experiments we studied whether hydrogen bonds are of sufficient strength to act as control elements in photochemical reactions. To this end, chiral aldehyde **7** was irradiated at $\lambda \geq 300$ nm with dihydropyridone **8** (Bach et al. 1999). The reaction was expected to occur with high diastereoselectivity if the pyridone was pre-coordinated to the substrate by hydrogen bonds. Indeed, this proved to be the case. Oxetane **9** was isolated as a single diastereoisomer with intramolecular hydrogen bonds still present as evidenced by an X-ray crystallographic study (Scheme 3). Further indications for the hydrogen bonds being the decisive diastereocontrol elements were the observations that the reaction proceeded unselectively in polar solvents, and that the reaction of *N*-alkylated amides remained unselective (Bach et al. 2001a).

As a straightforward extension of the promising results obtained in diastereoselective reactions, we sought to employ the lactam binding motif of compound **7** also for enantioselective and possibly even for catalytic enantioselective transformations.

2 Stoichiometric Template-Based Enantioselectivity

The diastereoselective reaction using aldehyde **7** clearly showed that the lactam motif was well capable of sterically fixing a second, prochiral lactam in a defined fashion closely to the nearby benzaldehyde moiety. Replacing this photochemically active backbone of **7** by a photochemically inactive, yet bulky shield should enable the complexation of a prochiral lactam substrate in a chemically unreactive though sterically well-defined chiral environment, and lead to the efficient face differentiation between the two chiral planes of the substrate by simple sterical shielding. In other words, such a complexing agent could reasonably be seen as a non-covalently attached chiral auxiliary binding the substrate in a sterically defined fashion by hydrogen bonds.

2.1 Properties and Synthesis of the Template

The synthesis of the as yet most successful, photochemically inactive template is shown in Scheme 4. Starting from the known precursor **10** (Stack et al. 1992) derived from Kemp's triacid, the introduction of a bulky, photochemically unreactive shield into the backbone was achieved by condensation of the acid chloride with a suited aminoalcohol to give the oxazole **11**. Selective reduction of imide **11** using NaBH$_4$ and Et$_3$SiH gave the corresponding lactam *rac*-**12** as a racemic mixture of enantiomers. These could be easily separated after conversion to their *N*-menthyloxycarbonyl derivatives by simple flash-chromatography on silica gel. *N*-defunctionalisation with trifluoroacetic acid finally regenerated both enantiomers of (+)-**12** and (–)-**12** in pure form (Bach et al. 2001b).

Complexing agents (+)-**12** and (–)-**12** proved to be excellently suited for the purpose of enantioselective photochemical reactions in solution. As expected from our experience with compound **7**, the lactam motif enables a strong and sterically defined complexation of other substrate amides and lactams. In addition, the tetrahydronaphthalene shield in the backbone is bulky enough to provide an efficient shielding of one enantiotopic face of the substrate while being photochemically unreactive at wavelengths >300 nm and not interfering with the hydrogen bonding of the substrate. Similar templates with a smaller, bicyclic shield or an

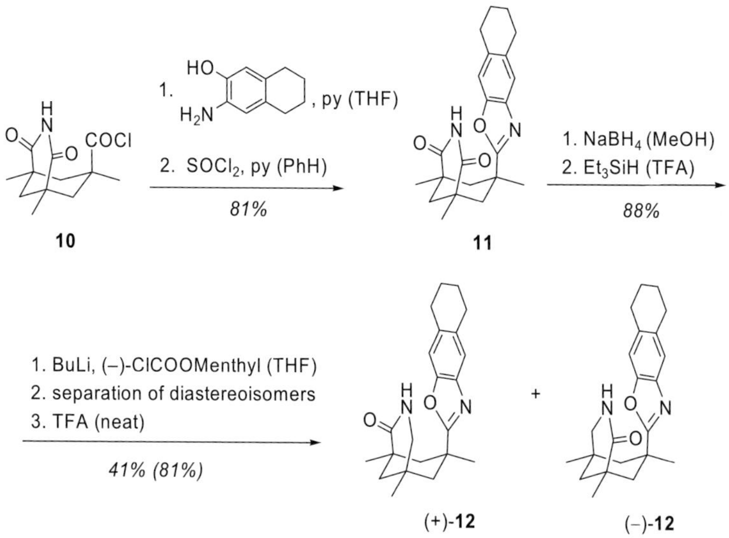

Scheme 4.

aromatic naphthyl shield either provided no acceptable differentiation of the prochiral planes or turned out to be unstable under the irradiation conditions due to unfavourable long wavelength absorptions.

With the face differentiation provided by the templates (+)-**12** and (−)-**12** being virtually complete, the remaining decisive factor for the extent of chirality transfer is the amount of substrate actually being bound to the template when the reaction takes place. While (+)-**12** or (−)-**12** cannot form hydrogen bond-mediated dimers with themselves, if they are enantiomerically pure, dimerisation of the substrate (K_{dim}) is always to be considered as a competing process to host–guest complexation (K_{ass}). To favour hydrogen-bond mediated complexations, reactions using templates (+)-**12** or (−)-**12** are generally performed in non-polar solvents (e.g. toluene) at low temperature (e.g. −60 °C) with greater than a one-fold excess of the template. While the first two conditions are beneficial for hydrogen bonding in general, the excess of template serves to ensure maximum substrate complexation in contrast to substrate dimerisation. However, even in cases where substrate dimerisation is considerably stronger than complexation to the template (i.e. $K_{dim} > K_{ass}$), complexation is *always* favoured over dimerisation enthalpically. High *ee* values could thus be achieved even in reactions using substrates with unfavourable dimerisation behaviour (Selig and Bach 2006). Although the use of an excess of template may seem uneconomical at first, it is important to note that templates (+)-**12** and (–)-**12**, being photochemically unreactive themselves, can be recovered from the reaction mixture generally in yields between 80% and 99% and used repeatedly.

2.2 [2+2]-Photocycloaddition Reactions

As the synthetically most useful and most frequently used photochemical reactions known, [2+2]-photocycloadditions were conducted enantioselectively in the presence of templates (+)-**12** and (–)-**12**. 2(1*H*)-Quinolones proved to be excellent substrates for this reaction, as they possess a lactam motif for binding to the template and are well known for their excellent suitability for both intra- and intermolecular [2+2]-photocycloaddition reactions. Initially, 4-alkoxyquinolones were used in both intramolecular (Bach et al. 2000b) and intermolecular reactions (Bach and Bergmann 2000), giving enantiomeric excesses between 80% and 98% *ee* (Bach et al. 2002a). More recently, templates (+)-**12** and (–)-**12** proved to be of general applicability also for structurally more complex quinolones such as **5** or **13**. In all cases, the template had no observable effect on yields and diastereomeric ratios when compared to the corresponding racemic reactions. For example, the reaction depicted in Scheme 2 provided a 78% yield of **6** with 93% *ee* when conducted in the presence of 2.3 equivalents of (+)-**12** (Brandes et al. 2004). A more complex example of an intermolecular [2+2]-photocycloaddition using the cyclic terpene tulipaline (**14**) resulted in the formation of a spirocyclic cyclobutane **15**, which was further converted into the tetracyclic lactam **16** (Scheme 5) (Selig and Bach 2006).

2.3 Other Cycloaddition Reactions

While the chiral complexing agents (+)-**12** and (–)-**12** proved to be generally suitable for a wide range of enantioselective [2+2]-photocycloaddition reactions on the *c*-bond of 2(1*H*)-quinolones, their applicabil-

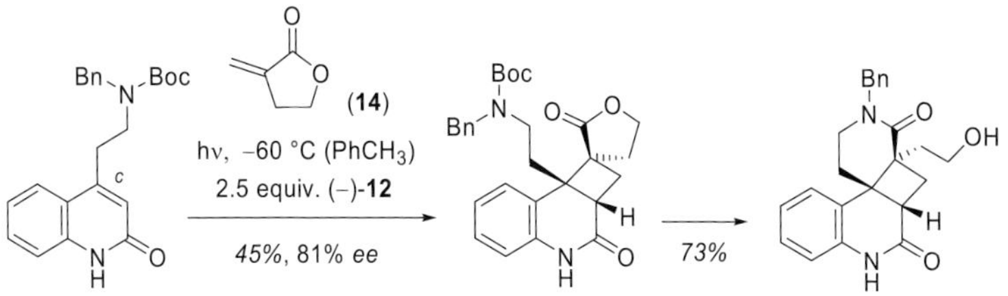

13 15 16

Scheme 5.

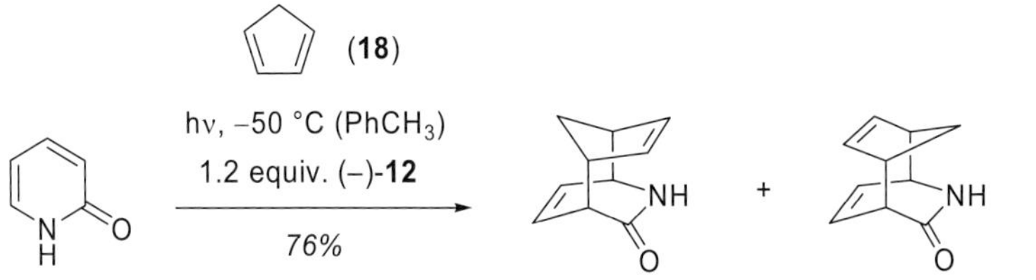

17 d.r.*(exo/endo)* = 3/2 *exo*-**19** 84% *ee* *endo*-**19** 87% *ee*

Scheme 6.

ity is by no means restricted to these reactions. Other photochemically induced cycloaddition reactions successfully performed enantioselectively include, for example, the [4+4]-photocycloaddition of pyridone (**17**) and cyclopentadiene (**18**) (Scheme 6) to give the diastereomeric products *exo*-**19** and *endo*-**19** (Bach et al. 2001c).

The Diels-Alder [4+2]-cycloaddition reaction of the photochemically generated *ortho*-quinodimethane from substrate **20** and acrylonitrile resulted in tricyclic product **21** (Scheme 7) (Grosch et al. 2003; Grosch et al. 2004). In an analogous fashion the reactive diene could be trapped by methyl acrylate or dimethyl fumarate. It was shown that the association constant of the corresponding products to the template was much lower than that of the substrates, an observation that is in line with an increasing *ee* upon increasing reaction time. This fact was also responsible for high enantioselectivities even at higher irradiation temperature. The pressure dependence of the reactions was studied and it was found that despite an increased association the enantioselectivity of the re-

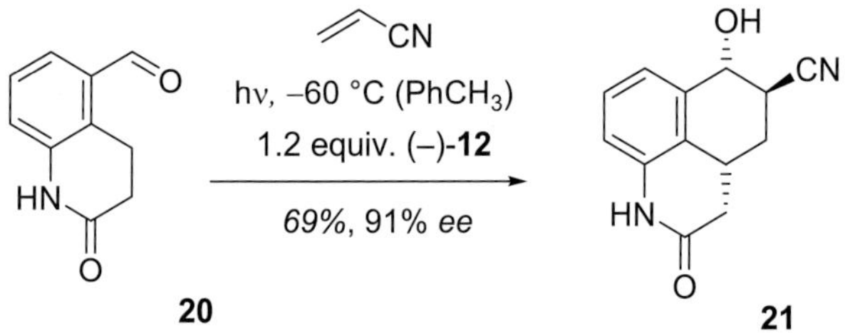

Scheme 7.

action decreased with increasing pressure. At 25 °C the enantiomeric excess for the enantioselective reaction **20**→**21** went down from 68% *ee* at 0.1 MPa to 58% *ee* at 350 MPa. This surprising behaviour was explained by different activation volumes for the diastereomeric transition states leading to **21** and its enantiomer.

As illustrated by the two examples above, the use of a 2.5-fold excess of complexing agent established for the [2+2]-photocycloaddition reactions is not always necessary to achieve high enantiomeric excesses of the products. As the chirality transfer is limited only by the amount of host–guest complexation, suitable—strongly binding—substrates can result in product *ee*′s of >80% *ee* even when using as little as 1.2 equivalents of the template.

2.4 Photocyclisation Reactions

Further types of photochemical reactions suitable for the induction of enantioselectivity by chiral templates (+)-**12** and (−)-**12** are 4π- and 6π-electrocyclisation reactions (Scheme 8) (Bach et al. 2001c, 2003).

As shown below, the enantiomeric excess achieved in the 6π-cyclisation reaction of amide **22** to the tetrahydrophenanthridinone **23** remained below 60% *ee*. While this is still an impressive value for a photochemical reaction in solution, the observed enantiomeric excess is clearly inferior to the values achieved in the different photochemically induced cycloaddition reactions presented in Sect. 2.2 and Sect. 2.3. As pointed out previously, complete chirality transfer from the template to the substrate is only possible if a complete complexation of the sub-

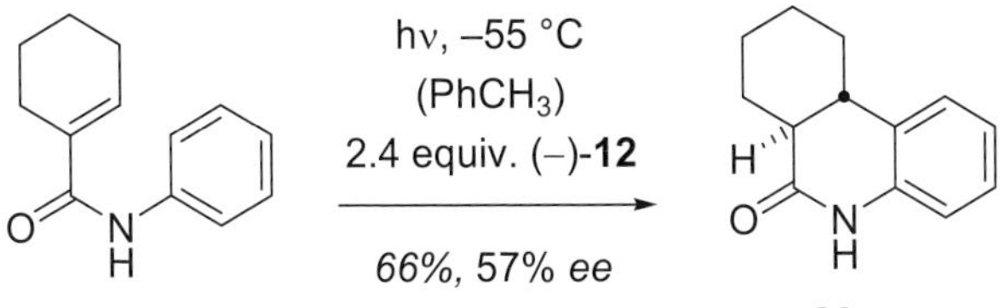

Scheme 8.

$$22 \xrightarrow[\text{66\%, 57\% ee}]{\substack{h\nu,\ -55\ °C \\ (PhCH_3) \\ 2.4\ \text{equiv.}\ (-)\text{-}12}} 23$$

strate is achieved. In contrast to all examples shown earlier, substrate **22** does not incorporate its hydrogen bonding amide functionality into a (lactam) ring. The additional conformational flexibility of the open chain amide **22** obviously represents a major hindrance to hydrogen-bond mediated complexation to the template. On the other hand, sterically constrained cyclic amides, that is, lactams, as used for almost all kinds of photochemical cycloaddition reactions do not impair hydrogen bonding to the template, thus allowing high ratios of chirality transfers with only a moderate excess of complexing agent (see Sect. 2.3). Consequently, lactams were the substrates of choice for the extension of the methodology of template-induced enantioselectivity from photochemical to radical reactions.

3 Radical Reactions

Radical reactions have been recognised only recently for the construction of enantiomerically pure compounds (Renaud and Sibi 2001; Zimmerman and Sibi 2006). In addition to substrate- or auxiliary-induced diastereoselective radical reactions, and in addition to the use of chiral Lewis acids, chiral hydrogen atom donors or chiral transition metal complexes, template molecules can be used to generate a chiral environment and induce chirality to the substrate. With the chiral complexing agent **12**, enantioselective radical reactions were achieved with enantiomeric excesses up to 99% *ee*.

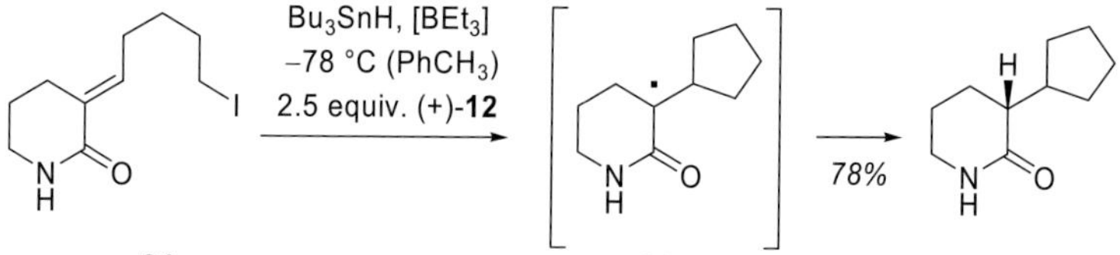

Scheme 9.

3.1 Enantioselective Hydrogen Abstraction

A first example shows the enantioselective reductive radical cyclisation reaction of 3-(5′-iodopentylidene)-piperidin-2-one (**24**) (Scheme 9). After the primary cyclisation step the hydrogen abstraction leads to the formation of a stereogenic centre. The complexing agent (+)-**12** was used as source of chirality (Aechtner et al. 2004; Dressel et al. 2006).

The radical precursor **24** was synthesised in a five-step procedure starting from commercially available pentane-1,5-diol with an overall yield of 15%. The radical reaction conditions were optimised for the synthesis of racemic product (up to 87% yield) and then adapted to the enantioselective reaction. In these studies triethyl borane was the ideal choice for the initiation of the radical reaction at low temperature. Toluene as a nonpolar solvent was found to be best suited to afford a high association between the complexing agent (+)-**12** and the substrate **24**. An excess of 2.5 equivalents of (+)-**12** was used in all experiments. The recovery rate of the chiral template was shown to be excellent and consistently exceeded 90%. Even at room temperature an enantiomerically enriched product with 38% *ee* could be obtained, and at lower temperatures the enantioselectivity could be increased even further. At –10 °C and –78 °C an enantiomeric excess of 40% *ee* and 84% *ee* was achieved. The amount of triethyl borane for initiation had a large effect on the product formation. Triethyl borane increased the polarity of the solution and was troublesome with regard to high enantioselectivity, but decreasing its load resulted in exceedingly long reaction times. Hence, UV initiation was tested as a means for radical initiation but had no beneficial effect on the formation of **26**. In all experiments the *exo*-radical cyclisation product was formed exclusively. This regioselectivity can be explained by the excellent overlap of the interacting π-orbitals of the terminal radical and the *exo* carbon atom of the olefin in the transition state, and the stabilisation of the newly formed radical by the carbonyl group in α-position. The hydrogen abstraction step is crucial for inducing enantioselectivity. The face differentiation occurs supposedly in the complex of the intermediate radical **25** and the template (+)-**12**. A job plot ^{1}H NMR analysis confirmed the assumed 1:1 stoichiometry between complexing agent (+)-**12** and substrate **24**.

3.2 Cyclisation Reactions

Another substrate class for reductive radical cyclisation reactions, which was studied in our laboratories, are 4-(4′-iodoalkyl)quinolones (e.g. **27**; Scheme 10). High enantioselectivities could be achieved even at 0 °C (up to 99% *ee*) or at ambient temperature (up to 96% *ee*). Furthermore, an unexpected chirality multiplication was observed with low catalyst loadings (Dressel and Bach 2006). 4-(4′-Iodobutyl)quinolone (**27**) can be synthesised from 4-methylquinolone in three steps by alkylation with 3-*tert*-butyldimethylsilyl(TBDMS)oxy-1-iodopropane, deprotection of the alcohol and iodo-dehydroxylation in an overall yield of 31%.

Radical reactions of substrate **27** were initially conducted at low temperature in toluene with triethyl borane as initiator and tributylstannane as reducing agent. Under these conditions no conversion was detected. Increasing the temperature to ambient temperature led to 99% yield and a diastereomeric ratio (d.r.) of 47/53 in favour of the *cis* compound *cis*-**29**. Reactions in the presence of the chiral complexing agent

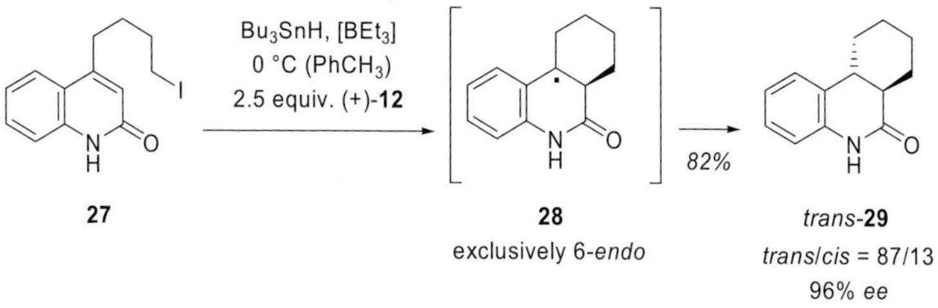

Scheme 10.

(+)-**12** (2.5 equivalents) resulted in high *ee* values both at 25 °C (d.r. = 63/37) and at 0 °C (d.r. = 87/13) for the predominant *trans*-diastereoisomer *trans*-**29** (80% *ee* and 96% *ee*). Changing the solvent to trifluorotoluene at 0 °C increased the enantioselectivity to 99% *ee* while the diastereomeric ratio remained unchanged (d.r. = 88/12). The chiral complexing agent (+)-**12** could be recovered in yields over 90% in all cases. By reducing the amount of chiral complexing agent at 0 °C to catalytic amounts, a chirality multiplication could be detected. With 0.1 equivalents of (+)-**12** a chirality turnover could be achieved resulting in an *ee* of 55%. The diastereomeric ratio dropped to an almost 1:1 mixture, the reaction mixture being heterogeneous throughout the course of the reaction. We assume that the chiral complexing agent (+)-**12** can dissolve the substrate and that the radical reaction proceeds under homogeneous conditions. This allows faster reaction rates and the substrates can pass through more than one catalytic cycle until full conversion is achieved.

The model in Scheme 11 explains the regioselectivity, enantioselectivity and diastereoselectivity of the reaction. In the radical cyclisation step the *endo* radical **28** is formed exclusively due to the high stability of the resulting benzyl radical. The approach of the alkyl radical in the complex **27**·(+)-**12** occurs from the sterically unhindered *re* face whereas the attack of the radical from the *si* face is blocked by the tetrahydronaphthalene shield. For the same reason the hydrogen abstraction step in the complex **28**·(+)-**12** takes place at the same face to form product *trans*-**29** predominantly with high enantioselectivity.

Introducing two geminal methyl groups into the butyl side chain changed the regioselectivity of the reaction (Scheme 12). Without chiral complexing agent the regioisomeric ratio **31/32** was 65/35, whereas in the presence of template (+)-**12** the *exo*-product **31** was the exclusive product. The *exo*-regioselectivity can be explained by enhanced stereoelectronic factors, which favour the chair-type transition state of the *exo*-cyclisation. The increase of regioselectivity in the presence of the chiral complexing agent is not fully understood, but it is shown by models that the interaction between 1′-H of the alkyl chain of substrate **30** and the tetrahydronaphthalene of the chiral complexing agent (+)-**12** is higher in the transition state of the *endo*-cyclisation.

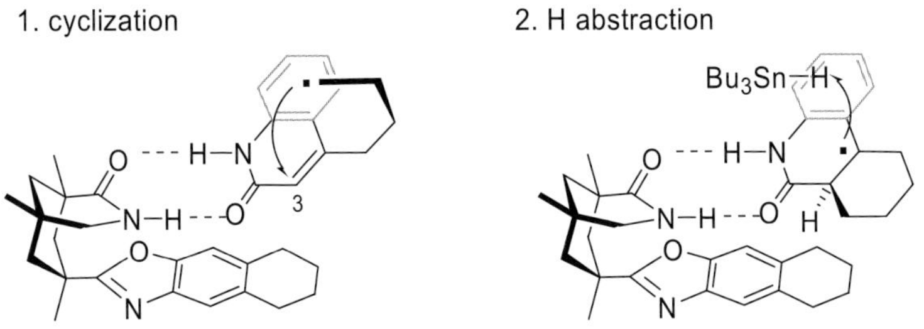

1. cyclization 2. H abstraction

27·(+)-12   **28·(+)-12**

Scheme 11.

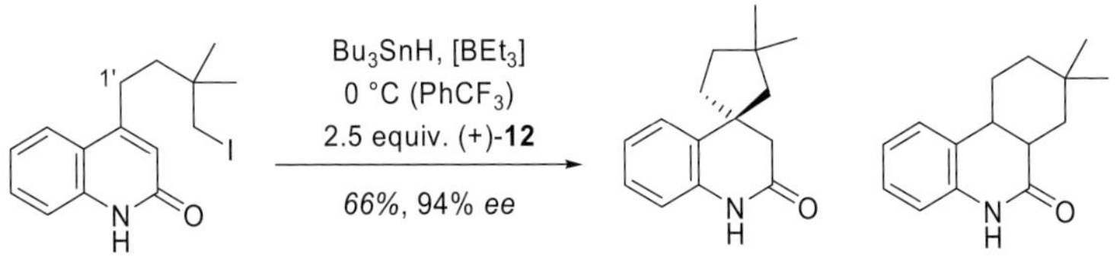

30 → **31** (66%, 94% ee) **32**

Bu$_3$SnH, [BEt$_3$]
0 °C (PhCF$_3$)
2.5 equiv. (+)-**12**

Scheme 12.

3.3 Norrish–Yang Cyclisation

Upon irradiation of carbonyl compounds the photoexcited intermediates can abstract hydrogen atoms either inter- or intramolecularly. The newly formed C–C bond results from the recombination of the two generated radicals.

In the case of an intramolecular reaction the 1,n-biradical can undergo a Norrish–Yang cyclisation reaction to build up an n-membered ring (Yang and Yang 1958). For 1,4-biradicals the Norrish type II cleavage reaction is detected as a side reaction. In Scheme 13 the enantioselective Norrish–Yang Cyclisation of *N*-(3-oxo-3-phenylpropyl)imidazolidin-2-one (**33**) is shown (Bach et al. 2001d, 2002b). The precursor for this reaction can easily be synthesised from readily available monoacetylated imidazolidinone followed by *N*-alkylation with 3-bromopropiophenone and subsequent hydrolysis of the *N*-acetyl protection group to yield imidazolidinone **33**. Upon irradiation at a wavelength of

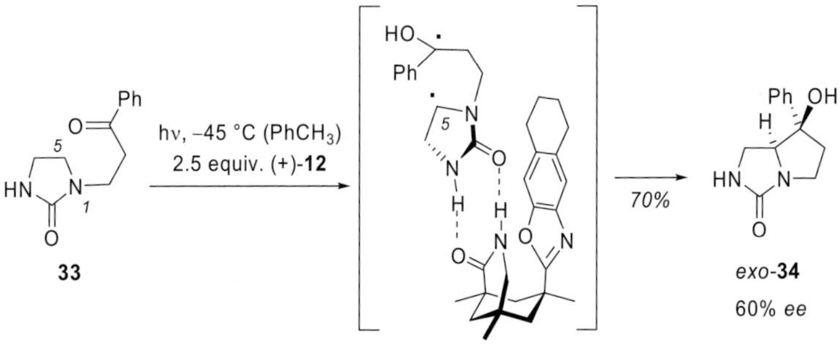

Scheme 13.

$\lambda \geq 300$ nm the carbonyl group of substrate **33** is excited followed by δ-hydrogen abstraction from the imidazolidinone ring (Scheme 13). After radical recombination, two new stereogenic centres are formed giving rise to four possible stereoisomeric bicyclic products. In the presence of chiral template (+)-**12**, stereoisomer **34** was predominantly formed.

The diastereoselectivity arises from the side differentiation of the prostereogenic hydroxybenzyl radical. In toluene, the thermodynamically more stable *exo*-product is mainly formed [d.r.(*exo/endo*) = 88/12] whereas in ^{t}BuOH the *endo*-product is favoured [d.r.(*exo/endo*) = 39/61]. The change in diastereoselectivity can be explained by the increased bulk of the hydroxyl group due to solvent association in ^{t}BuOH. In the presence of the chiral complexing agent (+)-**12** an enantioselective reaction with up to 60% *ee* was achieved. Substrate **33** binds to (+)-**12** with two hydrogen bonds. In this complex one side of the imidazolidinone is—as discussed previously for other substrates—sterically blocked by the tetrahydronaphtalene shield of (+)-**12**. The attack of the hydroxybenzyl radical can therefore occur only from the unhindered *re* face. To achieve good enantioselectivities it is essential that most of the substrate is bound to the chiral complexing agent (+)-**12**. This goal was accomplished by using 2.5 equivalents of (+)-**12** resulting in an enantiomeric excess of 60% *ee* at –45 °C whereas the reaction with 1.0 equivalent only yielded a 37% *ee*. As in previous cases, higher *ee* values could be obtained at lower temperature (60% *ee* at −45 °C compared to 5% *ee* at 30 °C).

4 Sensitised Reactions

Both energy transfer and electron transfer from a photoexcited compound to a given substrate are distance dependent. This property allows one to delineate—at least on paper—a catalytic cycle for a sensitised process with an appropriately modified template (Scheme 14). If the passive tetrahydronaphthalene shield in **12** is replaced by a photoactive moiety, this part of the compound can, after excitation, facilitate an energy or electron transfer significantly faster at a bound than at an unbound substrate.

The second property of the photoactive moiety is identical to the previously used shield, that is it must be bulky and rigid enough to guarantee the desired enantioface differentiation. After the reaction, the binding site at the sensitiser can be occupied by another substrate molecule unless product dissociation is disfavoured. In most photochemical processes the product is more space demanding than the substrate so that the equilibrium is shifted favouring substrate association. From our experience there are three major constraints to the ideal picture depicted in Scheme 14. First, intermolecular sensitisation becomes increasingly important if the substrate–template association is not sufficiently strong. Second, most substrates are not UV transparent but have a significant absorption band, frequently at shorter wavelengths than the sensitiser. Direct excitation can compete with sensitisation. Third, the lifetime of the sensitiser is limited by intra- and intermolecular decomposition processes. Commonly used carbonyl sensitisers absorb light with an energy content of 300–400 kJ Es^{-1}. If this energy is not quickly distributed bond cleavage reactions are unavoidable.

4.1 Possible Templates and Substrates

The ideal substrate for a sensitised reaction should absorb light at a wavelength, which is at least 50 nm shorter than the absorption maximum of the sensitiser. It should coordinate with an association constant, which is significantly higher than its dimerisation constant, and it

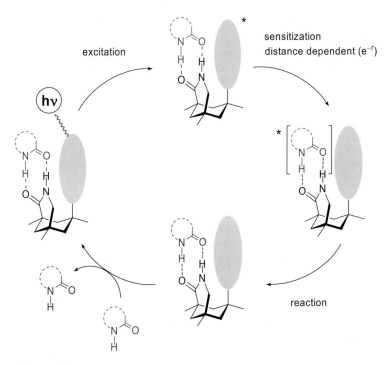

Scheme 14.

should generate a product which neither coordinates to the template nor shows any photochemical instability. Judging from their photophysical properties many combinations of a carbonyl compound and an olefin appear suited for an application in enantioselective sensitised reactions. Moreover, carbonyl compounds are very efficient photoelectron acceptors, that is they exhibit a significant reduction potential upon excitation and can effectively induce electron transfer processes. Given these two attractive modes of application, we concentrated our initial work on the synthesis of templates in which the tetrahydronaphthalene was replaced by an adequate carbonyl chromophore. While it was tempting to attach such a chromophore by an ester or amid linkage, we had to realise that the resulting sensitisers are not effective due to insufficient complexation and due to a lack of enantioface differentiation. The search for a catalyst/substrate combination which was suited to demonstrate the general principle of enantioselective photochemical reaction eventually led to the development of a catalytic photoinduced electron transfer (PET) reaction, which is discussed in the next section.

4.2 PET-Catalysed Reaction

Benzophenones have been described as useful sensitisers for PET catalysed conjugate addition reactions of α-amino alkyl radicals to enones (Bertrand et al. 2000). We tried to modify this reaction and synthesised the pyrrolidinylethyl-substituted quinolone **35** from the known bromide (Bauer et al. 2005). Upon electron transfer from the pyrrolidine to a given acceptor, a radical cyclisation occurs (Scheme 15), which after electron and proton transfer generates a pyrrolizidine. We found 4,4′-dimethoxybenzophenone to be a suitable catalyst for this reaction. Remarkably, the reaction proceeded with excellent simple diastereoselectivity and a single diastereoisomeric product *rac*-**36** was obtained. With 10 mol% of the catalyst, a chemical yield of 71% was achieved.

In order to install a benzophenone at the bicyclic scaffold we relied on the previously used oxazole linkage. To this end, the known aminohydroxybenzophenone **37** (Aichaoui et al. 1990) was coupled to the free acid *rac*-**38**, which is available from Kemp's triacid in five synthetic steps. Remarkably, an *O*-acylation instead of the expected *N*-acylation was observed resulting in ester *rac*-**39**. As a consequence, oxazole formation was less straightforward but could eventually be achieved under more forceful conditions. The reaction sequence led to the racemic benzophenone *rac*-**40**, i.e. to a 1/1 mixture of the enantiomers (+)-**40** and (−)-**40**, which was separated by chiral HPLC (Daicel Chiralpak AD). It is important to mention that a separation of enantiomers at an earlier stage is not sensible. While carboxylic acid **38** can be obtained in enantiomerically pure form, racemisation occurs upon activation, presumably due to a bridged symmetrical intermediate (Kirby et al. 1998) (Scheme 16).

The assignment of the absolute configuration to the individual enantiomers of **40** was conducted by a molecular recognition experiment (Bauer and Bach 2004) with the related 3-aza-bicyclo[3.3.1]nonan-2-

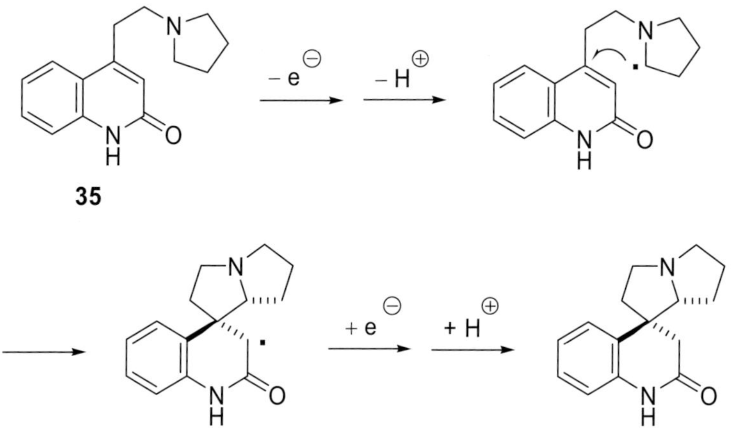

Scheme 15.

ones (+)-**12** and (−)-**12**, the absolute configuration of which had earlier been proven by anomalous X-ray diffraction. If (+)-**12** and (−)-**12** are added separately to samples of a given 3-aza-bicyclo[3.3.1]nonan-2-one enantiomer there is the possibility of a homo- or a heterochiral association. Provided that the unknown enantiomer possesses a configuration identical to either (+)-**12** or (−)-**12** (homochiral case) there is essentially no association indicated by an insignificant chemical shift change in the ^{1}H-NMR spectrum for the respective NH proton. Contrary to that, in the heterochiral situation, a remarkable and clearly evident ^{1}H-NMR shift change results due to strong association.

Using the single enantiomers (+)-**40** and (−)-**40** as chiral catalysts, the reaction to pyrrolizidines **36** and *ent*-**36** could be conducted enantioselectively. With 5 mol% of compound (−)-**40** product **36** was obtained in 61% yield and with an enantioselectivity of 20% *ee*. The reactions were conducted at −60 °C in toluene as the solvent with a substrate concentration of 4 mM. Increasing the amount of catalyst resulted in an improved enantioselectivity and in a decrease of reaction time. With 30 mol% (−)-**40** product **36** was obtained in 70% *ee* and in a chemical

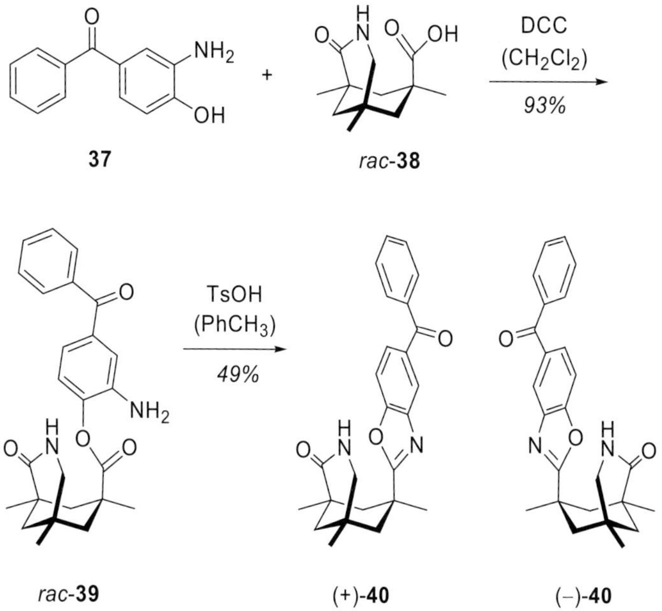

Enantioselective Photochemistry

Scheme 16.

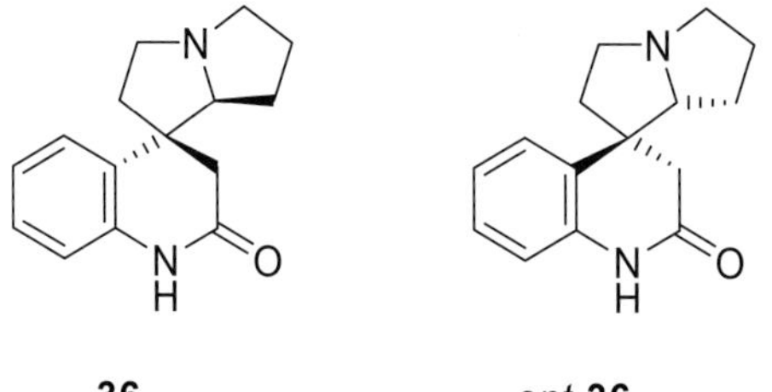

Scheme 17.

yield of 64%. Expectedly, the exchange of the catalyst configuration—the use of (+)-**40**—led to formation of the opposite enantiomer *ent*-**36**. The determination of the absolute product configuration turned out to be difficult because **36** and many derivatives thereof were not crystalline.

Eventually, the configuration proof was based on a comparison of calculated and measured CD spectra (Scheme 17). The absolute configuration is in line with a coordination of the substrate to the template and a subsequent cyclisation of the α-aminoalkyl radical to the quinolone core from the most accessible face, that opposite to the shield.

5 Summary

In summary, the utility of hydrogen bonds for controlling the absolute configuration in photomediated reactions has been demonstrated. Despite the simple binding motif, a number of reactions have been shown to proceed enantioselectively (up to 99% *ee*). The complexing agent, which is used in stoichiometric or superstoichiometric amounts, can be repeatedly recovered and allows consequently a multiplication of chirality. Chirality turnover in a single reaction vessel has so far been observed for compound (+)-**12** in the radical cyclisation **27**→*trans*-**29** and for sensitiser (–)-**40** in the PET-catalysed reaction **35**→**36**. Further work is directed towards application of this methodology in more complex syntheses and towards a development of more efficient catalytic transformations.

References

Achenbach TV, Slater EP, Brummerhop H, Bach T, Müller R (2000) Inhibition of cyclin-dependent kinase activity and induction of apoptosis by preussin in human tumour cells. Antimicrobiol Agents Chemother 44:2794–2801

Aechtner T, Dressel M, Bach T (2004) Hydrogen bond mediated enantioselectivity of radical reactions. Angew Chem 116:5974–5976, Angew Chem Int Ed 43:5849–5851

Aichaoui H, Lesieur I, Hénichart JP (1990) Unequivocal preparation of 4- and 5-acyl-2-aminophenols. Synthesis 679–680

Bach T, Bergmann H (2000) Enantioselective intermolecular [2+2]-photocycloaddition reactions of alkenes and a 2-quinolone in solution. J Am Chem Soc 122:11525–11526

Bach T, Brummerhop H (1998) Unprecedented facial diastereoselectivity in the Paternò-Büchi reaction of a chiral dihydropyrrol—A short total synthesis of (+)-preussin. Angew Chem 110:3577–3579, Angew Chem Int Ed 37:3400–3402

Bach T, Bergmann H, Harms K (1999) High facial diastereoselectivity in the photocycloaddition of a chiral aromatic aldehyde and an enamide induced by intermolecular hydrogen bonding. J Am Chem Soc 121:10650–10651

Bach T, Brummerhop H, Harms K (2000a) The synthesis of (+)-preussin and related pyrrolidinols by diastereoselective Paternò-Büchi reactions of chiral 2-substituted 2,3-dihydropyrroles. Chem Eur J 6:3838–3848

Bach T, Bergmann H, Harms K (2000b) Enantioselective intramolecular [2+2]-photocycloaddition reactions in solution. Angew Chem 112:2391–2393, Angew Chem Int Ed 39:2302–2304

Bach T, Bergmann H, Brummerhop H, Lewis W, Harms K (2001a) The [2+2] photocycloaddition of aromatic aldehydes and ketones to 3,4-dihydro-2-pyridones. Regioselectivity, diastereoselectivity and reductive ring opening of the product oxetanes. Chem Eur J 7:4512–4521

Bach T, Bergmann H, Grosch B, Harms K, Herdtweck E (2001b) Synthesis of enantiomerically pure 1,5,7-trimethyl-3-azabicyclo[3.3.1]nonan-2-ones as chiral host compounds for enantioselective photochemical reactions in solution. Synthesis 1395–1405

Bach T, Bergmann H, Harms K (2001c) Enantioselective photochemical reactions of 2-pyridones in solution. Org Lett 3:601–603

Bach T, Aechtner T, Neumüller B (2001d) Intermolecular hydrogen binding of a chiral host and a prochiral imidazolidinone: Enantioselective Norrish–Yang cyclisation in solution. Chem Commun 607–608

Bach T, Bergmann H, Grosch B, Harms K (2002a) Highly enantioselective intra- and intermolecular [2+2]-photocycloaddition reactions of 2-quinolones mediated by a chiral lactam host: Host-guest interactions, product configuration, and the origin of the stereoselectivity in solution. J Am Chem Soc 124:7982–7990

Bach T, Aechtner T, Neumüller B (2002b) Enantioselective Norrish–Yang cyclisation reactions of N-(ω-oxo-ω-phenylalkyl)-substituted imidazolidinones in solution and in the solid state. Chem Eur J 8:2464–2475

Bach T, Grosch B, Strassner T, Herdtweck E (2003) Enantioselective [6π]-photocyclisation reaction of an acrylanilide mediated by a chiral host. Interplay between enantioselective ring closure and enantioselective protonation. J Org Chem 68:1107–1116

Basler B, Brandes S, Spiegel A, Bach T (2005) Total syntheses of kelsoene and preussin. Top Curr Chem 243:1–42

Bauer A, Bach T (2004) Assignment of the absolute configuration of 7-substituted 3-azabicyclo[3.1.1]nonan-2-ones by NMR titration experiments. Tetrahedron Asymmetry 15:3799–3803

Bauer A, Westkämper F, Grimme S, Bach T (2005) Catalytic enantioselective reactions driven by photoinduced electron transfer. Nature 436:1139–1140

Berkessel A, Gröger H (2005) Asymmetric organocatalysis. Wiley-VCH, Weinheim

Bertrand S, Hoffmann N, Pete JP (2000) Highly efficient and stereoselective radical addition of tertiary amines to electron-deficient alkenes—application to the enantioselective synthesis of necine bases. Eur J Org Chem 2227–2238

Brandes S, Selig P, Bach T (2004) Stereoselective intra- and intermolecular [2+2]-photocycloaddition reactions of 4-(2′-aminoethyl)quinolones. Synlett 2588–2590

Dressel M, Bach T (2006) Chirality multiplication and efficient chirality transfer in *exo*- and *endo*-radical cyclisation reactions of 4-(4′-iodobutyl)quinolones. Org Lett 8:3145–3147

Dressel M, Aechtner T, Bach T (2006) Enantioselectivity and diastereoselectivity in reductive radical cyclisation reactions of 3-(ω-iodoalkylidene)-piperidin-2-ones. Synthesis 2206–2214

Grosch B, Orlebar CN, Herdtweck E, Massa W, Bach T (2003) Highly enantioselective Diels-Alder reaction of a photochemically generated *o*-quinodimethane with olefins. Angew Chem 115:3822–3824, Angew Chem Int Ed 42:3693–3696

Grosch B, Bach T (2004) Template-induced enantioselective photochemical reactions in solution. In: Inoue Y, Ramamurthy V (eds) Molecular and supramolcular photochemistry, vol 11: Chiral Photochemistry. Dekker, New York, pp 315–340

Grosch B, Orlebar CN, Herdtweck E, Kaneda M, Wada T, Inoue Y, Bach T (2004) Enantioselective [4+2]-cycloaddition reaction of a photochemically generated *o*-quinodimethane: Mechanistic details, association studies, and pressure effects. Chem Eur J 10:2179–2189

Kirby AJ, Komarov IV, Wothers PD, Feeder N (1998) The most twisted amide: structure and reactions. Angew Chem 110:830–831, Angew Chem Int Ed 37:785–786

Ojima I (ed) (2000) Catalytic asymmetric synthesis. Wiley-VCH, Weinheim

Rau H (2004) Direct asymmetric photochemistry with circularily polarised light. In: Inoue Y, Ramamurthy V (eds) Molecular and Supramolcular Photochemistry, vol 11: Chiral Photochemistry. Dekker, New York, pp 1–44

Renaud P, Sibi MP (eds) (2001) Radicals in organic synthesis. VCH, Weinheim

Selig P, Bach T (2006) Photochemistry of 4-(aminoethyl)quinolones: Enantioselective synthesis of tetracyclic tetrahydro-1aH-pyrido[4′,3′:2,3]cyclobuta[1,2-*c*]quinoline-2,11(3H,8H)-diones by intra- and intermolecular [2+2]-photocycloaddition reactions in solution. J Org Chem 71:5662–5673

Stack JG, Curran DP, Geib SV, Rebek J, Ballester P (1992) A new chiral auxiliary for asymmetric thermal reactions: High stereocontrol in radical addition, allylation, and annulation reactions. J Am Chem Soc 114:7007–7018

Yang NC, Yang DDH (1958) Photochemical reactions of ketones in solution. J Am Chem Soc 80:2913–2914

Zimmerman J, Sibi MP (2006) Enantioselective radical reactions. Top Curr Chem 263:107–162

Organocatalysis by Hydrogen Bonding Networks

A. Berkessel

Department of Chemistry, University of Cologne, Greinstraße 4, 50939 Cologne, Germany

email: *berkessel@uni-koeln.de*

1	Introduction	282
2	Case Studies	283
2.1	Epoxidation and Baeyer–Villiger Oxidation with Hydrogen Peroxide in Fluorinated Alcohol Solvents	283
2.2	Peptide-Catalyzed Asymmetric Epoxidation of Enones: On the Mechanism of the Juliá–Colonna Reaction	287
2.3	Enantiomerically Pure α- and β-Amino Acids from the Dynamic Kinetic Resolution of Azlactones and the Kinetic Resolution of Oxazinones	290
3	Epilogue	294
References		295

Abstract. In biological systems, hydrogen bonding is used extensively for molecular recognition, substrate binding, orientation and activation. In organocatalysis, multiple hydrogen bonding by man-made catalysts can effect remarkable accelerations and selectivities as well. The lecture presents four examples of non-enzymatic (but in some cases enzyme-like!) catalysis effected by hydrogen bonding networks: epoxidation of olefins and Baeyer–Villiger oxidation of ketones with H_2O_2 in fluorinated alcohol solvents; peptide-catalyzed asymmetric epoxidation of enones by H_2O_2; dynamic kinetic resolution of azlactones, affording enantiomerically pure α-amino acids; and kinetic resolution of oxazinones, affording enantiomerically pure β-amino acids. All four types of transformations are of preparative value, and their mechanisms are discussed.

1 Introduction

The term organocatalysis refers to the use of low-molecular weight and metal-free catalysts for the selective, in most cases enantioselective, transformation of organic substrates (Berkessel and Gröger 2005). The mechanisms by which metal-free enzymes (more than half of all enzymes known to date do not contain catalytically active metals) effect dramatic rate accelerations have been a major field of research in bioorganic chemistry for decades, dating back to the first half of the twentieth century (Langenbeck 1949). In many instances, organocatalysts can be considered as 'minimal versions' of metal-free enzymes, and the mechanisms and the categorizations of enzymatic catalysis apply to the action of organocatalysts as well. In both cases, the accelerations observed depend on the typical interactions of organic molecules with one another. A general distinction can be made between processes that involve the *formation of covalent adducts* between catalyst and substrate(s) within the catalytic cycle, and processes that rely on *non-covalent* interactions such as hydrogen bonding or the formation of ion pairs. The former type of interaction has been termed 'covalent catalysis', whereas the latter situation is usually called 'noncovalent catalysis'. The formation of covalent substrate–catalyst adducts may occur e.g. by single-step *Lewis*-acid – *Lewis*-base interaction, or by multi-step reactions such as the formation of enamines from aldehydes and secondary amines. The catalysis of aldol reactions by the formation of enamines is a striking example of common mechanisms in enzymatic catalysis and organocatalysis. In class I aldolases, lysine provides the catalytically active amine group, whereas typical organocatalysts for this purpose are secondary amines, the most simple one being proline (List and Seayad 2005). Noncovalent catalysis in many instances rests on the formation of hydrogen-bonded adducts between substrate and catalyst, or on protonation/deprotonation processes. Phase-transfer catalysis (PTC) by organic phase-transfer catalysts falls into the category 'noncovalent catalysis'. The latter is mechanistically unique because PTC promotes reactivity, not only by altering the chemical properties of the reactants, but also involves a transport phenomenon.

This chapter discusses four cases of organocatalysis effected by multiple hydrogen bonding. It should be pointed out at the very beginning that the crucial feature is the reduction of activation energy by effective hydrogen bonding to the transition state of a reaction, whereas substrate binding alone is not sufficient.

2 Case Studies

2.1 Epoxidation and Baeyer–Villiger Oxidation with Hydrogen Peroxide in Fluorinated Alcohol Solvents

The epoxidation of olefins with hydrogen peroxide is dramatically accelerated in fluorinated alcohols as solvents (Berkessel and Andreae 2001; Neumann and Neimann 2000, Sheldon et al. 2001) For example, in 1,1,1,3,3,3-hexafluoro-2-propanol (HFIP) as solvent, Z-cyclooctene is fully converted to the corresponding epoxide by 50% aqueous hydrogen peroxide at 60 °C within minutes. In 1,4-dioxane as solvent, only a few percent of product are formed under the same conditions. In fact, the acceleration in HFIP is in the range of 10^5-fold (Berkessel and Adrio 2004). A thorough kinetic analysis revealed that the fluorinated alcohol solvent had a kinetic order of 2–3, indicating the involvement of 2–3 molecules of HFIP (Berkessel and Adrio 2006; Berkessel et al. 2006). Both the olefin substrate and the oxidant showed first order dependence of rate on concentration. An investigation of the aggregation behavior of HFIP, by determination of its hitherto unknown crystal structure, proved the existence of H-bonded and endless helices, as shown in Fig. 1. DFT calculations on the H-bond donor ability of HFIP aggregates indicated that the clustering of the HFIP monomers leads to cooperatively enhanced H-bond donor ability (Berkessel et al. 2006).

According to our quantum chemical analysis, the oxygen transfer from hydrogen peroxide to the substrate olefin, in the presence of HFIP, proceeds through a single transition state. Hydrogen bonding by the fluorinated alcohol solvent significantly reduces the activation barrier (Berkessel and Adrio 2006; for a DFT-study involving only one HFIP molecule see Neumann et al. 2003). We were able to localize transition states comprising up to four molecules of HFIP. However, as the activation entropy becomes more and more unfavorable the more molecules of HFIP are involved, the overall optimal free energy of activation is found for 2–3 HFIPs (Fig. 2) which coincides with our kinetic results

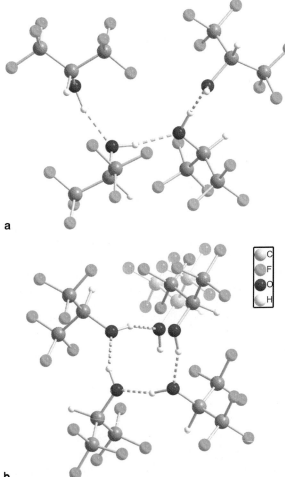

Fig. 1. Single-crystal X-ray structures of HFIP: (**a**) View perpendicular to the helix axis; (**b**) view along the helix structure

(see above). Figure 3 shows the suggested optimal transition state structures for oxygen transfer from hydrogen peroxide to *Z*-2-butene in the presence of two (Fig. 3a) and three (Fig. 3b) HFIP molecules. In sum-

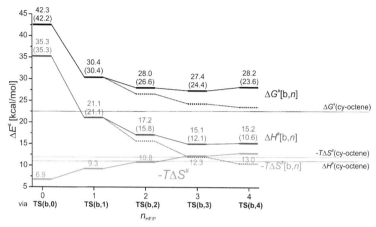

Fig. 2. Activation parameters vs number of HFIP molecules for the epoxidation of Z-butene within a solution model at 298K (RB3LYP/6–311++G(d,p)//RB3LYP/6–31+G(d,p)). *Dashed lines* refer to values in parentheses which include a correction for the dispersion interaction from a BSSE-corrected MP2/6–31+G(2d,p)//B3LYP/6–31+G(d,p) single-point calculation on the corresponding initial aggregates. ΔG (cyclooctene) and $-T\Delta S$ (cyclooctene) correspond to the experimentally determined activation parameters for the epoxidation of Z-cyclooctene at 298 K

mary, the acceleration of olefin epoxidation in fluorinated alcohols such as HFIP rests on the specific interaction of the solvent with the substrates by multiple H-bonding.

In fluorinated alcohol solvents, nonstrained ketones such as cyclohexanone (**1**) undergo oxidation to lactones in the presence of hydrogen peroxide and catalytic amounts of Brønsted acids (Berkessel and Andreae 2001; Berkessel et al. 2002). Unlike the classical Baeyer–Villiger reaction, ketone oxidation with H_2O_2 in e.g. HFIP proceeds via a spiro-bisperoxide **2** intermediate (Scheme 1). In contrast to other solvents, the acid-catalyzed rearrangement of the spiro-bisperoxide **2** to two equivalents of the product lactone **3** proceeds rapidly and cleanly in HFIP. Preliminary calculations indicate active participation of the fluorinated alcohol solvent in the rate-determining step also in this case.

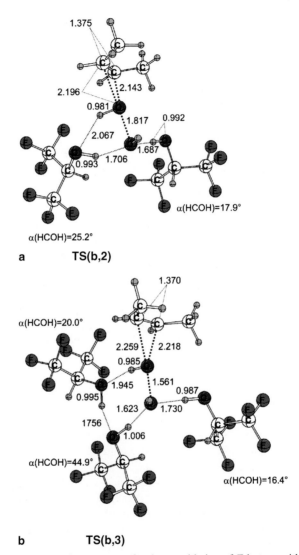

Fig. 3. Stationary-point structures for the epoxidation of Z-butene with hydrogen peroxide in the presence of two (**a**) and three (**b**) molecules of HFIP, optimized at RB3LYP/6–31+G(d,p) (selected bond lengths in Å)

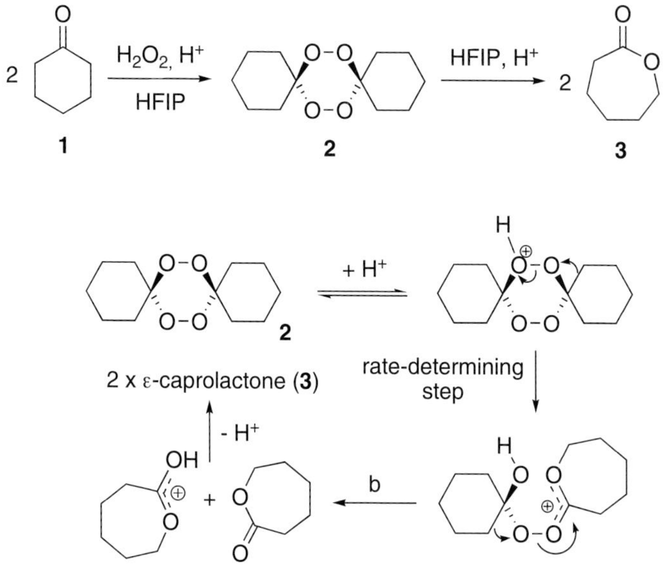

Scheme 1. *Top*: formation of the spiro-bisperoxide **2** from cyclohexanone (**1**) and hydrogen peroxide. *Bottom*: suggested mechanism of the acid-catalyzed rearrangement of the spiro-bisperoxide **2** to two molecules of the product lactone **3**

2.2 Peptide-Catalyzed Asymmetric Epoxidation of Enones: On the Mechanism of the Juliá–Colonna Reaction

In the 1980s, Juliá and Colonna discovered that the Weitz–Scheffer epoxidation of enones such as chalcone (**4**, Scheme 2) by alkaline hydrogen peroxide is catalyzed in a highly enantioselective fashion by poly-amino acids such as poly-alanine or poly-leucine (Juliá et al. 1980, 1982). The poly-amino acids used for the Juliá–Colonna epoxidation are statistical mixtures, the maximum length distribution being around 20–25 mers (Roberts et al. 1997). The most fundamental question to be addressed refers to the minimal structural element (i.e. the minimal peptide length) required for catalytic activity and enantioselectivity. To tackle this question, we have synthesized the whole series of *L*-leucine oligomers from 1- to 20-mer on a solid support (Berkessel

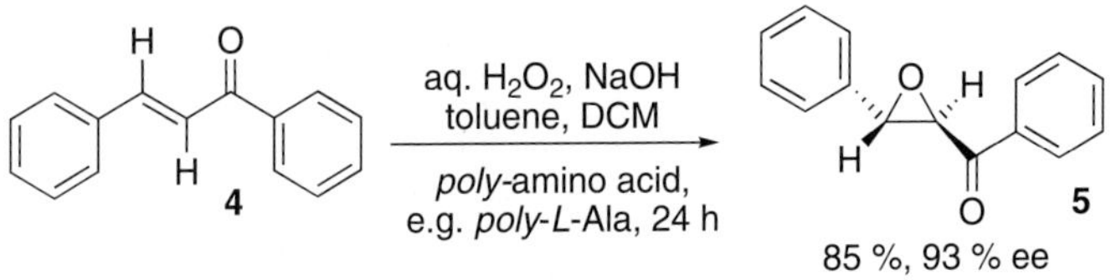

Scheme 2. Epoxidation of chalcone (**4**) in the presence of a peptide catalyst

et al. 2001). The result of the subsequent screening of the monodisperse and solid-phase bound peptides for activity and selectivity in the epoxidation of chalcone (**4**, Scheme 2) is shown in Fig. 4. Full enantioselection ($\geq 95\%$ ee of the epoxide **5**) is achieved already with the 4- to 5-mers, and no further improvement results from using longer peptide chains (green columns). However, the catalytic efficiency (measured as chemical yield after 24 h reaction time) increases with the increasing chain length of the peptide catalyst (blue columns). From the sum of the experimental data it was concluded that the three non-intra-helical NH bonds present at the N terminus play a crucial role in the catalytic mechanism. Based on the results of molecular modeling studies, it is suggest that the carbonyl oxygen atom of the enone substrate forms two H-bonds to the NH groups of the amino acids at the N terminus, and that the hydroperoxide nucleophile forms a third hydrogen bond. Consequently, the sense of helicity of the peptide catalyst determines the sense of induction in the epoxidation reaction. Figure 5 (left) illustrates the binding of chalcone to the N terminus of the peptide, Fig. 5 (right) shows the binding of the β-hydroperoxyenolate intermediate to the peptide's N terminus.

According to this model, the action of the peptide catalysts used in the Juliá–Colonna epoxidation bears a lot of similarity to that of enzymes, in particular the binding/activation and proper orientation of the substrates which ultimately effects the excellent enantioselectivities in the overall process. In fact, the H-bonding motif discovered as the catalytically active site also acts as the oxy-anion hole in serine esterases and is known to bind/stabilize a variety of fully or partially negatively charged entities (Milner-White and Watson 2002a,b), the β-hydroperoxyenolate, in the present case. Studies by Roberts, Kelly,

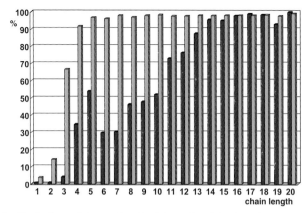

Fig. 4. Catalytic activity and selectivity of solid phase-bound L-leucine oligomers. *Green columns*: enantiomeric excess of the product epoxide **5**. *Blue columns*: conversion of the substrate chalcone (**4**) after 24 h

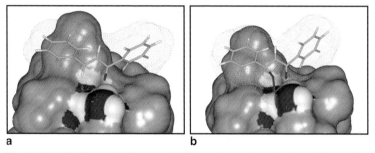

Fig. 5. *Left*: binding of chalcone to the N terminus of a L-leucine helix. *Right*: binding of the β-hydroperoxyenolate intermediate to the peptide's N terminus

Blackmond et al. are in accord with our suggestion (Berkessel et al. 2001, 2006) of an H-bonding controlled reaction at the N terminus of the helical peptide (Kelly and Roberts 2004; Blackmond et al. 2005). In other words, the naturally evolved oxyanion hole motif serves as the basis of catalysis in a purely 'man-made' reaction.

2.3 Enantiomerically Pure α- and β-Amino Acids from the Dynamic Kinetic Resolution of Azlactones and the Kinetic Resolution of Oxazinones

Azlactones (**6**) are cyclic five-membered ring N-analogs of anhydrides, derived from α-amino acids (Scheme 3). Oxazinones (**7**) are the six-membered ring homologs derived from β-amino acids (Scheme 3). Both azlactones and oxazinones can be ring-opened with nucleophiles such as alcohols (affording N-acyl amino acid esters) or amines (affording N-acyl amino acid amides). Both heterocycles **6** and **7** are accessible from the corresponding N-acyl amino acids by treatment with condensing agents, such as thionyl chloride or DCC.

Azlactones are configurationally labile, implying potential for dynamic kinetic resolution (DKR), e.g. in the catalytic alcoholytic ring-opening (Scheme 4). We reasoned that organocatalysts suitable for this transformation might again be based on hydrogen bonding: we hoped that a pseudo-Lewis acidic—Brønsted basic bifunctional organocatalyst of the general formula **8** (Scheme 4) would simultaneously activate the azlactone towards nucleophilic attack by H-bonding to the azlactone carbonyl oxygen atom, and activation/steering of the alcohol nucleophile by H-bonding to the tertiary amine base.

Indeed, this concept proved successful and after optimization of the catalyst structure and of the reaction conditions, a number of azlactones **6** could be transformed to highly enantiomerically enriched N-acyl amino acid esters **9** of high enantiomeric purity (Berkessel et al. 2005, 2006). Some of the results are summarized in Scheme 5.

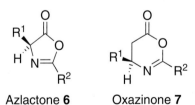

Azlactone 6   Oxazinone 7

Scheme 3. Azlactones (**6**) and oxazinones (**7**) – cyclic and N-analogous amino acid anhydrides

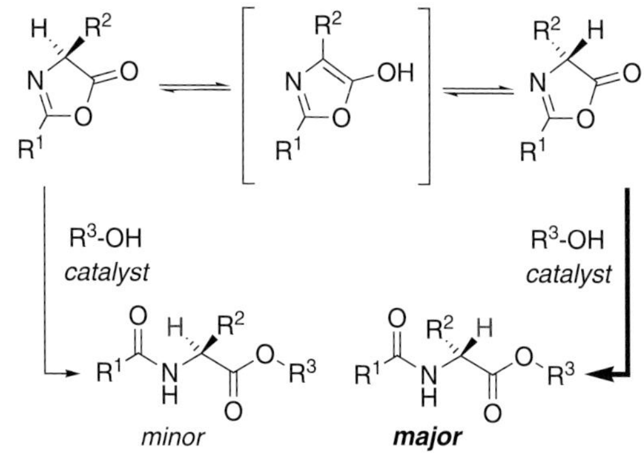

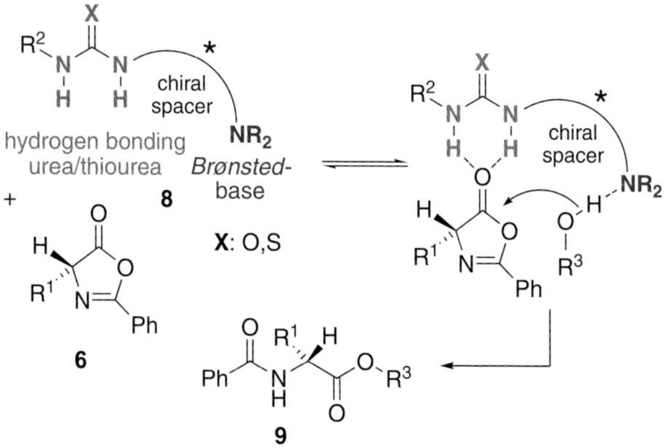

Scheme 4. Dynamic kinetic resolution of azlactones **6** by alcoholytic ring-opening, effected by the bifunctional organocatalysts **8**

Oxazinones of the general structure **7** are configurationally stable. Consequently, alcoholytic ring opening with a chiral organocatalyst was expected to afford kinetic resolution (KR), but no DKR (Scheme 6). In fact, a number of oxazinones could be resolved kinetically and with excellent selectivity (Scheme 7; Berkessel et al. 2005). Figure 6 shows a typical time course for the kinetic resolution of a β^3-amino acid derived oxazinone.

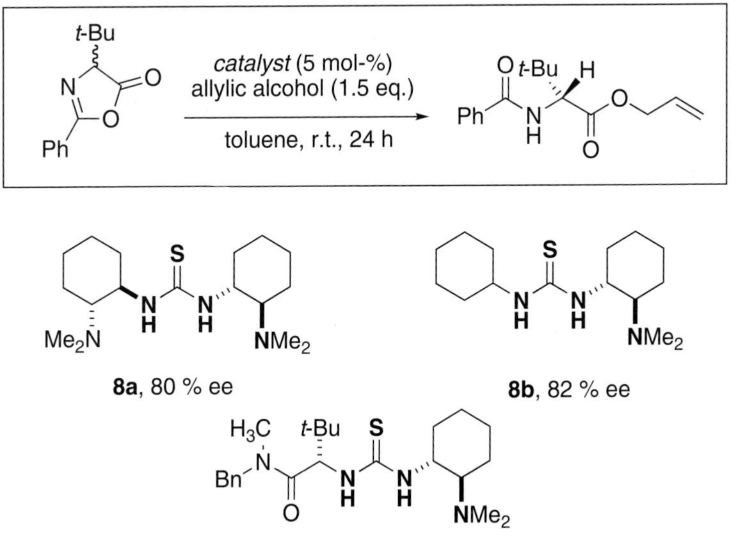

8a, 80 % ee

8b, 82 % ee

8c, 95 % ee

Scheme 5. Some examples for the dynamic kinetic resolution of azlactones

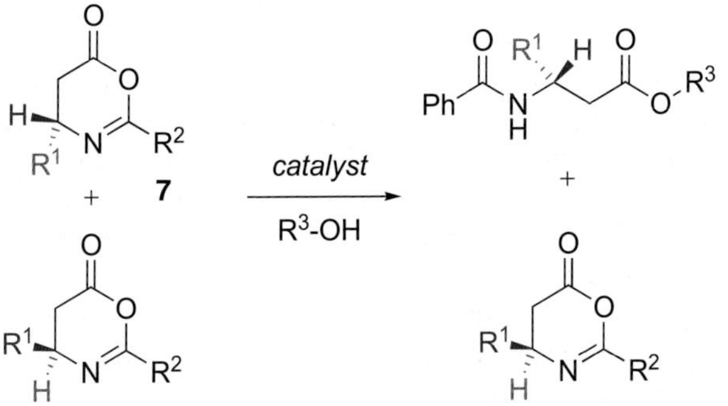

ent-**7**

Scheme 6. Kinetic resolution of oxazinones **7** by alcoholytic ring opening

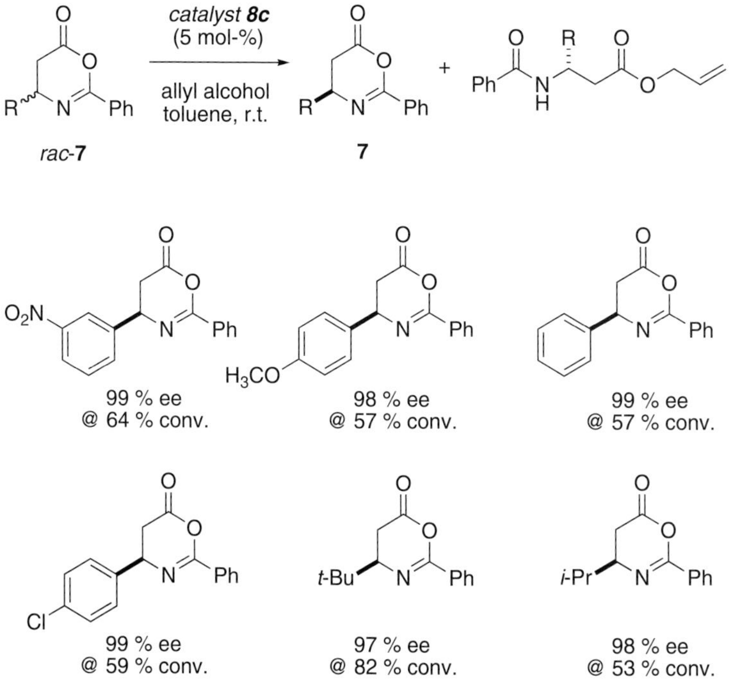

Scheme 7. Kinetic resolution of some 4-aryl- and 4-alkyl-substituted oxazinones

It was mentioned in the beginning that azlactones and oxazinones are activated amino acid derivatives. In the KR discussed here, the remaining oxazinone enantiomer can be reacted further, in the crude reaction mixture, with nucleophiles. For example, treatment with resin-bound and N-terminally nonprotected peptides results in coupling with a β-amino acid. Heating of the homogeneous crude reaction mixture (typically toluene as solvent) with dilute aqueous hydrochloric acid results in hydrolysis of the unreacted oxazinone enantiomer and precipitation of the corresponding N-acyl β-amino acid. The latter can be isolated in excellent enantiomeric purity by simple filtration. The filtrate contains the β-amino acid ester of opposite configuration (Berkessel et al. 2005).

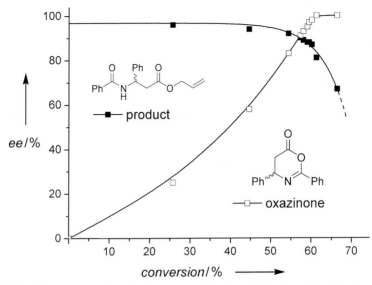

Fig. 6. Time course of a typical oxazinone kinetic resolution by alcoholytic ring opening. The organocatalyst used was **8c** from Scheme 5

3 Epilogue

Multiple and specific hydrogen bonding has been recognized as a highly efficient motif not only in enzymatic catalysis, but nowadays also in organocatalysis (Jacobsen and Taylor 2006). In the first part of the lecture, multiple hydrogen bonding within supramolecular aggregates of fluoroalcohols formed the basis for epoxidation catalysis with hydrogen peroxide as oxygen donor. There appears to be no direct analog for this type of catalysis among natural systems. For the Juliá–Colonna reaction, it appears that a catalytic motif evolved for amide and ester hydrolysis, namely the oxy-anion hole, can be 'side-tracked' to accelerate other reactions involving intermediates/transition states with a negatively charged oxygen atom. The same holds for the ring openings of azlactones and oxazinones using thiourea catalysts. In these cases, negative charge on the carbonyl oxygen atom is stabilized by hydrogen bonding to the thiourea. In other words, both the Juliá–Colonna reaction and the azlactone/oxazinone opening involve a naturally evolved catalytic principle. It is tempting to speculate which other types of enzymatic rate acceleration by hydrogen bonding might be suitable for adaptation to organocatalysis. Along the same lines, one may wonder whether organocatalysts might eventually substitute for enzymes—potentially in a living cell.

Acknowledgements. This work was supported financially by the EU (Research Training Networks 'The Discovery of New Molecular Catalysts through Combinatorial Chemistry: Activity and Selectivity from Diversity', COMBICAT, RTN-CT-2000–00014 and '(R)Evolutionary Catalysis', REVCAT, MRTN-CT-2006–038566), the Deutsche Forschungsgemeinschaft (Priority Program 'Organocatalysis', SPP 1179), the Fonds der Chemischen Industrie, and by the BASF AG, Ludwigshafen. Generous gifts of amino acids by the Degussa AG, Hanau, are gratefully acknowledged.

References

Berkessel A, Adrio JA (2004) Kinetic studies of olefin epoxidation with hydrogen peroxide in 1,1,1,3,3,3-hexafluoro-2-propanol reveal a crucial catalytic role for solvent clusters. Adv Synth Catal 346:275–280

Berkessel A, Adrio JA (2006) Dramatic acceleration of olefin epoxidation in fluorinated alcohols: activation of hydrogen peroxide by multiple H-bond networks. J Am Chem Soc 128:13412–13420

Berkessel A, Adrio JA, Hüttenhain D, Neudörfl JM (2006a) Unveiling the "booster effect" of fluorinated alcohol solvents: aggregation-induced conformational changes, and cooperatively enhanced H-bonding. J Am Chem Soc 128:8421–8426

Berkessel A, Andreae MRM (2001) Efficient catalytic methods for the Baeyer-Villiger oxidation and epoxidation with hydrogen peroxide. Tetrahedron Lett 42:2293–2295

Berkessel A, Andreae MRM, Schmickler H, Lex J (2002) Baeyer–Villiger oxidations with hydrogen peroxide in fluorinated alcohols: lactone formation by a nonclassical mechanism. Angew Chem Int Ed 41:4481–4484

Berkessel A, Cleemann F, Mukherjee S (2005a) Kinetic resolution of oxazinones: an organocatalytic approach to enantiomerically pure beta-amino acids. Angew Chem Int Ed 44:7466–7469

Berkessel A, Cleemann F, Mukherjee S, Müller TN, Lex J (2005b) Highly efficient dynamic kinetic resolution of azlactones by urea-based bifunctional organocatalysts. Angew Chem Int Ed 44:807–811

Berkessel A, Gasch N, Glaubitz K, Koch C (2001) Highly enantioselective enone epoxidation catalyzed by short solid phase-bound peptides. Org Lett 3:3839–3842

Berkessel A, Gröger H (2005) Asymmetric organocatalysis—from biomimetic concepts to applications in asymmetric synthesis. Wiley-VCH, Weinheim

Berkessel A, Koch B, Toniolo C, Rainaldi M, Broxterman QB, Kaptein B (2006b) Asymmetric enone epoxidation by short solid-phase bound peptides: further evidence for catalyst helicity and catalytic activity of individual peptide strands. Biopolymers: Pept Sci 84:90–96

Berkessel A, Mukherjee S, Cleemann F, Müller TN, Lex J (2005c) Second-generation organocatalysts for the highly enantioselective dynamic kinetic resolution of azlactones. Chem Commun 2005:1898–1900

Berkessel A, Mukherjee S, Müller TN, Cleemann F, Roland K, Brandenburg M, Neudörfl JM (2006c) Structural optimization of thiourea-based bifunctional organocatalysts for the highly enantioselective dynamic kinetic resolution of azlactones. Org Biomol Chem 4:4319–4330

Blackmond DG, Mathew SP, Gunathilagan S, Roberts SM (2005) Mechanistic insights from reaction progress kinetic analysis of the polypeptide-catalyzed epoxidation of chalcone. Org Lett 7:4847–4850

Jacobsen EN, Taylor MS (2006) Asymmetric catalysis by chiral hydrogen bond donors. Angew Chem Int Ed 45:1520–1543

Juliá S, Guixer J, Masana J, Rocas J, Colonna S, Annunziata R, Molinari H (1982) Synthetic enzymes. Part 2. Catalytic asymmetric epoxidation by means of polyamino-acids in a triphase system. J Chem Soc Perkin Trans 1:1317–1324

Juliá S, Masana J, Vega JC (1980) Synthetic Enzymes. Highly stereoselective epoxidation of chalcone in a triphasic toluene–water–poly[(S)-alanine] system. Angew Chem Int Ed Engl 19:929–931

Kelly DR, Roberts SM (2004) The mechanism of the polyleucine catalyzed asymmetric epoxidation. Chem Commun 2004:2018–2020

Langenbeck W (1949) Die organischen Katalysatoren und ihre Beziehungen zu den Fermenten (Organic Catalysts and their Relation to the Enzymes) 2nd ed. Springer, Berlin

List B, Seayad J (2005) Asymmetric organocatalysis. Org Biomol Chem 3:719–724

Milner-White EJ, Watson JD (2002a) A novel main-chain anion-binding site in proteins: the nest. A particular combination of Φ, Ψ values in successive residues gives rise to anion-binding sites that occur commonly and are found often at functionally important regions. J Mol Biol 315:171–182

Milner-White EJ, Watson JD (2002b) The conformations of polypeptide chains where the main-chain parts of successive residues are enantiomeric. Their occurence in cation and anion-binding regions of proteins. J Mol Biol 315:183–191

Neumann R, Neimann K (2000) Electrophilic activation of hydrogen peroxide: selective oxidation reactions in perfluorinated alcohol solvents. Org Lett 2:2861–2863

Neumann R, Shaik S, de Visser SP, Kaneti J (2003) Fluorinated alcohols enable olefin epoxidationby H_2O_2: template catalysis. J Org Chem 68:2903–2912

Roberts SM, Bentley PA, Kroutil W, Littlechild JA (1997) Preparation of polyamino acid catalysts for use in Juliá asymmetric epoxidation. Chirality 9:198–202

Sheldon RA, van Vliet MCA, Arends IWCE (2001) Fluorinated alcohols: effective solvents for uncatalysed epoxidations with aqueous hydrogen peroxide. Synlett 2001:248–250

Recoverable, Soluble Polymer-Supported Organic Catalysts

M. Benaglia (✉)

Dipartimento di Chimica Organica e Industriale, Universita' degli Studi di Milano, Via C. Golgi 19, 20133 Milan, Italy
email: *maurizio.benaglia@unimi.it*

1	Introduction	300
2	General Considerations on the Immobilization Process	302
3	Achiral Organic Catalysts	307
3.1	Oxidation Catalysts	307
3.2	Phase-Transfer Catalysts	308
4	Chiral Organic Catalysts	311
4.1	Proline-Derived Catalysts	311
4.2	MacMillan's Catalyst	314
5	Conclusion	316
References		318

Abstract. Enantioselective organic catalysis is currently the subject of an intense research activity aimed to develop 'metal-free' alternatives to established 'metal-based' catalytic processes. In the case of chiral catalysts, which are often expensive, or obtained after a complex synthesis, the immobilization on solid support represents an attractive methodology that allows the recovery and possibly the recycling of the catalytic species. Among the different solid matrixes employed, soluble polymers recently emerged as very convenient supports for the synthesis of a variety of small organic molecules, ligands and catalysts. This chapter highlights the more recent results obtained by our group in the field of poly(ethylene glycol) (PEG)-supported chiral and achiral organic catalysts. A few considerations on the methodologies, the future and the problems related to the immobilization of chiral organic catalysts are also briefly discussed.

1 Introduction

Catalysts are used in the production of a large variety of chemicals and fuels, as demonstrated by the fact that catalyst-based manufacturing accounts for about 60% of chemical products and 90% of processes (Senkan 2001). These numbers will likely increase in the future, considering all the advantages of a catalytic process: it requires only small amount of a 'smart' molecule to produce a large quantity of the desired compound; the catalyst usually allows operation under mild reaction conditions; also the economic benefits of an efficient catalytic process are enormous since it is less capital-intensive, has lower operating costs, produces products of higher purity and fewer by-products. In addition, catalysts provide important environmental benefits.

These undoubted positive features are very attractive also for the preparation of chiral compounds. Demand for enantiomerically pure compounds is continuously increasing, not only for use in pharnaceuticals but also in other fields such as agrochemicals, flavour and aroma chemicals, and speciality materials. Recent, strict government regulations that require the individual evaluation of all the possible stereoisomers of a compound and the commercialization of a chiral product only as single enantiomer have called for further improvements in the stereoselective synthesis of chiral compounds. In this context it is surprising how relatively few enantioselective catalytic reactions are used on an industrial scale today (Blaser 2003). This is even more difficult to understand if one thinks about the impressive progress made over the last few years in the field of enantioselective catalysts, where hundreds of catalytic transformations of great chemical and stereochemical efficiency have been developed.

So the obvious question is this: Why has the application of enantioselective catalysis to the fine chemicals industry, of potentially great economic and environmental interest, not been widely pursued on large scale? It is true that different issues must be addressed: firstly there is the cost of the chiral catalyst, but other problems must also be considered, such as general applicability; many of the very selective catalysts have been developed for reactions with selected model substrates but not tested on differently functionalized molecules. In addition, for many catalysts little information is available on catalyst selectivity, activity, and productivity. The stability of the catalyst and the possibility of an easy separation and maybe recycling are also important aspects to be considered for an industrial asymmetric catalytic process.

In this context immobilization of the catalytic species on a solid support may represent a solution to some of the problems; not only the recovery and the possible recycling of a catalyst may be investigated and successfully realized through its immobilization, but also the studies of other issues such as stability, structural characterization and catalytic behaviour may be better conducted on a supported version of the enantioselective catalyst.

These general considerations are true also for organic catalysts which have recently been the subject of incredibly intense research activity (Dalko and Moisan 2004). Today it is safe to state that a very high number of fundamental organic reactions that once required the use of metal-based enantiopure catalysts can now be performed, at an equal level of chemical and stereochemical efficiency, by using substoichiometric amounts of structurally simple organic molecules. In this context, the term 'organic' is synonymous of 'metal-free' with all the advantages of performing a reaction under metal-free conditions. These advantages might include, *inter alia*, the possibility of working in wet solvents and under an aerobic atmosphere, dealing with a stable and robust catalyst and avoiding the problem of a (possibly) expensive and toxic metal leaching into the organic product.

Furthermore, as mentioned before, organic catalysts may be ready immobilized on a support with the aim of facilitating catalyst recovery and recycling (Benaglia et al. 2003).

It is reasonable to suppose that a simple organic compound will be less affected by the connection to a support than a more structurally complex and somehow more 'delicate' enzyme. Metal-free catalysts are also more readily amenable than organometallic catalysts to anchoring on a support in order to easily separate the product from the catalyst, and to recover and recycle the latter. Indeed, it has repeatedly been shown that the use of a metal-based catalyst immobilized on a support is often problematic because of possible, extensive metal leaching and requires catalyst regeneration by metal replenishment before recycling.

2 General Considerations on the Immobilization Process

The main goals of the immobilization of a catalyst on a support are the simplification of the reaction work-up, and the recovery and hopefully the recycling of the precious chiral catalyst. But besides the recovery and the recycle of the catalyst, other reasons may lead to realize a supported version of a catalytic species.

Catalyst instability can be a problem that may be tackled by developing an immobilized catalyst. Organic catalysts do exist that slowly decompose under the conditions necessary for their reaction and release trace amounts of by-products that must be separated from the products. For instance, in photooxigenation reactions catalysed by porphyrin the release of highly coloured materials derived from the catalysts made product purification a real problem. This problem can be solved by immobilization of the catalyst because the decomposed material is also supported and can thus be removed from the reaction medium during the process of catalyst recovery.

Among the different methodologies developed for the oxidation of unsaturated compounds with singlet oxygen, dye-sensitized photooxidation of triplet oxygen offers the advantage of producing by-product-free reaction mixtures and uses a catalytic amount of a substance that promotes oxygen to its singlet state. However, the presence of the sensitizer, or its decomposition derivatives, can contaminate the reaction products and make the purification step really troublesome. A polymer-bound catalyst can offer the possibility of simple recovery and recycling, and easy product purification and isolation. In this context, soluble polymers have recently been the subject of intense research activity. By carrying out the reaction in homogeneous solution a higher chemical and stereochemical efficiency could be achieved than by using insoluble polymer.

Among the soluble polymeric matrixes used, poly(ethylene glycol)s (PEG) are the most successful (Janda et al. 2002). These polymers with $M_w > 2,000$ Da are readily functionalized, commercially available, inexpensive supports that feature convenient solubility properties: they are soluble in many common organic solvents and insoluble in a few other solvents, such as diethyl ether, hexanes, t-butyl-methyl ether. Therefore,

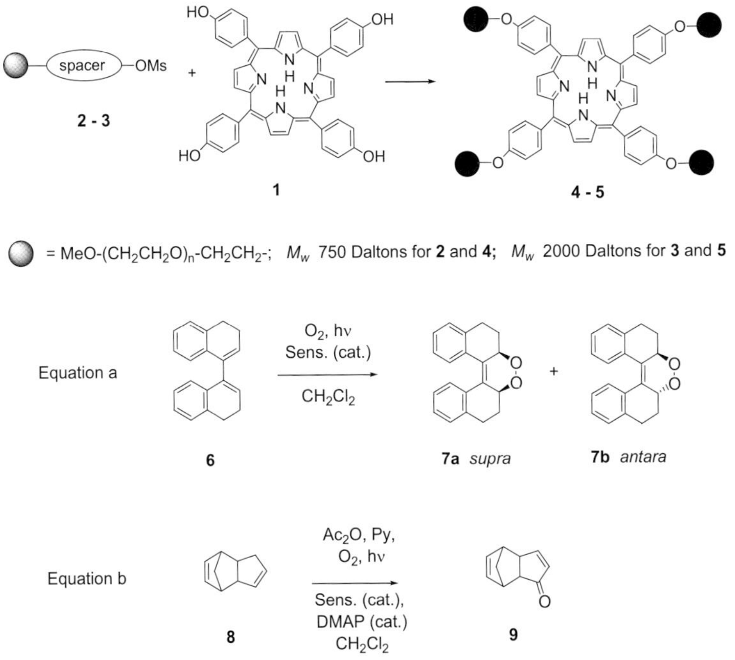

Recoverable, Soluble Polymer-Supported Organic Catalysts 303

Scheme 1.

○ = MeO-(CH$_2$CH$_2$O)$_n$-CH$_2$CH$_2$-; M_w 750 Daltons for **2** and **4**; M_w 2000 Daltons for **3** and **5**

Equation a: **6** → **7a** *supra* + **7b** *antara* (O$_2$, hv, Sens. (cat.), CH$_2$Cl$_2$)

Equation b: **8** → **9** (Ac$_2$O, Py, O$_2$, hv, Sens. (cat.), DMAP (cat.), CH$_2$Cl$_2$)

the choice of proper solvent systems makes it possible to run a reaction under homogeneous catalysis conditions (where the PEG-supported catalyst is expected to perform at its best) and then to recover the catalyst under heterogeneous conditions, as if it were bound to an insoluble matrix.

On the basis of our experience in the PEG-supported synthesis of small organic molecules, the mesylates **2** and **3**, M_W 750 and 2,000 respectively, were prepared in three steps and 95% overall yield from the commercially available monomethylether of PEG (MeOPEG) and attached to the commercially available, 5,10,15,20-tetrakis-(4-hydroxyphenyl)-porphyrin **1** (Scheme 1).

Unfortunately when the MeOPEG with M_W = 750 was used, the product **4** (obtained in 93% yield) could not be purified by precipitation from diethyl ether, but it was recovered as a very thick oil. However, by using the mesylate **3** of MeOPEG$_{2000}$ the expected PEG-supported porphyrin **5** was obtained as a solid readily precipitated with diethyl ether in 87% yield as a pure compound. The irradiation of a 0.01 M methylene chloride solution of bisdialine **6** with a 100-W halogen lamp, in the presence of 3 mol% of PEG supported tetrahydroxyphenyl-porphyrin (PEG-TPP, **5**) as sensitizer gave a 82/18 mixture of *supra* and *antara* diastereoisomeric endoperoxides **7** in quantitative yield after 1 h (equation a, Scheme 1). The polymer-bound catalyst not only showed the same activity of the non-supported species, but it greatly simplified isolation of the product (Fabris et al. 2002). At the end of the reaction the solvent was concentrated in a vacuum and diethyl ether was added to the PEG-supported porphyrin which was then recovered quantitatively by filtration. From this concentrated filtrate the endoperoxides were easily isolated by crystallization from ethanol. The one-pot oxidation of olefin to $α,β$-unsaturated ketones was applied to convert dicyclopentadiene **8** into the corresponding dicyclopentadienone **9**, a useful starting material for the preparation of enantiopure diols (equation b, Scheme 1). Once again, the soluble polymer-bound porphyrin **5** favourably compares with the non-supported catalyst. It is worth mentioning that it is possible to run the reaction on gram scale, using an even smaller amount of sensitizer; the oxidation of 13 mL. of dicyclopentadiene, promoted by 220 mg of PEG–TPP **5** (0.025% mol equivalent), gave 11 g of ketone **9**, isolated, after filtration of the supported catalyst, by simple evaporation of the organic solvent as analytically pure compound requiring nofurther purification.

Recycling of the PEG-bound porphyrin **5** was also studied. The catalyst was reused a second time in the photooxidation of the bisdialine to give the product **7** in 97% yield and 80/20 diastereoisomeric ratio after a 1-h reaction. The recovered PEG–TPP was recycled a third time without showing any decrease in the catalytic activity; the PEG-supported sensitizer was recycled six times with no appreciable loss of chemical or stereochemical efficiency.

Immobilization of an organic catalyst can be used also to facilitate the process of catalyst's optimization. Surprisingly, there is just a single example of the application of this methodology reported to date and so it deserves a special mention (Jacobsen and Sigman 1998).

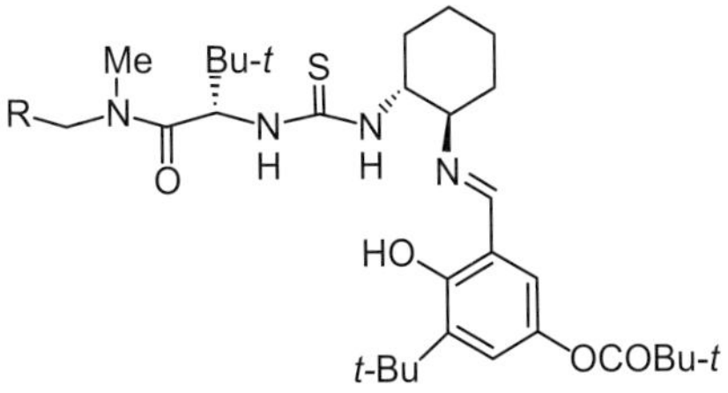

Scheme 2.

10a R = 1% cross-linked polystyrene
10b R = H

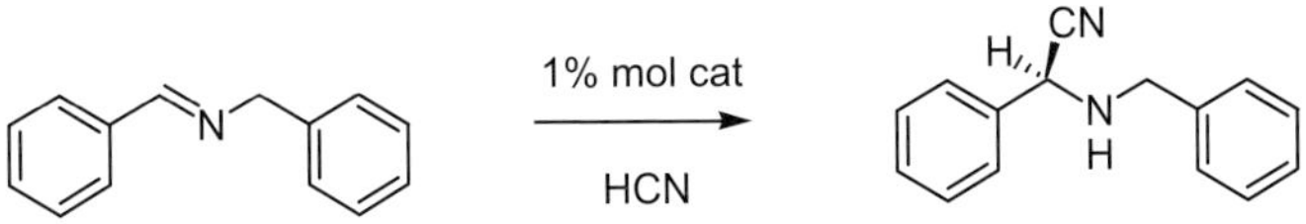

Jacobsen developed a fully organic catalyst for the Strecker reaction using a thiourea based chiral Brã̧nsted acid that turned out to be extremely chemically active, stereoselective, and with broad application. The optimization of the catalyst structure was realized through a series of modifications of the salen-based structure carried out on an insoluble polystyrene support, and using the principles of combinatorial chemistry for finding out the best aminoacid, diamine, and diamine-amino acid linker combination. The screening of three successive libraries led to the identification of the supported thiourea catalyst **10a** as the single best form from which the non-supported counterpart **10b** was derived.

At a loading as low as 1 mol%, **10b** promoted the hydrocyanation of *N*-allyl or -benzyl imines derived from aromatic and aliphatic aldehydes and of some ketones in very high yield and almost complete stereoselectivity (see Scheme 2). It is interesting to note that the soluble and the resin-bound catalysts performed equally well. Recovery and recycling of the supported catalyst was shown to occur without any erosion of chemical or stereochemical efficiency over ten reaction cycles.

Given these excellent results, it is surprising that this approach has not been used more extensively for chiral organic catalyst discovery. The success of this methodology is even more significant if one considers that catalysts **10** are among the few chiral organic catalysts to be currently used at the industrial level.

The preparation of supported catalysts for use in environmentally friendly or green solvents, as part of the drive towards developing more Green Chemistry, is also becoming more widespread. In this context PEG-supported enantiomerically pure bisoxazolines have been prepared and used in combination with Cu(II) salts as the catalyst for the Mukaiyama aldol condensation between the trimethylsilyl keteneacetal of methyl isobutyrate and various aldehydes carried out in water as the only reaction solvent. Enantiomeric excesses similar to those obtained under the same conditions with non-supported ligands were observed. The reaction proceeded at higher yields when electron-poor aldehydes were used as the substrates. The high solubility of the ligand in water allowed a very convenient catalyst-recycling procedure involving simple removal of the reaction product by extraction in diethylether and addition of fresh reagents to the catalyst-containing aqueous solution. The chemical and stereochemical efficiency of the catalyst was only marginally eroded over its use in three reaction cycles.

However, PEG supported metal-free catalysts have also been shown to perform well in water. For example the synthesis of a PEG-supported TEMPO (2,2,6,6-tetramethyl-piperidine-1-oxyl), and its use as a highly efficient, recoverable and recyclable catalyst in oxidation reactions was described (Pozzi et al. 2004).

TEMPO-catalysed oxidation of alcohols to carbonyl compounds with buffered aqueous NaOCl has found broad application, even in large-scale operations. Indeed, this selective methodology involves the use of safe, inexpensive inorganic reagents under mild reaction conditions. A new supported catalyst PEG–TEMPO **11**, soluble in organic solvents such as CH_2Cl_2 and AcOH, but insoluble in ethers and hexanes, was prepared and proved to be an effective catalyst for the selective oxidation of 1-octanol with various stoichiometric oxidants. When **11** was used at 1 mol% as a catalyst in combination with KBr (10 mol%) a slight excess of buffered bleach (pH = 8.6) as the terminal oxidant, partial over-oxidation of 1-octanol to octanoic acid was observed (91% yield in

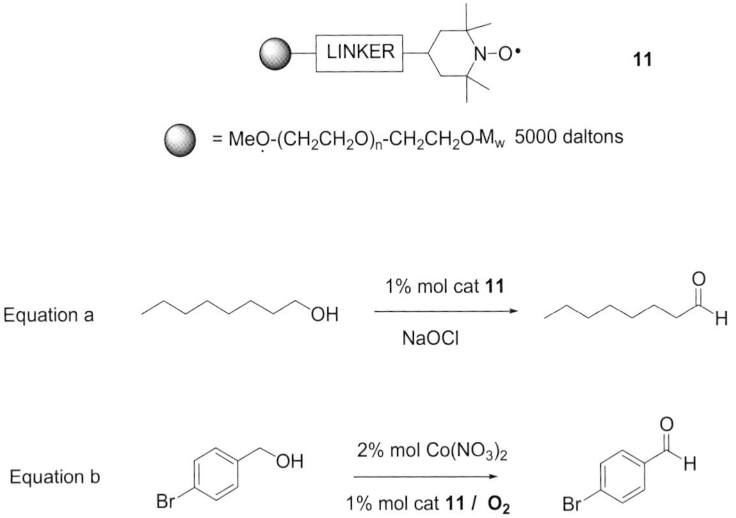

Scheme 3.

octanal). This can be avoided either by using a stoichiometric amount of NaOCl or, more conveniently, working under bromide-free conditions (equation a, Scheme 3). Although slightly decreased, the oxidation rate remains high even in the absence of KBr and the aldehyde was obtained in 95% yield after only 30 min.

3 Achiral Organic Catalysts

3.1 Oxidation Catalysts

The use of oxoammonium ions such as those derived from TEMPO in combination with inexpensive, safe, and easy-to-handle terminal oxidants in the conversion of alcohols into aldehydes, ketones, and carboxylic acids is a significant example of how it is possible to develop safer and greener chemistry, by avoiding the use of environmentally-unfriendly or toxic metals. However, separation of the products from TEMPO can be problematic, especially when the reactions are run on a large scale; immobilization on a solid support may offer a solution to this problem.

As mentioned before a PEG-supported TEMPO proved to be very efficient in the oxidation of 1-octanol to octanal not only with sodium hypochlorite, but also in combination with different terminal oxidants such as bis(acetoxy)iodobenzene and trichloroisocyanuric acid. This reaction could be extended to acyclic and cyclic primary and secondary alcohols with excellent results. It is remarkable that the PEG-supported TEMPO maintained the good selectivity for primary vs secondary benzylic alcohol oxidation typical of non-supported TEMPO.

Catalyst recovery exploited the well known different solubility of PEG in solvents of different polarity. In this case, where a PEG of relatively high M_W was used, addition of diethylether to the reaction mixture induced the precipitation of compound **11**. Subsequent filtration allowed recovery of the supported reagent with less than 10% weight loss for each recovery. Catalyst recycling was shown to be possible for seven reaction cycles in the oxidation of 1-octanol, that occurred in undiminished conversion and selectivity under the same reaction conditions.

The beneficial effect of the spacer was also demonstrated in a subsequent work, where spacer-containing and spacer-lacking pre-catalysts (1–5 mol%) were used for the oxidation of primary and secondary alcohols using oxygen as the terminal oxidant in the presence of 2 mol% of $Co(NO_3)_2$ hydrate and $Mn(NO_3)_2$ hydrate as co-catalyst (Pozzi 2005). For instance, under these conditions oxidation of 1-octanol and cyclooctanol occurred in quantitative yield with **1** and in only about 65% yield with **12** (Scheme 3). Recycling of pre-catalyst **1** was demonstrated for the oxidation of 4-bromobenzyl alcohol for six reaction cycles occurring in slowly decreasing yields (>99% first cycle and 74% sixth cycle) (equation b, Scheme 3).

3.2 Phase-Transfer Catalysts

The multifaceted applications of phase-transfer catalysts (PTC) in organic synthesis contributed decisively to the establishment of organic catalysts as useful preparative tools. Polymer-supported PTC was examined extensively but it was noted that the catalytic activity of the insoluble polystyrene-supported catalysts was strongly reduced in comparison with that of their non-supported soluble counterparts. Since the 1960s soluble polymeric supports have been envisaged as possible alternatives to their insoluble counterparts for catalyst immobilization.

A quaternary ammonium salt was easily synthesized on a modified MeOPEG, and this supported catalyst was shown to be an efficient and recoverable promoter of several reactions carried out under PTC conditions. Catalyst **13** showed a catalytic activity that was similar to, or even better than that of the non-supported catalysts (Benaglia et al. 2000).

Remarkably, the benzylation of phenol and pyrrole required only 0.01 equivalents of catalyst **13** to occur in ~95% yield. Dichloromethane, in which the catalyst is readily soluble, was found to be the organic solvent of choice, but the reaction could satisfactorily be carried out also in the absence of organic solvent. Generally the use of solid/liquid conditions led to higher yields than those observed under liquid/liquid conditions. The PEG-supported ammonium salt could be recovered by precipitation and filtration, and recycled three times to run the same or even a different reaction without any appreciable loss of catalytic activity. It is also worth mentioning that compound **13** compares favourably as catalyst to other quaternary ammonium salts immobilized on insoluble polystyrene supports. The use of these catalysts generally required higher reaction temperatures and/or longer reaction times than those used here. In addition, solid supported phase-transfer catalysts required a preliminary, long conditioning time (up to 15 h) to ensure bead swelling and optimum accessibility of substrate and reagent to the catalytic site. Finally, the high stirring rate necessary with these catalysts resulted in extensive mechanical degradation of the polymer beads, that were difficult to recover by filtration. It seems possible that the catalyst benefits from the involvement of the support, because the polyethylenoxy chain of PEG can complex the alkaline cation of the base helping the transfer of the HO^- counterion in the organic phase.

In order to increase the number and a proper spatial arrangement of the catalytic sites the loading expansion of PEG was carried out exploiting the principles of dendrimer chemistry and led to the synthesis of the PEG-supported tetrakis ammonium salt **14** (Tocco et al. 2002). This catalyst displayed a higher catalytic efficiency than **13**, while retaining the solubility properties peculiar of the PEG support, that allowed simple catalyst recovery and recycling by precipitation and filtra-

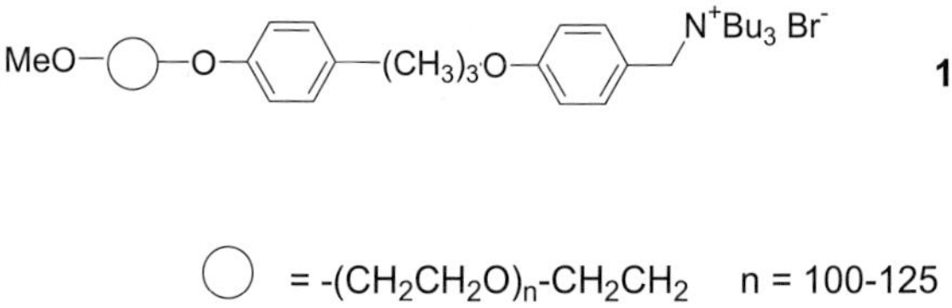

Scheme 4.

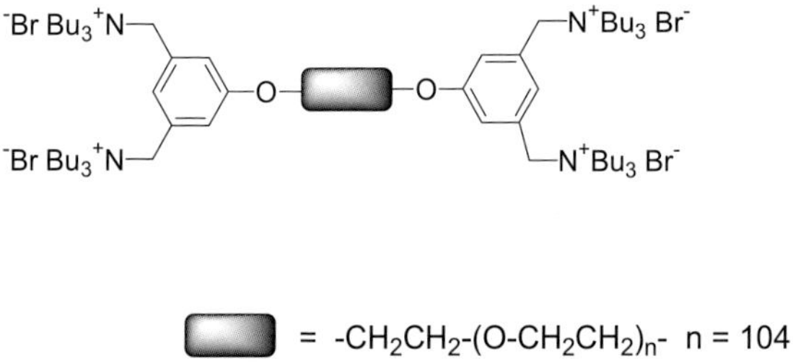

MeO—◯—O—⟨benzene⟩—(CH₃)₃O—⟨benzene⟩—CH₂N⁺Bu₃ Br⁻   **13**

◯ = -(CH₂CH₂O)$_n$-CH₂CH₂   n = 100-125

⁻Br Bu₃⁺N—⟨benzene⟩—O—▭—O—⟨benzene⟩—N⁺Bu₃ Br⁻   **14**
(with additional ⁻Br Bu₃⁺N— and —N⁺Bu₃ Br⁻ substituents on the benzene rings)

▭ = -CH₂CH₂-(O-CH₂CH₂)$_n$-   n = 104

tion (Scheme 4). The catalyst, that featured four quaternary ammonium groups located at the termini of the polymer backbone, displayed a remarkable efficiency in promoting different reactions carried out under PTC conditions at low catalyst concentrations (1–3 mol%) for short reaction times and under mild conditions. The polymer support allowed easy recovery and recycling of the catalyst.

In an attempt to develop a PEG-supported version of a chiral phase-transfer catalyst the *Cinchona* alkaloid-derived ammonium salt **15** used by Corey and Lygo in the stereoselective alkylation of amino acid precursors was immobilized on a modified PEG similar to that used in the case of **13**. The behaviour of the catalyst obtained **16**, however, fell short of the expectations (Danelli et al. 2003). Indeed, while this catalyst (10 mol%) showed good catalytic activity promoting the benzylation of the benzophenone imine derived from *tert*-butyl glycinate in 92% yield (solid CsOH, DCM, −78 to 23 °C, 22 h), the observed *ee* was only 30%. Even if this was increased to 64% by maintaining the reaction temperature at −78 °C and prolonging the reaction time to 60 h, the top level of stereoselectivity obtained with the non-supported catalyst could not be matched. PEG was considered to be responsible, at least in part, for these results because of the following effects. By increasing the polarity around the catalyst, PEG prevents the formation of a tight ion pair between the enolate and the chiral ammonium salt, the formation of which is regarded as crucial for high stereocontrol. Moreover, PEG enhances the solubility of the inorganic cation in the organic phase leading to a competing non-stereoselective alkylation occurring on the achiral cesium enolate. To check the validity of these hypotheses control experiments were carried out by performing the reaction with the non-supported catalyst in the presence of the bis-methylether of PEG_{2000}. The observed *ee* was 65%, a value that was in good agreement with that observed with catalyst **16** but largely inferior to the >90% *ee* easily achieved with the non-supported catalyst (Scheme 5).

4 Chiral Organic Catalysts

An organic catalyst has been recently defined as 'organic compound of relatively low molecular weight and simple structure capable of promoting a given transformation in substoichiometric quantity' (Benaglia et al. 2003). In this context chiral organic catalysts may be seen as a simplified version of enzymes, from which they are conceptually derived and to which they are often compared. Even if they rarely display the remarkable selectivity peculiar of enzymes, in general metal-free catalysts are more stable than bio-catalysts and show a larger field of application under a variety of conditions unsustainable by enzymes; furthermore organic catalysts may be readily immobilized on a support with the aim of facilitating catalyst recovery and recycling.

4.1 Proline-Derived Catalysts

Enantioselective catalysis promoted by enantiomerically pure amines (aminocatalysis) is the subject of considerable interest due to the ubiquitous presence and ready availability of these compounds in the chiral pool. In this context amino acids have always played a key role. One of the most successful and versatile chiral organic catalysts, proline, was

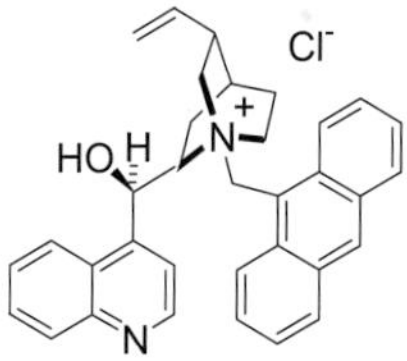

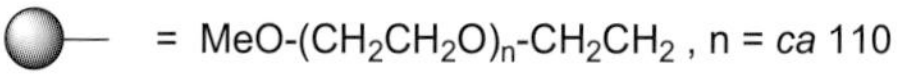

Scheme 5.

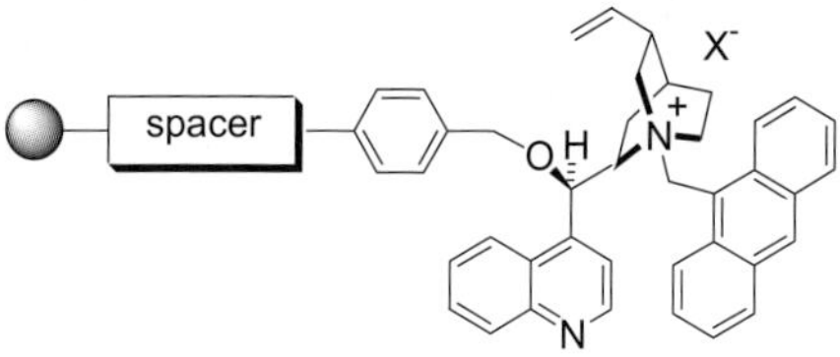

immobilized very soon after the first seminal works of List and Barbas, on both soluble and insoluble supports (Cozzi 2006). Our group developed the first soluble polymer-supported version of proline that was tested successfully tested in the adirect aldol condensation of ketones with aldehydes (Benaglia et al. 2001).

The immobilization of (2*S*,4*R*)-4-hydroxyproline on PEG$_{5000}$ monomethylether (MeOPEG) by means of a succinate spacer created, at very good yield, the supported catalyst **17**.

In the presence of 0.25–0.35 mol equivalents of the catalyst **17**, acetone reacted with enolizable and nonenolizable aldehydes in dimethyl fluoride at room temperature to afford the aldol product in yields (up to 80%) and *ee* (up to >98%), comparable with those obtained using nonsupported proline derivatives as catalysts (that did, however, give faster

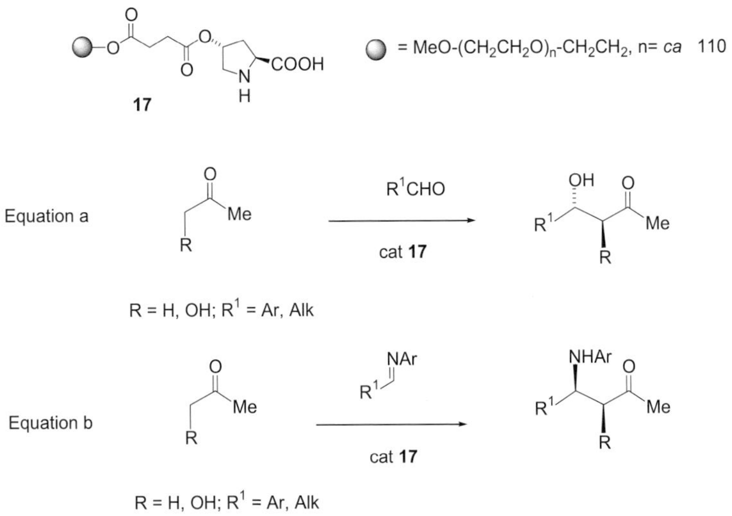

Scheme 6.

Equation a: Aldol reaction of a ketone (R-CH$_2$-C(O)-Me) with R^{1}CHO catalyzed by cat **17** gives the β-hydroxyketone. R = H, OH; R^1 = Ar, Alk

Equation b: Mannich reaction of a ketone (R-CH$_2$-C(O)-Me) with imine R^{1}CH=NAr catalyzed by cat **17** gives the β-aminoketone. R = H, OH; R^1 = Ar, Alk

Catalyst **17**: PEG-supported proline (succinate linker), where ● = MeO-(CH$_2$CH$_2$O)$_n$-CH$_2$CH$_2$, n = ca 110.

reactions). The use of PEG-supported proline was extended also to the condensation of hydroxyacetone with cyclohexane–carboxyaldehyde that afforded the corresponding anti-α,β-dihydroxyketone in 48% yield and 96% ee (*anti/syn* ratio >20/1) (equation in Scheme 6).

While the relatively low cost of many amino acids does not seem to justify the preparation of supported catalysts derived from them, other reasons may drive the immobilization of chiral catalysts, such as those mentioned above and the possibility of experimenting with different solubility properties, easy separation of the products from the catalysts and the catalyst's recyclability. The immobilization of these compounds on a support can also be seen as an attempt to develop a minimalistic version of an enzyme, with the amino acid playing the role of the enzyme's active site and the polymer that of an oversimplified peptide backbone not directly involved in catalytic activity.

The catalyst **17** was used also in Mannich condensations; the reaction of a ketone with imines afforded β-aminoketones in good yield and high enantiomeric excess (*ee*) (Puglisi et al. 2002). Extension of the PEG–

Proline-promoted condensation to hydroxyacetone as the aldol donor gave access to synthetically relevant *syn*-α-hydroxy-β²-aminoketones, that were obtained in moderate to good yields, and good to high diastereo- and enantioselectivities (equation b, Scheme 6). Exploiting its solubility properties, the PEG–Pro catalyst was easily recovered and recycled to promote all of the above-mentioned reactions that occurred in slowly diminishing yields but virtually unchanged enantioselectivities.

4.2 MacMillan's Catalyst

Another chiral organic catalyst of major success is MacMillan's catalyst, **18**, that has found widespread use in a number of relevant processes (MacMillan 2000). Immobilized versions of these catalysts for the enantioselective Diels–Alder cycloaddition of dienes with unsaturated aldehydes were developed on soluble (Benaglia et al. 2002) and insoluble supports (equation a, Scheme 7) (Benaglia 2006).

The PEG-supported imidazolidinone **19** (Scheme 7) was later used in 1,3-dipolar cycloadditions (Puglisi et al. 2004). By reacting *N*-benzyl-*C*-phenylnitrone with acrolein, it was shown that the outcome of the reaction was strongly dependent on the nature of the acid employed to generate the catalyst, and that only the use of HBF_4 as in the case of **19** allowed reproducible results to be obtained. Under the best reaction conditions (20 mol% of catalyst, DCM, –20 °C, 120 h) the product was obtained in 71% yield as a 85:15 *trans/cis* mixture of isomers, the *trans* isomer having 87% *ee* (equation b, Scheme 7).

A comparison of the results obtained with catalyst **19** with those obtained by MacMillan indicates that the major difference between the PEG-supported and the non-supported catalyst resides in the chemical rather than in the stereochemical efficiency. Indeed, while the supported catalyst gave *trans/cis* ratios almost identical to and *ee* only 3%–6% lower than those obtained with the non-supported catalyst, the difference in chemical yields was larger, ranging from 9% to 27%.

The PEG-supported catalyst was recycled twice to afford the product with constant level of diastereo- and enantio-selectivity but in chemical yields diminishing from 71% to 38%. In order to explain this behaviour several experiments were carried out. First, after each recovery the supported catalyst was examined by ^1H NMR that showed degradation,

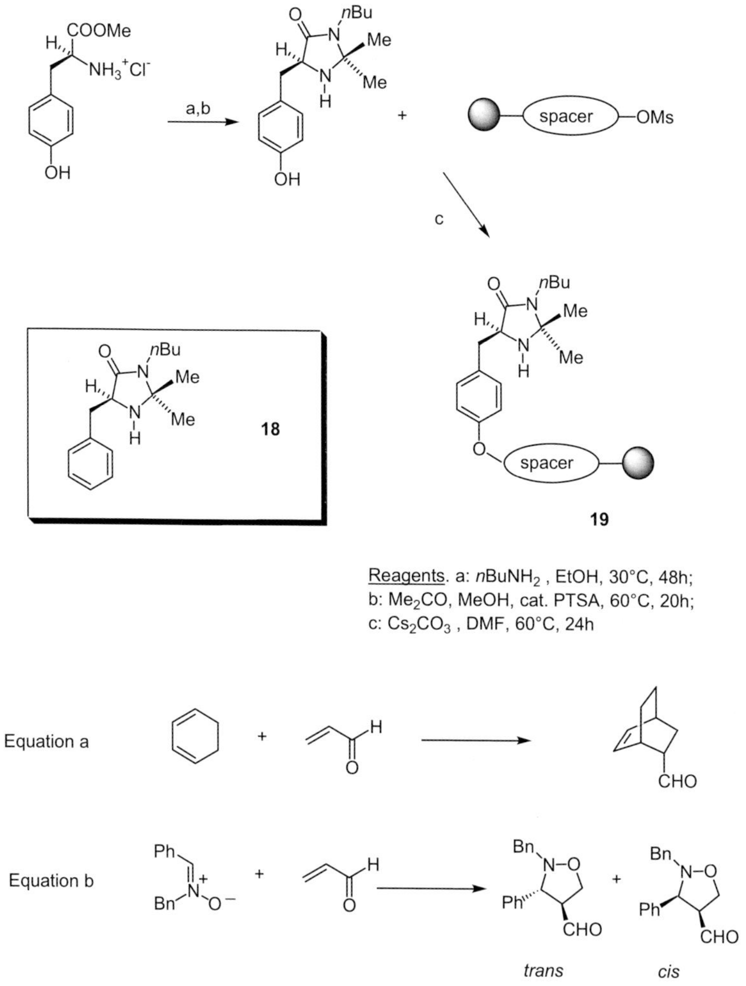

Reagents. a: nBuNH$_2$, EtOH, 30°C, 48h;
b: Me$_2$CO, MeOH, cat. PTSA, 60°C, 20h;
c: Cs$_2$CO$_3$, DMF, 60°C, 24h

Scheme 7.

increasing after each cycle, probably due to an imidazolidinone ring opening process. The triflate salt of the non-supported imidazolidinone was kept for 120 h at 24 °C in a 95/5 CD$_3$CN/D$_2$O mixture both in the presence and in the absence of the bismethyl ether of PEG_{2000} and in neither case was degradation observed by NMR, thus suggesting that the polymer is not playing a leading role in provoking the instability of the supported catalysts.

These findings seemed to point to catalyst degradation induced by the reagents, and indeed further NMR analysis showed extensive catalyst degradation in the presence of acrolein, less degradation with crotonaldehyde, and essentially no degradation with cinnamaldehyde (nitrone did not exert any effect on the catalyst stability). In agreement with the results of these experiments, we observed also that the non-supported catalyst showed a marked instability and decrease in chemical efficiency when recycled.

5 Conclusion

On the basis of our experience in the field of supported organic catalysts a few general considerations may be attempted.

The case of the PEG-imidazolidinone **19** is a clear example how a judicious choice of the catalyst to be immobilized is important. Possible candidates to be supported must be catalysts of great versatility and with a wide tolerance of structurally different substrates, possibly with a high catalytic efficiency. However, the stability of a metal-free catalyst under the reaction conditions of its standard applications is also a crucial issue to be carefully considered; therefore, it is a good policy to make a preliminary investigation on the real stability of the catalyst before its immobilization. Furthermore, only catalysts really versatile in scope (that is, catalysts capable of promoting more than one type of reaction, or tolerating a real variety of substrates in a given transformation) are worth considering as possible candidates for immobilization.

Another point of obvious interest is the choice of the support, which is crucial as several features of the support may influence the catalyst's behaviour at every level. The solubility properties are the most important, and the decision to develop either a homogeneous or a heterogeneous catalytic system is the first to be made in designing immobilization. While the homogenous catalytic system is expected to be more reactive, stereochemically more efficient (because it operates in solution exactly like the non-supported system) and more reliable in reproducibility than a heterogenously supported catalyst, the latter is normally believed to be more stable, easy to handle and more simply recovered and recycled. An ideal support does not exist, but probably the best immobilization technique must be selected for each specific catalytic system to be supported. In this sense, significant progress could derive from the development of interdisciplinary expertise with the contributions of organic, polymer and material chemists.

In the context of catalyst separation and recycling, it must be noted that a system where a catalyst must not be removed from the reaction vessel is very attractive. An example comes from continuous flow methods, when the immobilized catalyst permanently resides in the reactor where it transforms the entering starting materials into the exiting products. The retention of the catalyst inside the reaction vessel can be achieved by different techniques ranging from ultrafiltration through a M_W-selective membrane to immobilization on a silica gel column.

Another point of discussion is the presence of a linker. If it is true that the support should exert the minimum effect on the catalyst, it seems obvious that the longer the distance between the catalyst and the support, the higher the chances of the supported catalyst to mimic the behaviour of its non-supported analogue. Following this idea an appropriate spacer (Montanari et al. 1979) was often introduced to separate the catalytically active site from the support; the methodology has been applied also to soluble supports, even if it is worth mentioning that has been demonstrated how the principle of maximum separation between the catalyst and the support is likely to be overestimated.

In conclusion, it is clear now that asymmetric organocatalysis has achieved the same relevance to stereoselective synthesis as organometallic catalysis, and it can be safely anticipated that novel *fully organic* methods will soon be developed to perform an even wider variety of reactions. In this context, the development of immobilized chiral organic catalysts will play an important role in contributing to further expand the applicability of organic catalysts and even to help to discover new chiral organic catalytic species. The immobilization of chiral organic catalysts represents a relatively new field of research, in great expansion and open to the interdisciplinary contributions of organic and material chemists.

References

Benaglia M (2006) Recoverable and recyclable chiral organic catalysts. New J Chem 30:1525–1533

Benaglia M, Annunziata R, Cinquini M, Cozzi F, Tocco G (2000) A poly(ethylene glycol)-supported quaternary ammonium salt: an efficient, recoverable, and recyclable phase-transfer catalyst. Org Lett 2:1737–1739

Benaglia M, Celentano G, Cozzi F (2001) enantioselective aldol condensation catalyzed by poly(ethylene glicol)-supported proline. Adv Synth Catal 343:171–173

Benaglia M, Celentano G, Cinquini M, Cozzi F, Puglisi A (2002) Poly(ethylene glycol)-supported chiral imidazolidin-4-one: an efficient organic catalyst for the enantioselective diels-alder cycloaddition. Adv Synth Catal 344:149–152

Benaglia M, Puglisi A, Cozzi F (2003) Polymer supported organic catalysts. Chem Rev 103:3401–3429

Blaser HU (2003) Enantioselective catalysis in fine chemicals production. Chem Comm 293–297

Cozzi F (2006) Immobilization of organic catalysts: when, why, and how. Adv Synth Catal 348:1367–1390

Dalko PI, Moisan L (2004) In the golden age of organocatalysis. Angew Chem Int Ed 35:5138–5175

Danelli T, Annunziata R, Benaglia M, Cinquini M, Cozzi F, Tocco G (2003) Immobilization of catalysts derived from *Cinchona* alkaloids on modified poly(ethylene glycol). Tetrahedron: Asymmetry 14:461–467

Fabris F, Benaglia M, Danelli T, Sperandio D, Pozzi G (2002) Poly(ethylene glycol)-supported tetrahydroxyphenyl porphyrin: a convenient, recyclable catalyst for photooxidation reactions. Org Lett 4:4229–4242

Jacobsen EN, Sigman MS (1998) Schiff base catalysts for the asymmetric Strecker reaction identified and optimized from parallel synthetic libraries. J Am Chem Soc 120:4901–4902

Janda KD, Dickerson TJ, Reed NN (2002) Soluble polymers as scaffolds for recoverable catalysts and reagents. Chem Rev 102:3325–3244

MacMillan DWC, Ahrendt KA, Borths CJ (2000) New strategies for organic catalysis: the first highly enantioselective organocatalytic diels-Alder reaction. J Am Chem Soc 122:4243–4244

Pozzi G, Cavazzini M, Quici S, Benaglia M, Dell' Anna G (2004) Poly(ethylene glycol)-supported TEMPO: an efficient, recoverable, metal-free catalyst for the selective oxidation of alcohols. Org Lett 6:441–443

Pozzi G, Quici S, Benaglia M, Puglisi A, Holczknecht O (2005) Aerobic oxidation of alcohols to carbonyl compounds mediated by poly(ethylene glycol)-supported TEMPO radicals. Tetrahedron 61:12058–12064

Puglisi A, Benaglia M, Cinquini M, Cozzi F (2002) Poly(ethylene glycol)-supported proline: a versatile catalyst for the enantioselective aldol and iminoaldol reactions. Adv Synth Catal 344:533–542

Puglisi A, Benaglia M, Celentano G, Cinquini M, Cozzi F (2004) Enantioselective 1,3-dipolar cycloadditions of unsaturated aldehydes promoted by a poly(ethylene glycol)-supported organic catalyst. Eur J Org Chem 567–573

Senkan S (2001) Combinatorial heterogeneous catalysis–a new path in an old field. Angew Chem, Int Ed 40:312–337

Tocco G, Benaglia M, Cinquini M, Cozzi F (2002) Synthesis of a poly(ethylene glycol)-tetrakis ammonium salt: a recyclable phase-transfer catalyst of improved catalytic efficiency. Tetrahedron Lett 43:3391–3393

Controlling the Selectivity and Stability of Proteins by New Strategies in Directed Evolution: The Case of Organocatalytic Enzymes

M.T. Reetz(✉)

Max-Planck-Institut für Kohlenforschung, Kaiser-Wilhelm-Platz 1, 45470 Mülheim/Ruhr, Germany
email: *reetz@mpi-muelheim.mpg.de*

1	Introduction	322
2	Directed Evolution of Enantioselective Lipases	325
3	Directed Evolution of Enantioselective Baeyer-Villigerases	329
4	Directed Evolution of Enantioselective Epoxide Hydrolases	331
5	Directed Evolution of Hyperthermostable Enzymes	336
6	Conclusions	337
	References	337

Abstract. The directed evolution of functional enzymes as catalysts in organic reactions has emerged as a powerful method of protein engineering. This includes the directed evolution of enantioselective enzymes as pioneered by the author. In recent years the challenges in this new area of asymmetric catalysis has shifted to solving the problem of probing protein sequence space more efficiently than before. Iterative saturation mutagenesis (ISM) is one way of addressing this crucial question. This chapter reviews the concept of ISM and its application in controlling the enantioselectivity and thermostability of enzymes, specifically those that have an organocatalytic mechanism. Illustrative examples include the directed evolution of lipases, Baeyer-Villigerases and epoxide hydrolases.

1 Introduction

Due to economic and ecological factors, catalytic processes in the production of fine chemicals are gaining in importance, especially in the area of asymmetric catalysis (Collins et al. 1997; Breuer et al. 2004). Accordingly, the practicing chemist has three major options: transition metal catalysts (Jacobsen et al. 1999), organocatalysts (Berkessel and Gröger 2004) or enzymes (Drauz and Waldmann 2002; Liese et al. 2006). All of them have advantages and disadvantages, which means that a given type of catalysis cannot be expected to provide general solutions to all problems of relevance in academic and industrial laboratories. Therefore, research in all three approaches needs to be intensified.

The renaissance of organocatalysis since 2000 is truly impressive (Berkessel and Gröger 2004; Seayad and List 2005) (see also the other chapters in this monograph). Much of the design of organocatalysts for a variety of different reaction types is inspired by the knowledge that has accumulated regarding the mechanisms of enzyme catalysis. It has been estimated that about 40% of all enzymes are metalloenzymes, while the majority (60%) unfold their catalytic power in the absence of transition metals. Thus, the latter can be considered to be organocatalytic enzymes. Examples are Type I aldolases (which served as models for numerous proline-catalysed organocatalytic reactions), lipases, epoxide hydrolases and flavin-dependent monooxygenases (Collins et al. 1997; Breuer et al. 2004; Drauz and Waldmann 2002; Liese et al. 2006). The use of these enzymes in synthetic organic chemistry is well documented, and numerous industrial processes leading to the production of chiral or achiral products are known.

Hundreds of examples of the successful use of enzymes in asymmetric catalysis have been reported, including numerous industrial processes (Collins et al. 1997; Breuer et al. 2004; Drauz and Waldmann 2002; Liese et al. 2006). Nevertheless, the traditional limitations of enzymes as catalysts in synthetic organic chemistry revolve around limited substrate acceptance, poor enantioseletivity in many cases and/or insufficient thermostability under operating conditions. Various approaches to solving these problems have been suggested, and numerous successful examples are known which add to the power of enzyme catalysis. These include immobilization, use of additives, post-translational chemical modification, site-directed mutagenesis and directed evolution. Site-directed mutagenesis is based on rational design, requiring structural and mechanistic knowledge of the enzyme under study as well as sound predictive power, which makes general use difficult (Cedrone et al. 2000). For this reason an alternative genetic approach, namely directed evolution, has provided completely new perspectives in the quest to optimize and tune existing enzymes according to the needs of chemists and biotechnologists (Arnold and Georgiou 2003; Brakmann and Schwienhorst 2004). Directed evolution of functional enzymes is based on the appropriate combination of random gene mutagenesis, expression and high-throughput screening (or selection). Darwinism in the test tube, as it is sometimes called, was suggested and strived for many decades ago, but it was not until the 1980s and 1990s that molecular biologists developed efficient gene mutagenesis methods needed for this type of protein engineering. These techniques include error-prone polymerase chain reaction (epPCR), saturation mutagenesis and DNA shuffling (Arnold and Georgiou 2003; Brakmann and Schwienhorst 2004). In particular epPCR is used most often, three to eight such cycles being common. Typically, an epPCR library of 1,000–10,000 clones is generated, and following expression and screening the best mutant (hit) is identified and sequenced. The mutant gene of this hit is subsequently used as a template for performing another cycle of epPCR, and so on. These and other procedures were increasingly used in the 1990s to improve the thermostability of enzymes and/or stability against organic solvents. In 1995 we launched a project with the aim of controlling the enantioselectivity of enzymes by directed evolution (Reetz et al. 1997). Figure 1 illustrates the concept, which constitutes a fundamentally new approach to asymmetric catalysis.

The challenges in putting this concept into practice revolve around: (1) the problem of probing protein sequence space efficiently; and (2) the necessity of developing high-throughput screening systems for determining the enantiomeric purity of thousands of samples per day. In the present overview the emphasis is on the development of efficient methods for probing protein sequence space, whereas reviews regarding *ee*-assays have appeared elsewhere (Reetz 2004). In order to illustrate the essential concepts here, organocatalytic enzymes are chosen,

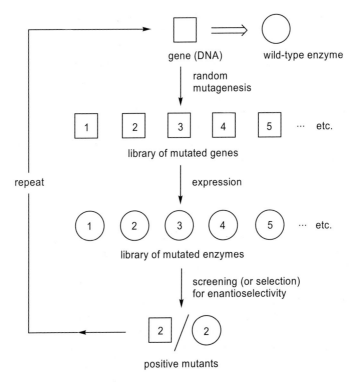

Fig. 1. Strategy for directed evolution of an enantioselective enzyme

namely lipases, Baeyer-Villigerases and epoxide hydrolases. The problem of protein sequence space can be illuminated as follows. When considering a mutagenesis method which introduces one amino acid exchange randomly in a protein composed of 300 amino acids, it can be calculated that there are 5,700 different mutants theoretically possible. When the mutation rate is increased leading to an average of two or three simultaneous exchanges, the number of possible mutants increases to 16 million or 30 billion, respectively.

2   Directed Evolution of Enantioselective Lipases

Lipases catalyse the hydrolysis of carboxylic acid esters (or esterification/transesterification in organic solvents) (Schmid and Verger 1998). Enantioselectivity is relevant in kinetic resolution of racemates or in desymmetrization (e. g., of meso-substrates). The mechanism is organocatalytic, in which a catalytic triad (Asp, His, Ser) initiates a proton shuttle. The activated serine attacks the ester function with formation of the so-called oxyanion in the rate-determining step (Fig. 2). Subsequently the acyl enzyme intermediate is hydrolysed via a similar mechanism. The respective protein environment ensures high reaction rate (Pauling postulate). Among other interactions, it is H-bond stabilization of the oxyanion which is crucial. This is the fundamental difference between enzyme catalysis and synthetic organocatalysis in which only the solvent surrounds transition states or intermediates.

Hundreds of impressive examples of enantioselective lipase-catalysed reactions are known, including industrial processes as in the case of the BASF method of chiral amine production (Collins et al. 1997; Breuer et al. 2004; Schmid and Verger 1998). However, the classical problem of substrate acceptance or lack of enantioselectivity (or both) persists. We were able to meet this challenge in model studies regarding the hydrolytic kinetic resolution of the ester *rac*-**1** with formation of carboxylic acid **2**, catalysed by the lipase from *Pseudomonas aeruginosa*. The wild-type (WT) lipase is only slightly (*S*)-selective, the selectivity factor amounting to a mere $E = 1.1$ (Scheme 1).

Proof-of-principle of the concept of directed evolution of enantioselective enzymes was provided in 1997 in a study describing four consecutive cycles of epPCR at low mutation rate, leading to $E = 11.3$ (Reetz et al. 1997). This was the first example of directed evolution of an enantioselective enzyme. Nevertheless, such an *E*-value is not yet practical, and therefore subsequent studies were undertaken which included epPCR at higher mutation rate, saturation mutagenesis at the identified hot spots, and DNA shuffling. The total efforts amounted to the production and screening of 40,000 clones. The best mutant of these important exercises in probing protein sequence space showed an *E*-value of 51 in the kinetic resolution of the model reaction (Reetz et al. 2001). Figure 3 summarizes this work (Reetz 2004). Moreover, it was possible to invert

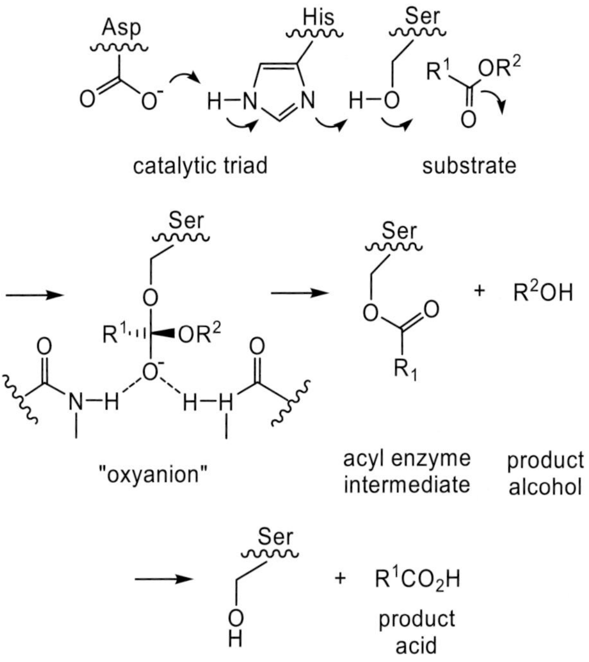

Fig. 2. Mechanism of the lipase-catalysed hydrolysis of esters

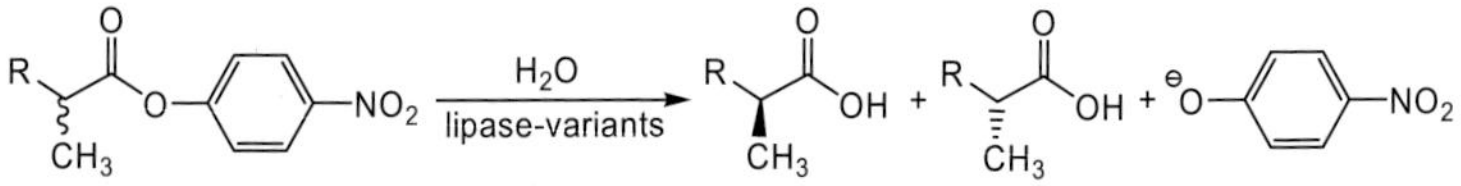

rac-**1** (R = *n*-C_8H_{17}) (*S*)-**2** (*R*)-**2** **3**

Scheme 1.

Fig. 3. Schematic summary of the directed evolution of enantioselective lipase-variants originating from the WT PAL used as catalysts in the hydrolytic kinetic resolution of ester *rac*-**1**. *CMCM* = Combinatorial multiple-cassette mutagenesis

Directed Evolution

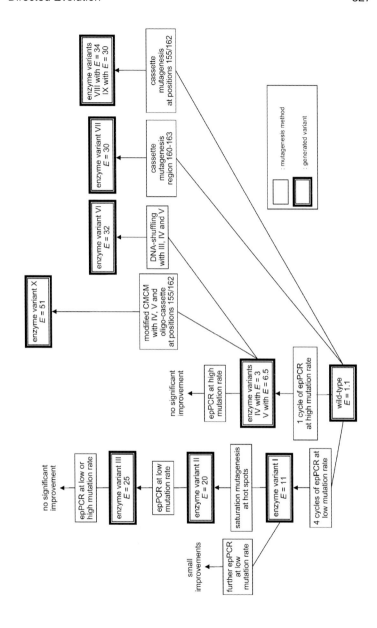

the sense of enantioselectivity, that is, to evolve (*R*)-selective mutants (Zha et al. 2001).

The best mutant, enzyme variant X in Fig. 3, has six mutations, five of them being remote. In view of Emil Fischer's lock-and-key model (or induced fit), remote mutations may appear as a surprise. In a detailed MM/QM study performed in collaboration with the theoretician Walter Thiel, we were able to illuminate the source of enhanced enantioselectivity (Bocola et al. 2004; Reetz et al. 2007). Only two of the six mutations are crucial, one occurring on the enzyme surface, the other next to the binding pocket. A relay mechanism was postulated as shown in Fig. 4. For details the reader is referred to the original publications (Bocola et al. 2004; Reetz et al. 2007).

Based on the theoretical predictions, we went back to the laboratory and prepared by site-specific mutagenesis some of the relevant double- and triple-mutants in a deconvolution process (Reetz et al. 2007). Indeed, a double mutant with an *E*-value of 63 was discovered. Thus, the intertwinement of experiment and theory not only leads to an understanding of the results, it also points the way to even better mutants. Much can be learned from directed evolution, provided a sound theoretical analysis is performed.

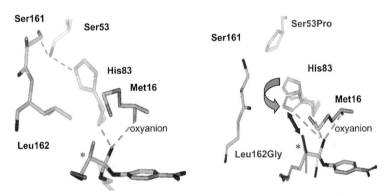

Fig. 4. The oxyanions originating from *rac*-**1** in the WT PAL (*left*) and mutant X (*right*). In the case of mutant X, additional stabilization of the oxyanion by His83 is possible with (*S*)-**1** (methyl group = *green line*), but not with (*R*)-**1** (methyl group = *purple line*)

3 Directed Evolution of Enantioselective Baeyer-Villigerases

Baeyer-Villiger reactions (BV) involve the interaction of ketones with per-acids or alkyl hydroperoxides, the products being esters or lactones (Krow 1993). The process is catalysed by acids, bases or transition metal salts. In the latter case several chiral catalysts for enantioselective BV-reactions have been reported, but applications are restricted to strained ketones (mainly cyclobutanone derivatives) (Bolm et al. 2004). Moreover, an organocatalytic system based on the use of a synthetic chiral flavin-type compound has been devised (*ee* up to 60%) (Murahashi et al. 2002). This interesting work was inspired by previous reports regarding the mechanism of cyclohexanone monooxygenase (e.g. CHMO from *Acinetobacter* sp. NCIMB 9871), a flavin-dependent enzyme (Flitsch and Grogan 2002). Accordingly, oxygen from air reacts with the reduced form of enzyme-bound flavin with formation of an alkylhydroperoxide which initiates the BV reaction. The oxidized flavin then has to be reduced by co-factor NADPH (Fig. 5). Thus, one practical option is to use whole cells as the catalytic machine of CHMO-catalysed BV reactions. Indeed, a number of successful enantioselective BV reactions (kinetic resolution of racemic ketones or desymmetrization) have been reported (Flitsch and Grogan 2002).

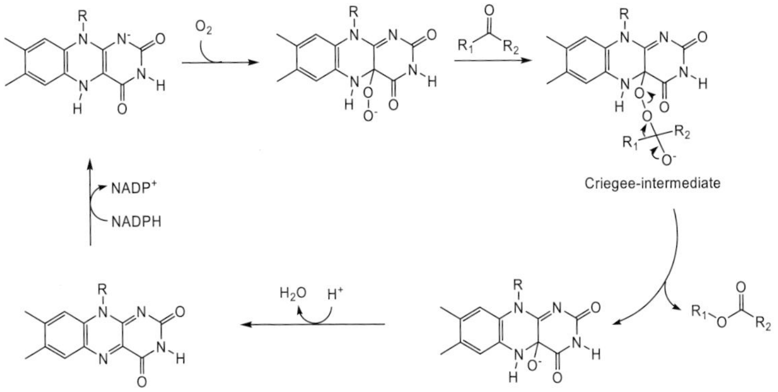

Fig. 5. Mechanism of CHMO-catalysed Baeyer-Villiger reaction

Naturally, the WT of any enzyme has limitations regarding substrate scope (acceptance) and the degree of enantioselectivity. For example, the *ee* of the desymmetrization of 4-hydroxycyclohexanone **4** catalysed by the WT-CHMO amounts to only 9% in favour of (*S*)-**5**. We applied our previous experience with the lipases using epPCR and were able to evolve a mutant showing *ee* = 90% (*S*) (Reetz et al. 2004a). Reversal of enantioselectivity was also possible. One of the mutants also displayed fairly large substrate scope, a number of cyclic and bicyclic ketones undergoing desymmetrization with *ee* > 95% (Mihovilovic et al. 2006). No synthetic catalysts are currently available that allow such transformations (Scheme 2).

In a related study, we also evolved CHMO-mutants which catalyse the sulfoxidation of thio-ethers such as **6** (Reetz et al. 2004b) (Scheme 3).

These results and the previous ones regarding the lipase are impressive because, *inter alia*, no synthetic catalysts are available which match the efficiency and enantioselectivity of the enzyme mutants described herein. Indeed, the concepts that we proposed, including some of the *ee*-assays developed in our laboratories (Reetz et al. 1997), have been applied successfully by other academic and industrial groups (Reetz 2006). Nevertheless, we were not fully content with the traditional use of epPCR, saturation mutagenesis and DNA shuffling as tools. Therefore, another direction of our research beginning in 2003 was the development of more efficient ways to probe protein sequence space. A principally new strategy was developed, initially using an epoxide hydrolase (Sect. 4).

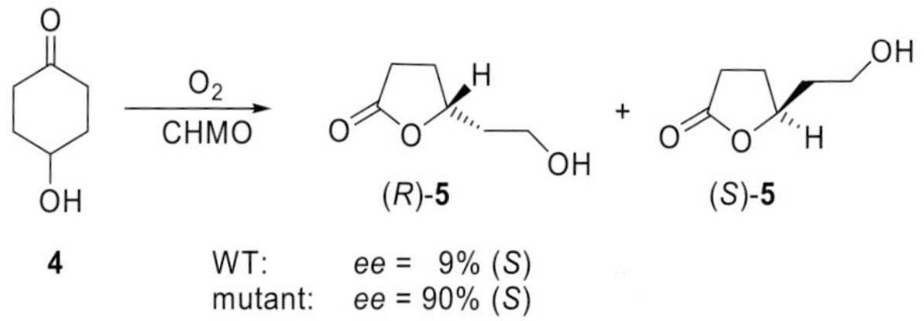

Scheme 2.

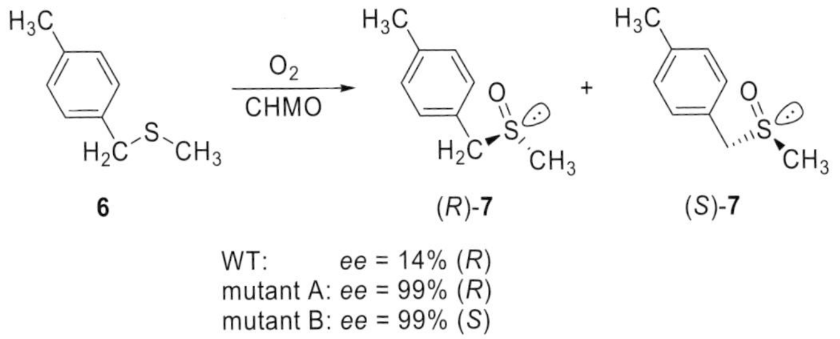

Scheme 3.

4 Directed Evolution of Enantioselective Epoxide Hydrolases

In the years 2000–2004 we were studying the directed evolution of enantioselective epoxide hydrolases (EH), specifically the kinetic resolution of glycidyl phenyl ether (*rac*-**8**) catalyzed by the EH from *Aspergillus niger* (ANEH) (Reetz et al. 2004c). The WT-ANEH shows an *E*-value of only 4.6 in slight favour of (*S*)-**9** (Scheme 4).

Upon applying the traditional mutagenesis methods, especially epPCR at various mutation rates, the results were somewhat disappointing. After screening a total of 20,000 clones, the best mutant showed an *E*-value of only 11.8 (Reetz et al. 2004c). We speculated as to why the ANEH is so difficult to evolve, perhaps because the binding pocket is an unusually narrow tunnel as demonstrated by the X-ray structure (Zou et al. 2000). Mechanistically it was also known that two tyrosines bind and activate appropriate epoxides, an aspartate then attacking nucleophilically with formation of a covalent intermediate which is subsequently hydrolysed (Fig. 6) (Archelas and Furstoss 1998). This again shows that nature is a master in devising organocatalytic processes!

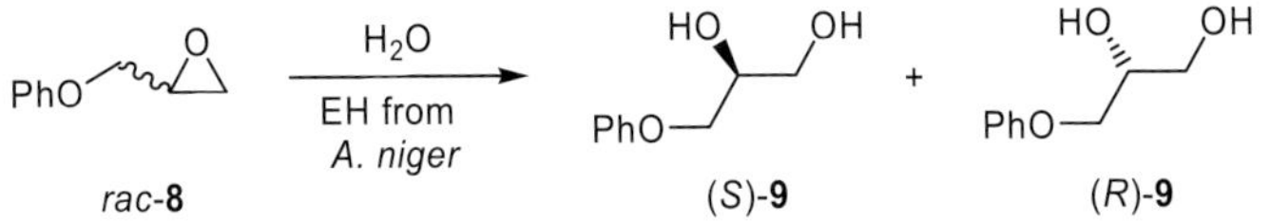

Scheme 4.

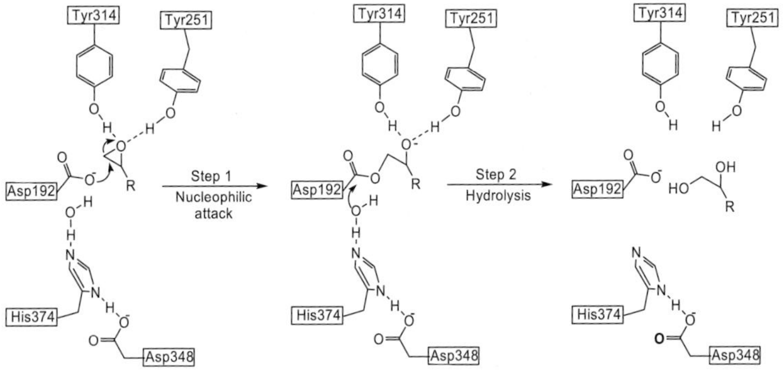

Fig. 6. Mechanism of ANEH-catalysed hydrolytic reactions of epoxides

In the quest to devise a new and hopefully more efficient method for probing protein sequence space, which is crucial in the area of directed evolution, we proposed the concept of iterative saturation mutagenesis (ISM) (Reetz et al. 2006a). It is based on a Cartesian view of the enzyme structure, specifically by performing iterative cycles of saturation mutagenesis at rationally chosen sites (Reetz et al. 2006a; Reetz and Carballeira 2007). A given site may be composed of one, two or three amino acid positions. Randomization at the chosen sites generates small focused libraries of mutants. Following screening for some property of interest–for example enantioselectivity, substrate acceptance (rate) or thermostability–the gene of the respective best mutant is used as a template for performing further saturation mutagenesis experiments at the other sites. Figure 7 illustrates the case of four sites (Reetz et al. 2006a). The initial identification of the appropriate sites is crucial for the success of ISM. The basis for choosing these sites depends upon the nature of the catalytic property to be improved.

Convergence is reached after generating and screening 64 focused libraries prepared by saturation mutagenesis. However, it is not at all necessary to explore all upward pathways in the fitness landscape. It is also clear that the hits produced by the process of ISM are not likely to be evolved by conventional strategies such as repeating cycles of epPCR or DNA shuffling which address the whole gene (and thus enzyme),

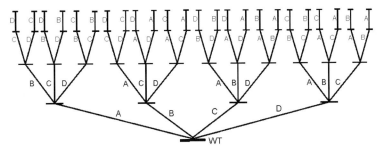

Fig. 7. ISM using four sites A, B, C and D, each site in a given upward pathway being visited only once

simply on statistical grounds. Each new cycle of ISM maximizes the probability of obtaining additive and/or cooperative effects of newly introduced mutations in a defined region of the enzyme. We have demonstrated several times the enormous benefits of conducting the search in protein sequence space by ISM. These studies include the drastic improvement of thermostability of a lipase (Reetz et al. 2006c) as well as the enhancement of enantioselectivity of an epoxide hydrolase (ANEH) (Reetz et al. 2006a) as reviewed herein and of a hybrid catalyst in which an achiral diphosphine/Rh-complex is anchored to a protein host (Reetz et al. 2006b).

In the case of ANEH, the challenge to evolve high enantioselectivity in the kinetic resolution of epoxide *rac*-**8** was nearly overwhelming, because our previous attempts were not very successful (Reetz et al. 2004c). The first step in applying ISM was to find and apply a criterion for choosing appropriate sites for saturation mutagenesis. For this purpose our previously developed Combinatorial Active-Site Saturation Test (CAST) appeared ideally suited (Reetz et al. 2005). It had been developed to solve the long-standing problem of limited substrate range of enzymes. CAST involves the systematic formation of focused libraries (based on saturation mutagenesis) around the complete binding pocket. This distinguishes it from previously reported focused libraries in which only one or two sites were considered (Arnold and Georgiou 2003; Brakmann and Schwienhorst 2004). CASTing thus requires structural knowledge, either the X-ray structure of the enzyme or a homology model, which is then carefully analysed in the region around the binding pocket. All those sites A, B, C, etc. harbouring one, two or three amino acids having side chains next to the binding pocket are chosen for saturation mutagenesis (Reetz et al. 2005). A site comprising a single amino acid gives rise to 20 different mutants, those having two or three amino acids may result in 400 or 8,000 different mutants, respectively. In each case over-sampling is necessary if 95% coverage of the defined protein sequence space is to be ensured, namely about 150, 3,000 and 98,000 clones, respectively. However, such high coverage is not mandatory, and in the ANEH-project much lower numbers were screened (Reetz et al. 2006a).

The CAST analysis of ANEH suggested six sites A, B, C, D, E and F, some compromising two amino acid positions, other three (Fig. 8) (Reetz et al. 2006a). Thus, the ISM scheme is more extensive that the model illustration shown in Fig. 7.

Initially only one upward pathway in the limited and focused fitness landscape was considered, namely B–C–D–F–E. The results are

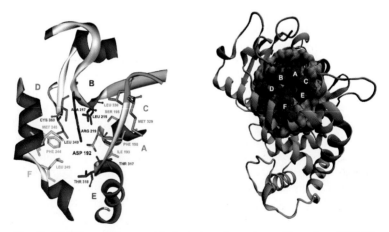

Fig. 8. CASTing of the epoxide hydrolase from *Aspergillus niger* (ANEH) based on the X-ray structure of the WT. *Left*: Defined randomization sites A–E; *Right*: Top view of tunnel-like binding pocket showing sites A–E (*blue*) and the catalytically active D192 (*red*)

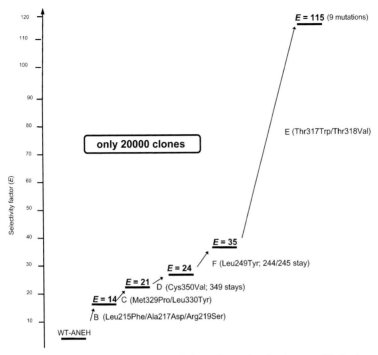

Fig. 9. Iterative CASTing in the evolution of enantioselective epoxide hydrolases as catalysts in the hydrolytic kinetic resolution of *rac*-**8**

remarkable in several respects (Fig. 9) (Reetz et al. 2006a). First, the final mutant, harbouring nine mutational changes solely around the binding pocket, leads to a selectivity factor of $E = 115$ (Fig. 9). Second, only 20,000 clones had to be produced and screened which happens to be equivalent to the number of clones screened in our first ANEH-project leading to an E-value of only 11.8 (Reetz et al. 2004c). Thus, ISM in the embodiment of iterative CASTing means rapid and efficient directed evolution. Industry in particular needs fast methods for efficient directed evolution. The computer program CASTER as an aid in designing saturation mutagenesis libraries is available from our website (Reetz and Carballeira 2007).

These results were obtained by performing saturation mutagenesis with simultaneous randomization of amino acids at the respective sites B, C, D, F and E (A was not considered). In doing so, NNK codon degeneracy was chosen, meaning all 20 proteinogenic amino acids as building blocks. However, different codon degeneracy is also possible, for example NDT, meaning only 12 amino acids as building blocks (Phe, Leu, Ile, Val, Tyr, His, Asn, Asp, Cys, Arg, Ser, Gly). Although such a restriction reduces structural diversity somewhat, the degree of over-sampling is reduced drastically (Reetz and Carballeira 2007). This strategic advantage was used in solving yet another major problem, namely the enantioselective ring-opening hydrolysis of *trans*-disubstituted epoxides (Reetz 2007). Such substrates are not at all accepted by ANEH. By applying restricted CASTing and employing NDT codon degeneracy, we succeeded in evolving highly active and enantioselective mutants for these difficult substrates.

5 Directed Evolution of Hyperthermostable Enzymes

Sufficient thermostability is a requirement for real applications of enzymes (Drauz and Waldmann 2002; Liese et al. 2006). Many approaches for increasing the thermostability of enzymes have been described, including directed evolution using the traditional techniques of epPCR, DNA shuffling, etc. (Arnold and Georgiou 2003; Brakmann and Schwienhorst 2004; Eijsink et al. 2005). We have shown that the concept of ISM is particularly well suited for improving this catalytic parameter significantly in a rapid manner (Reetz and Carballeira 2007; Reetz et al. 2006c). Obviously, the criterion for choosing the sites to be randomized by saturation mutagenesis is different. We developed the idea of using the B-factors of a given enzyme available from its X-ray structure. They reflect smearing of atomic electron densities around their equilibrium positions as a result of thermal motion. Thus, high B-factors indicate high flexibility. Since it was known that high rigidity is characteristic of hyperthermophilic enzymes (Radivojac et al. 2004), the goal was to increase rigidity at the flexible positions. Thus, we chose those sites of an enzyme for ISM which show the highest average B-factors. Again, a computer aid (B-FITTER) is available on our website for choosing and designing appropriate ISM libraries (Reetz and Carballeira 2007). We were able to improve the thermo-stability of a lipase from $T_{50}^{60} = 48\ °C$ to $T_{50}^{60} = 93\ °C$, which corresponds to an increase in thermostability of 45 °C, and has no precedence in the literature (Reetz and Carballeira 2007; Reetz et al. 2006c).

6 Conclusions

Directed evolution of functional enzymes has emerged as a powerful method of engineering their catalytic profiles. Traditional approaches such as a epPCR, saturation mutagenesis and DNA shuffling are successful. Nevertheless, even better strategies and methods are being sought. ISM appears to be particularly fast, reliable and efficient, requiring less molecular biological and screening work, while providing excellent results. ISM is useful in controlling three of the most important catalytic properties of an enzyme, namely range of substrate acceptance, enantioselectivity and thermostability. So far only a few enzyme types have been engineered with ISM, which means that more research is necessary. This approach is complementary to the development of synthetic organocatalysts as described in the other chapters of this monograph.

References

Archelas A, Furstoss R (1998) Epoxide hydrolases: New tools for the synthesis of fine organic chemicals. Trends Biotechnol 16:108–116

Arnold FH, Georgiou G (eds) (2003) Methods in molecular biology. Humana Press, Totowa, vol 230

Berkessel A, Gröger H (eds) (2004) Asymmetric organocatalysis. VCH, Weinheim

Bocola M, Otte N, Jaeger K-E, Reetz MT, Thiel W (2004) Learning from directed evolution: Theoretical investigations into cooperative mutations in lipase enantioselectivity. Chem Bio Chem 5:214–223

Bolm C, Palazzi C, Beckmann O (2004) Metal-catalyzed Baeyer-Villiger reactions. In: Beller M, Bolm C (eds) Transition metals for organic chemistry: building blocks and fine chemicals, vol 2. Wiley-VCH, Weinheim, pp 267–274

Brakmann S, Schwienhorst A (eds) (2004) Evolutionary methods in biotechnology (clever tricks for directed evolution). Wiley-VCH, Weinheim

Breuer M, Ditrich K, Habicher T, Hauer B, Keßer M, Stürmer R, Zelinski T (2004) Industrial methods for the production of optically active intermediates. Angew Chem Int Ed 43:788–824

Cedrone F, Ménez A, Quéméneur E (2000) Tailoring new enzyme functions by rational redesign. Curr Opin Struct Biol 10:405–410

Collins AN, Sheldrake GN, Crosby J (eds) (1997) Chirality in industry II: Developments in the commercial manufacture and applications of optically active compounds. Wiley, Chichester

Drauz K, Waldmann H (eds) (2002) Enzyme catalysis in organic synthesis: A comprehensive handbook. VCH, Weinheim, 2nd edn, vol I–III

Eijsink VGH, Gåseidnes S, Borchert TV, van den Burg B (2005) Directed evolution of enzyme stability. Biomol Eng 22:21–30

Flitsch S, Grogan G (2002) Baeyer-Villiger oxidations. In: Drauz K, Waldmann H (eds) Enzyme catalysis in organic synthesis. Wiley-VCH, Weinheim, vol 2, pp 1202–1245

Jacobsen EN, Pfaltz A, Yamamoto H (eds) (1999) Comprehensive asymmetric catalysis. Springer, Berlin, vol I–III

Krow GR (1993) The Baeyer-Villiger oxidation of ketones and aldehydes. Org React (New York) 43:251–798

Liese A, Seelbach K, Wandrey C (eds) (2006) Industrial biotransformations. Wiley-VCH, Weinheim, 2006

Mihovilovic MD, Rudroff F, Winninger A, Schneider T, Schulz F, Reetz MT (2006) Microbial Baeyer-Villiger oxidation: Stereopreference and substrate acceptance of cyclohexanone monooxygenase mutants prepared by directed evolution. Org Lett 8:1221–1224

Murahashi S-I, Ono S, Imada Y (2002) Asymmetric Baeyer-Villiger reaction with hydrogen peroxide catalyzed by a novel planar-chiral bisflavin. Angew Chem Int Ed 41:2366–2368

Radivojac P, Obradovic Z, Smith DK, Zhu G, Vucetic S, Brown CJ, Lawson JD, Dunker AK (2004) Protein flexibility and intrinsic disorder. Protein Sci 13:71–80

Reetz MT (2004a) Controlling the enantioselectivity of enzymes by directed evolution: Practical and theoretical ramifications. Proc Natl Acad Sci USA 101:5716–5722

Reetz MT (2004b) High-throughput screening of enantioselective industrial biocatalysts. In: Brakmann S, Schwienhorst A (eds) Evolutionary methods in biotechnology. Wiley-VCH, Weinheim, pp 113–141

Reetz MT (2006) Directed evolution of enantioselective enzymes as catalysts for organic synthesis. In: Gates BC, Knözinger H (eds) Advances in catalysis, vol 49. Elsevier, San Diego, pp 1–69

Reetz MT (2007) Directed evolution as a means to engineer enantioselective enzymes. In: Gotor V (ed) Asymmetric organic synthesis with enzymes. Wiley-VCH, Weinheim, in press

Reetz MT, Carballeira JD (2007) Iterative saturation mutagenesis (ISM) for rapid directed evolution of functional enzymes. Nat Protoc 2:891–903

Reetz MT, Zonta A, Schimossek K, Liebeton K, Jaeger K-E (1997) Creation of enantioselective biocatalysts for organic chemistry by in vitro evolution. Angew Chem Int Ed Engl 36:2830–2832

Reetz MT, Wilensek S, Zha D, Jaeger K-E (2001) Directed evolution of an enantioselective enzyme through combinatorial multiple cassette mutagenesis. Angew Chem Int Ed 40:3589–3591

Reetz MT, Brunner B, Schneider T, Schulz F, Clouthier CM, Kayser MM (2004a) Directed evolution as a method to create enantioselective cyclohexanone monooxygenases for catalysis in Baeyer-Villiger reactions. Angew Chem Int Ed 43:4075–4078

Reetz MT, Daligault F, Brunner B, Hinrichs H, Deege A (2004b) Directed evolution of cyclohexanone monooxygenases: Enantioselective biocatalysts for the oxidation of prochiral thioethers. Angew Chem Int Ed 43:4078–4081

Reetz MT, Torre C, Eipper A, Lohmer R, Hermes M, Brunner B, Maichele A, Bocola M, Arand M, Cronin A, Genzel Y, Archelas A, Furstoss R (2004c) Enhancing the enantioselectivity of an epoxide hydrolase by directed evolution. Org Lett 6:177–180

Reetz MT, Bocola M, Carballeira JD, Zha D, Vogel A (2005) Expanding the range of substrate acceptance of enzymes: Combinatorial active-site saturation test. Angew Chem Int Ed 44:4192–4196

Reetz MT, Wang L-W, Bocola M (2006a) Directed evolution of enantioselective enzymes: Iterative cycles of CASTing for probing protein-sequence space. Angew Chem Int Ed 45:1236–1241; Erratum 2494

Reetz MT, Peyralans JJ-P, Maichele A, Fu Y, Maywald M (2006b) Directed evolution of hybrid enzymes: Evolving enantioselectivity of an achiral Rh-complex anchored to a protein. Chem Commun (Cambridge UK) 4318–4320

Reetz MT, Carballeira JD, Vogel A (2006c) Iterative saturation mutagenesis on the basis of B factors as a strategy for increasing protein thermostability. Angew Chem Int Ed 45:7745–7751

Reetz MT, Puls M, Carballeira JD, Vogel A, Jaeger K-E, Eggert T, Thiel W, Bocola M, Otte N (2007) Learning from directed evolution: Further lessons from theoretical onvestigations into cooperative mutations in lipase enantioselectivity. ChemBioChem 8:106–112

Schmid RD, Verger R (1998) Lipases: Interfacial enzymes with attractive applications. Angew Chem Int Ed 37:1608–1633

Seayad J, List B (2005) Asymmetric organocatalysis. Org Biomol Chem 3:719–724

Zha D, Wilensek S, Hermes M, Jaeger K-E, Reetz MT (2001) Complete reversal of enantioselectivity of an enzyme-catalyzed reaction by directed evolution. Chem Commun (Cambridge UK) 2664–2665

Zou JY, Hallberg BM, Bergfors T, Oesch F, Arand M, Mowbray SL, Jones TA (2000) Structure of Aspergillus niger epoxide hydrolase at 1.8 resolution: Implications for the structure and function of the mammalian microsomal class of epoxide hydrolases. Structure (London) 8:111–122

Ernst Schering Foundation Symposium Proceedings

Editors: Günter Stock
Monika Lessl

Vol. 2006/1: Tissue-Specific Estrogen Action
Editors: K.S. Korach, T. Wintermantel

Vol. 2006/2: GPCRs: From Deorphanization
to Lead Structure Identification
Editors: H.R. Bourne, R. Horuk, J. Kuhnke, H. Michel

Vol. 2006/3: New Avenues to Efficient Chemical Synthesis
Editors: P.H. Seeberger, T. Blume

Vol. 2006/4: Immunotherapy in 2020
Editors: A. Radbruch, H.-D. Volk, K. Asadullah, W.-D. Doecke

Vol. 2006/5: Cancer Stem Cells
Editors: O.D. Wiestler, B. Haendler, D. Mumberg

Vol. 2007/1: Progestins and the Mammary Gland
Editors: O. Conneely, C. Otto

Vol. 2007/2: Organocatalysis
Editors: M.T. Reetz, B. List, S. Jaroch, H. Weinmann

Vol. 2007/3: Sparking Signals
Editors: G. Baier, B. Schraven, U. Zügel, A. von Bonin

Printing: Krips bv, Meppel, The Netherlands
Binding: Stürtz, Würzburg, Germany